Receptors and
Recognition

Series B Volume 17

Monoclonal Antibodies to Receptors

Probes for Receptor Structure and Function

Edited by
M. F. Greaves

Leukaemia Research Fund Centre,
Institute for Cancer Research,
London, U.K.

LONDON NEW YORK
CHAPMAN AND HALL

First published 1984 by
Chapman and Hall Ltd
11 New Fetter Lane, London EC4P 4EE

Published in the USA by
Chapman and Hall
733 Third Avenue, New York NY 10017

Printed in Great Britain at the
University Press, Cambridge

ISBN 0 412 25330 5

British Library Cataloguing in Publication Data

Monoclonal antibodies to receptors.——(Receptors and recognition. Series B; V.17)
1. Antibodies, Monoclonal 2. Cell receptors
I. Greaves, M.F. II. Series
596′.0293 QR186.85

ISBN 0-412-25330-5

Library of Congress Cataloging in Publication Data

Main entry under title:

Monoclonal antibodies to receptors.

(Receptors and recognition. Series B; v. 17)
Bibliography: p.
Includes index.
1. Antibodies, Monoclonal. 2. Cell receptors.
3. Hormone receptors. 4. Neurotransmitter receptors.
I. Greaves, M. F. (Melvyn F.), 1941– . II. Series.
[DNLM: 1. Antigen-antibody reactions. 2. Antibodies, Monoclonal. 3. Receptors, Immunologic.
W1 RE107MA v. 17/QW 570 A6246]
QR186.85.A58 1984 596′.087 84-4976
ISBN 0-412-25330-5

Receptors and Recognition

General Editors: P. Cuatrecasas and M.F. Greaves

About the series

Cellular Recognition – the process by which cells interact with, and respond to, molecular signals in their environment – plays a crucial role in virtually all important biological functions. These encompass fertilization, infectious interactions, embryonic development, the activity of the nervous system, the regulation of growth and metabolism by hormones and the immune response to foreign antigens. Although our knowledge of these systems has grown rapidly in recent years, it is clear that a full understanding of cellular recognition phenomena will require an integrated and multidisciplinary approach.

This series aims to expedite such an understanding by bringing together accounts by leading researchers of all biochemical, cellular and evolutionary aspects of recognition systems. This series will contain volumes of two types. First, there will be volumes containing about five reviews from different areas of the general subject written at a level suitable for all biologically oriented scientists (Receptors and Recognition, series A). Secondly, there will be more specialized volumes (Receptors and Recognition, series B), each of which will be devoted to just one particularly important area.

Advisory Editorial Board

Receptors and Recognition

Series A

Published

Volume 1 (1976)
M.F. Greaves (London), Cell Surface Receptors: A Biological Perspective
F. Macfarlane Burnet (Melbourne), The Evolution of Receptors and Recognition in the Immune System
K. Resch (Heidelberg), Membrane Associated Events in Lymphocyte Activation
K.N. Brown (London), Specificity in Host-Parasite Interaction

Volume 2 (1976)
D. Givol (Jerusalem), A Structural Basis for Molecular Recognition: The Antibody Case
B.D. Gomperts (London), Calcium and Cell Activation
M.A.B. de Sousa (New York), Cell Traffic
D. Lewis (London), Incompatibility in Flowering Plants
A. Levitski (Jerusalem), Catecholamine Receptors

Volume 3 (1977)
J. Lindstrom (Salk, California), Antibodies to Receptors for Acetylcholine and other Hormones
M. Crandall (Kentucky), Mating-Type Interaction in Micro-organisms
H. Furthmayr (New Haven), Erythrocyte Membrane Proteins
M. Silverman (Toronto), Specificity of Membrane Transport

Volume 4 (1977)
M. Sonenberg and A.S. Schneider (New York), Hormone Action at the Plasma Membrane: Biophysical Approaches
H. Metzger (NIH, Bethesda), The Cellular Receptor for IgE
T.P. Stossel (Boston), Endocytosis
A. Meager (Warwick) and R.C. Hughes (London), Virus Receptors
M.E. Eldefrawi and A.T. Eldefrawi (Baltimore), Acetylcholine Receptors

Volume 5 (1978)
P.A. Lehmann (Mexico), Stereoselective Molecular Recognition in Biology
A.G. Lee (Southampton, U.K.), Fluorescence and NMR Studies of Membranes
L.D. Kohn (NIH, Bethesda), Relationships in the Structure and Function of Receptors for Glycoprotein Hormones, Bacterial Toxins and Interferon

Volume 6 (1978)
J.N. Fain (Providence, Rhode Island), Cyclic Nucleotides
G.D. Eytan (Haifa) and B.J. Kanner (Jerusalem), Reconstitution of Biological Membranes
P.J. O'Brien (NIH, Bethesda), Rhodopsin: A Light-sensitive Membrane Glycoprotein

Index to Series A, Volumes 1–6

Series B

Published

The Specificity and Action of Animal Bacterial and Plant Toxins (B1)
edited by P. Cuatrecasas (Burroughs Wellcome, North Carolina)
Intercellular Junctions and Synapses (B2)
edited by J. Feldman (London), N.B. Gilula (Rockefeller University, New York) and J.D. Pitts (University of Glasgow)
Microbial Interactions (B3)
edited by J.L. Reissig (Long Island University, New York)
Specificity of Embryological Interactions (B4)
edited by D.R. Garrod (University of Southampton)
Taxis and Behavior (B5)
edited by G.L. Hazelbauer (University of Uppsala)
Bacterial Adherence (B6)
edited by E.H. Beachey (Veteran's Administration Hospital and University of Tennessee, Memphis, Tennessee)
Virus Receptors Part 1 Bacterial Viruses (B7)
edited by L.L. Randall and L. Philipson (University of Uppsala)
Virus Receptors Part 2 Animal Viruses (B8)
edited by K. Lonberg-Holm (Du Pont, Delaware) and L. Philipson (University of Uppsala)
Neurotransmitter Receptors Part 1 Amino Acids, Peptides and Benzodiazepines (B9)
edited by S.J. Enna (University of Texas at Houston) and H.I. Yamamura (University of Arizona)
Neurotransmitter Receptors Part 2 Biogenic Amines (B10)
edited by H.I. Yamamura (University of Arizona) and S.J. Enna (University of Texas at Houston)
Membrane Receptors: Methods for Purification and Characterization (B11)
edited by S. Jacobs and P. Cuatrecasas (Burroughs Wellcome, North Carolina)
Purinergic Receptors (B12)
edited by G. Burnstock (University College, London)
Receptor Regulation (B13)
edited by R.J. Lefkowitz (Duke University Medical Center, North Carolina)
Histocompatibility Antigens: Structure and Function (B14)
edited by P. Parham (Stanford University School of Medicine, California) and J. Strominger (Harvard University, Massachussetts)
Receptor-Mediated Endocytosis (B15)
edited by P. Cuatrecasas (Burroughs Wellcome, North Carolina) and T. Roth (University of Maryland, Baltimore County)
Genetic Analysis of the Cell Surface (B16)
edited by P. Goodfellow (Imperial Cancer Research Fund, London)

Contents

Contributors

Maurizio Bifulco, Laboratory of Biochemical Pharmacology, National Institutes of Health, Bethesda, Maryland, U.S.A.

Pedro Cuatrecasas, Wellcome Research Laboratories, Research Triangle Park, North Carolina, U.S.A.

Michele L. De Luca, Laboratory of Biochemical Pharmacology, National Institutes of Health, Bethesda, Maryland, U.S.A.

Joel M. Depper, Metabolism Branch, National Cancer Institute, Bethesda, Maryland, U.S.A.

Claire M. Fraser, Department of Molecular Immunology, Roswell Park Memorial Institute, Buffalo, New York, U.S.A.

Sara Fuchs, Department of Chemical Immunology, Weizmann Institute of Science, Rehovot, Israel.

Melvyn F. Greaves, Membrane Immunology Laboratory Leukaemia Research Fund Centre, Institute for Cancer Research, London, U.K.

Warner C. Greene, Metabolism Branch, National Cancer Institute, Bethesda, Maryland, U.S.A.

Evelyn F. Grollman, Laboratory of Biochemical Pharmacology, National Institutes of Health, Bethesda, Maryland, U.S.A.

Steven Jacobs, Wellcome Research Laboratories, Research Triangle Park, North Carolina, U.S.A.

Hana Kanety, Department of Chemical Immunology, Weizmann Institute of Science, Rehovot, Israel.

Leonard D. Kohn, Laboratory of Biochemical Pharmacology, National Institutes of Health, Bethesda, Maryland, U.S.A.

Frederick C. Kull Jr, Wellcome Research Laboratories, Research Triangle Park, North Carolina, U.S.A.

Irit Lax, Department of Chemical Immunology, Weizmann Institute of Science, Rehovot, Israel.

Warren J. Leonard, Metabolism Branch, National Cancer Institute, Bethesda, Maryland, U.S.A.

Tovia Libermann, Department of Chemical Immunology, Weizmann Institute of Science, Rehovot, Israel.

Daria Mochly-Rosen, Department of Chemical Immunology, Weizmann Institute of Science, Rehovot, Israel.

Bruno Moncharmont, Wellcome Research Laboratories, Research Triangle Park, North Carolina, U.S.A.

Roland A. Newman, Department of Cancer Biology, Salk Institute for Biological Studies, San Diego, California, U.S.A.

Indu Parikh, Wellcome Research Laboratories, Research Triangle Park, North Carolina, U.S.A.

Gordon D. Ross, Department of Medicine and Department of Microbiology – Immunology, University of North Carolina, Chapel Hill, North Carolina, U.S.A.

Joseph Schlessinger, Department of Chemical Immunology, Weizmann Institute of Science, Rehovot, Israel.

Claudio Schneider, Membrane Immunology Laboratory, Imperial Cancer Research Fund, Lincoln's Inn Fields, London, U.K.

Alain B. Schreiber, Institute of Biological Sciences, Syntex Research, Palo Alto, California, U.S.A.

Miry C. Souroujon, Department of Chemical Immunology, Weizmann Institute of Science, Rehovot, Israel.

A. Donny Strosberg, Molecular Immunology, Institut Jacques Monod, C.N.R.S. and University of Paris VII, Paris, France.

Cox Terhorst, Laboratory of Molecular Immunology, Dana Farber Cancer Institute, Harvard Medical School, Boston, Massachusetts, U.S.A.

Donatella Tombaccini, Laboratory of Biochemical Pharmacology, National Institutes of Health, Bethesda, Maryland, U.S.A.

Ian S. Trowbridge, Department of Cancer Biology, Salk Institute for Biological Studies, San Diego, California, U.S.A.

William A. Valente, Laboratory of Biochemical Pharmacology, National Institutes of Health, Bethesda, Maryland, U.S.A.

Thomas A. Waldmann, Metabolism Branch, National Cancer Institute, Bethesda, Maryland, U.S.A.

Yosef Yarden, Department of Chemical Immunology, Weizmann Institute of Science, Rehovot, Israel.

Preface

Receptor specific antibodies are excellent probes for a wide range of biological investigations on receptor structure and function. The hybridoma technology (Kohler and Milstein, 1975) has inevitably had a major impact on this field with most of the better known receptors now identified with monoclonal antibodies. This volume of the *Receptors and Recognition* series provides reviews of recent developments in this field and emphasizes in particular the new opportunities afforded by the judicious application of monoclonal reagents. It is assumed that most readers will be familiar with the now fairly routine methods of cell fusion, hybridoma cloning and selection for producing monoclonal antibodies and so few details of the basic technical procedures are described. Several good reviews on this topic are however available (see Galfre and Milstein, 1981; Goding, 1980; Yelton and Scharf, 1981; McMichael and Fabre, 1982).

By no means all vertebrate receptor species are discussed here; omissions include antibodies to low density lipoprotein receptors (Beisiegel *et al.*, 1981; Kita *et al.*, 1981), prolactin and growth hormone receptors (Friesen *et al.*, 1982; Simpson *et al.*, 1983) and the hepatocyte asialoglycoprotein receptor (Schwartz *et al.*, 1981; Harford *et al.*, 1982). Nevertheless the coverage is comprehensive and critical and the individual chapters provided illustrate vividly the rapid progress being made. In addition to monoclonal antibodies three other types of receptor antibodies are described in this volume: experimentally induced polyclonal antibodies, human autoantibodies, and anti-anti-hormone (= anti-idiotype) antibodies. The latter reagents are capable of recognizing hormone or neurotransmitter receptors via 'molecular mimicry' and are discussed in detail by Strosberg and Schreiber in Chapter 2, and also by Ross (Chapter 5), Kohn *et al.* (Chapter 9) and Fuchs *et al.* (Chapter 8) (see also Sege and Peterson, 1978; Shechter *et al.*, 1982; Wassermann *et al.*, 1982; Homcy *et al.*, 1982; Cleveland *et al.*, 1983).

I should like to thank all the authors contributing to this volume for providing up to date and critical reviews of high quality. Most of them were submitted on time! My thanks are also due to Jackie Needham for her expert secretarial assistance, and Richard Stileman of Chapman and Hall Publishers for his untiring efforts and enthusiastic support of this and other volumes in the *Receptors and Recognition* series.

REFERENCES

Beisiegel, U., Schneider, W.J., Goldstein, J.L., Anderson, R.G.W. and Brown, M.S. (1981), *J. Biol. Chem.*, **256,** 11923–11931.

Friesen, H.G., Shiu, R.P.C., Elsholtz, H., Simpson, S. and Hughes, J. (1982), *Ciba Found. Symp.*, **90,** 263–278.
Galfre, G. and Milstein, C. (1981), *Methods Enzymol.*, **73B,** 13–46.
Goding, J.W. (1980), *J. Immunol. Methods,* **39,** 285–308.
Harford, J., Lowe, M., Tsunoo, H. and Ashwell, G. (1982) *J. Biol. Chem.*, **257,** 12685–12690.
Homcy, C.J., Rockson, S.C. and Haber, E. (1982), *J. Clin. Invest.*, **69,** 1147–1154.
Kita, T., Beisiegel, U., Goldstein, J.L., Schneider, W.J. and Brown, M.S. (1981), *J. Biol. Chem.*, **256,** 4701–4703.
McMichael, A.J. and Fabre, J.W. (eds) (1982) *Monoclonal Antibodies in Clinical Medicine,* Academic Press, London.
Schwartz, A.L., Marshak-Rothstein, A., Rup, D. and Lodish, H.F (1981), *Proc. Natl. Acad. Sci. U.S.A.*, **78,** 3348–3352.
Sege, K. and Peterson, P.A. (1978), *Proc. Natl. Acad. Sci. U.S.A.*, **75,** 2443–2447.
Shechter, Y., Maron, R., Elias, D. and Cohen, T.R. (1982), *Science,* **216,** 542–544.
Simpson, J.S.A., Hughes, J.P. and Friesen, H.G. (1983), *Endocrinology,* **112,** 2137.
Wassermann, N.H., Penn, H.S., Freimuth, P.I., Treptow, N., Wentzel, S., Cleveland, W.L. and Erlanger, B.F. (1982) *Proc. Natl. Acad. Sci. U.S.A.* **79,** 4810–4814.
Yelton, D.E. and Scharf, M.D. (1981) *Ann. Rev. Biochem.*, **50,** 657–680.

1 Introduction

MELVYN F. GREAVES

Monoclonal Antibodies to Receptors: Probes for Receptor Structure and Function
(*Receptors and Recognition*, Series B, Volume 17)
Edited by M. F. Greaves
Published in 1984 by Chapman and Hall, 11 New Fetter Lane, London EC4P 4EE

1.1 DIVERSITY OF APPLICATIONS OF ANTI-RECEPTOR ANTIBODIES

Successful applications of monoclonal antibodies to receptor biology are remarkably diverse (Table 1.1) and are perhaps best exemplified by the many elegant studies on nicotinic acetylcholine receptors which range from gene cloning and detailed protein subunit anatomy to the immunopathology of myasthenia gravis.

In many of these studies it is clear that monoclonal antibodies have had distinct advantages. Since these reagents can be selected for high affinity and bind to a single epitope, they permit a fine anatomical dissection of a receptor molecule and its subunits (see for example chapters by Ross, Fraser, Kohn *et al.*, Fuchs *et al.*, and also Sanchez-Madrid *et al.*, 1983; Tzartos *et al.*, 1981; Gullick and Lindstrom, 1983) and enable functional effects of antibodies to be linked to a particular site. The species specificity of selected monoclonal antibodies has also been important for allocating receptor genes to particular chromosomes by somatic cell genetics (references in Table 1.1; see also Goodfellow and Solomon, 1982).

1.2 SPECIFICITY OF ANTIBODIES TO RECEPTORS

It is important to appreciate that different regions or subunits of receptors are not equipotent immunologically; as far as a mouse immune response (for example) is concerned, certain areas of receptor will be 'immunodominant' (see Tzartos *et al.*, 1981). In addition, the physical form of the receptor material used for both immunization and hybridoma screening has a marked influence on the spectrum of antibody specificities and affinities obtained. Thus antibodies raised and selected against isolated receptors or their subunits may preferentially recognize intracytoplasmic domains of receptor unavailable on the cell surface (cf. Froehner, 1981) or subunit determinants which are cryptic or unavailable in the complete, native structure. Also, since isolation of receptors usually involves denaturation, antibodies raised against purified receptors may not recognize conformation-dependent regions or may only do so with low affinity (Tzartos *et al.*, 1981). As discussed by Fuchs *et al.* (Chapter 8), antibodies raised against denatured versus non-denatured receptors do show the anticipated specificity differences.

The hybridoma screening or selection procedure itself introduces a strong bias into the apparent repertoire of antibody specificities. Thus selective screening of monoclonal antibodies by inhibition of ligand binding may create the impression that many or most antibodies raised against receptors recognize the receptor's binding site for hormone, neurotransmitter, etc., whilst in fact this is almost certainly not the case. Firstly, although antibody binding may prevent interaction of receptor with its natural ligand (e.g. acetylcholine) or

Table 1.1 Some examples of applications of monoclonal anti-receptor antibodies

	Receptors (R)	Reference (example)
1. Affinity purification of receptors	Trf R and others	See Chapter 13 by Schneider; also Jacobs and Cuatrecasas (1981)
2. Biosynthesis, processing and membrane insertion of receptors	ACh R	Anderson and Blobel (1981) Fuchs *et al.* (Chapter 8)
	Insulin R	Hedo *et al.* (1983) Chapter 11 by Jacobs *et al.*
	IL-2 R	Chapter 3 by Leonard *et al.*
	Trf R	Schneider, C. *et al.* (1982) Chapter 10 by Trowbridge and Newman
3. Biochemical characterization, subunit anatomy and inter-relationships	AChR	Tzartos *et al.* (1981, 1982) Gullick and Lindstrom (1983) Chapter 8 by Fuchs *et al.*
	β_2 Adr R	Chapter 6 by Fraser
	T cell Ags	Chapter 7 by Terhorst
	LDL R	Schneider, W.J. *et al.* (1982)
	C3b R	Chapter 5 by Ross Springer *et al.* (1982)
	Steroid H R	Chapter 4 by Moncharmont and Parikh
	Asialo GP R	Schwartz *et al.* (1981) Harford *et al.* (1982)
4. Receptor localization *in situ*	Trf R	Gatter *et al.* (1983)
	Steroid H R	Chapter 4 by Moncharmont and Parikh
	ACh R	Swanson *et al.* (1983)
5. Identification of 'traffic' pathways for receptor–ligand internalization and recycling	Trf R	Hopkins and Trowbridge (1983) Enns *et al.* (1983)
	Asialo GP R	Geuze *et al.* (1983)
	LDL R	Brown *et al.* (1983)
	EGF R	Chapter 12 by Schlessinger *et al.* (See also Cuatrecasas and Roth, 1983 and Brown *et al.* (1983) for review)

6. Functional studies of receptor–response coupling	TSH R	Chapter 9 by Kohn *et al.*
	EGF R	Chapter 12 by Schlessinger *et al.*
	Trf R	Chapter 10 by Trowbridge and Newman
	IL 2 R	Chapter 3 by Leonard *et al.*
	Insulin R	Chapter 11 by Jacobs *et al.*
	T cell Ags	Chapter 7 by Terhorst See also Chapter 2 by Strosberg and Schreiber
7. Genetic mapping of receptors	EGF R	Waterfield *et al.* (1982)
	Trf R	Goodfellow *et al.* (1982)
8. Molecular cloning of receptors	ACh R Trf R	See Chapter 13 by Schneider
9. Receptors, antibodies and disease pathology	TSH R	Chapter 9 by Kohn *et al.*
	ACh R	Chapter 8 by Fuchs *et al.* Tzartos *et al.* (1982)
	β_2 Adr R	Chapter 6 by Fraser; see also Chapter 2 by Strosberg and Schreiber for review.
	LDL R	Tolleshaug *et al.* (1982) Beisiegel *et al.* (1981)

TSH, thyroid-stimulating hormone/thyrotropin; ACh, acetylcholine; Trf, transferrin; IL-2, interleukin 2; β_2 Adr, adrenergic (receptor); T cell Ags, T lymphocyte, cell surface antigens; Steroid H, steroid hormone; EGF epidermal growth factor; Asialo GP, asialoglycoprotein; C3b, complement component(s) C3b (see Chapter 5 by Ross); LDL, low-density lipoprotein.

various agonists/antagonists (e.g. α-toxin) used as affinity probes for the binding site, the reverse is often not the case (see Fuchs *et al.*, Chapter 8). Assuming that the antibodies do not possess a much higher affinity for the receptor than the binding site specific probe (which seems very unlikely with, for example, the acetylcholine receptor), this observation suggests that immunoglobulins recognize epitopes close to but distinct from the receptor binding site(s) and inhibit by steric hindrance.

Secondly, there are theoretical reasons for anticipating a low or modest frequency of anti-binding site antibodies. Most of the receptors discussed in this volume have phylogenetically conserved combining site specificity; other, functionally less important, regions of the structure would be expected to have accumulated more sequence polymorphism and potential antigenicity. In accord with this expectation, when receptor antibodies are screened and selected independently of ligand-binding-inhibitory capacity, it is clear that usually only a small minority of antibodies against, for example, acetylcholine receptors (see Chapter 8 by Fuchs *et al.;* Chapter 2 by Strosberg and Schreiber; see also Tzartos *et al.*, 1982) and steroid receptors (see Chapter 4 by Moncharmont and Parikh) possess this activity and then only rare antibodies may actually bind to the binding site itself rather than in close proximity (see Fuchs *et al.*, Chapter 8). A word of caution here, however. The combining site of receptors is likely to be highly conformation-dependent and therefore monoclonal antibodies raised against denatured receptors are perhaps unlikely to recognize this region. Immunization with membrane-associated receptors (Watters and Maelicke, 1983) or the anti-idiotype strategy (see Chapter 2) may provide an immunological route to the combining sites themselves.

It is important therefore to appreciate the specificity bias introduced by immunization and screening procedures. This can obviously be exploited to advantage. For comprehensive analysis of receptors it is clearly preferable to collect a panel or 'library' of monoclonal antibodies with different specificities.

How confident can we be that the antibodies described are indeed specific for any particular receptor, particularly in those instances where the starting immunogen was an intact whole cell preparation rather than purified or enriched receptor material? Antibody binding should at least parallel hormone binding and responsiveness with respect to different cell types. In special circumstances mutant cells with a specific lesion in receptor may be available, e.g. LDL receptors in familial hypercholesterolemia (Beisiegel *et al.*, 1981). More significantly the antibody should either competitively block the 'natural' ligand binding to its receptor (e.g. asialoglycoprotein, Harford *et al.*, 1982; EGF, Schlessinger *et al.*, Chapter 12; T cell growth factor, see Chapter 3 by Leonard *et al.;* complement receptors, see Chapter 5 by Ross, and also Wright *et al.*, 1983), or co-precipitate receptor that has been affinity-labelled with its complementary ligand (e.g. insulin receptor antibodies, see Chapter 11; EGF receptor antibody, Waterfield *et al.*, 1982; transferrin receptor antibody, Sutherland *et al.*, 1981). An alternative strategy to the latter assay is to immunoprecipitate the putative receptor and directly test the complex for ligand binding (see Chapter 6 by Fraser on β_2-adrenergic receptors), or run the immunoprecipitate on a sodium dodecyl sulfate (SDS)-polyacrylamide gel and transfer by blotting to nitrocellulose paper (i.e. 'Western' blotting). The transferred protein (e.g. acetylcholine receptor; Tzartos *et al.*, 1982) can then be probed with iodinated ligand.

One additional point concerning the precise specificity of monoclonal antibodies is important. Since most cell surface receptors are glycosylated, monoclonal antibodies raised and selected against these structures may be anti-carbohydrate. Indeed such specificities may be quite common as indicated by recent studies on antibodies raised against whole cells but selected for binding to the EGF receptor (M. Waterfield and E. Mayes, personal communication). Some of the monoclonal anti-(TSH receptor) antibodies described by L.D. Kohn and colleagues (Chapter 9) react with glycolipid or glycoprotein plus glycolipid components of the TSH receptor and are presumably anti-carbohydrate. In other situations where cells are used as immunogens for monoclonal antibody production, cell surface carbohydrates are clearly immunodominant. This has been a striking feature of monoclonal antibodies against human cancer cells (Huang *et al.*, 1983a; Hansson *et al.*, 1983; Hakomori and Kannagi 1983) and myeloid cells (Gooi *et al.*, 1983; Huang *et al.* 1983b).

Such antibodies will not identify the nascent, newly synthesized receptor precursor protein but will bind to the mature receptor molecule. The anti-carbohydrate specificity may not be of significance for some studies and indeed could even be an advantage (e.g. for elution from affinity columns) but it could also lead to problems. In particular, such antibodies may well bind to other non-receptor molecules bearing the same or similar determinants (e.g. glycolipids) which would complicate interpretation of experiments in which antibodies are used to manipulate cell function via presumed binding to specific receptor proteins.

Many studies employ, as a screening assay, inhibition of ligand binding (see chapters by Fraser, Leonard *et al.*, Kohn *et al.*) which immediately identifies antibodies binding close to or at the ligand (e.g. hormone) binding site. For other purposes several different monoclonal antibodies or even a polyclonal reagent may be desirable. Thus, as described by Schneider in Chapter 13, monoclonal antibodies against the transferrin receptor were incapable of recognizing the nascent or newly synthesized receptor polypeptide in a cell-free translation system, but a suitable reagent could be prepared by immunizing rabbits with denatured (reduced, alkylated) receptor which had been affinity-purified with a monoclonal antibody. Alternatively, monoclonal antibodies can be made or selected that are less conformation-dependent by immunizing with denatured receptor or receptor subunits (Korman *et al.*, 1982; Tzartos *et al.*, 1981, 1982; see Fuchs *et al.*, Chapter 8). These antibodies will bind to specific polysome mRNA (Korman *et al.*, 1982) and recognize primary translation products (e.g. of acetylcholine receptor subunits – Giraudat *et al.*, 1982; Merlie *et al.*, 1983; Fuchs *et al.*, Chapter 8).

1.3 ANTIBODIES AND THE MOLECULAR CLONING OF RECEPTORS

Antibodies will be invaluable in the various approaches being applied now to the molecular cloning of receptors (see Chapter 13 by Schneider). With the exception of the acetylcholine receptor in electric organs of fish (see Fuchs *et al.*, Chapter 8), almost all vertebrate cell surface receptors for hormones, growth factors, neurotransmitters etc. are present at low density with correspondingly very low-abundance mRNA. This poses a considerable problem for receptor gene cloning. Antibodies specific for particular receptors or receptor subunits in the case of heterodimeric proteins are likely to be key reagents in resolving this difficulty. In addition to their use in isolating low-abundance, specific mRNA (polysomes) by affinity chromatography for cDNA cloning (Schneider, Chapter 13), they provide the tools for receptor protein isolation and sequencing (and hence synthesis of oligonucleotide probes, e.g. acetylcholine receptor, Noda *et al.*, 1982). Antibodies can also be used as probes for screening cloned cDNA by hybrid selection (e.g. transferrin receptor, Schneider *et al.*, 1983; acetylcholine receptor, Ballivet *et al.*, 1982; Giraudat *et al.*, 1982; Merlie *et al.*, 1983) as well as for identifying the receptor protein product of genomic DNA which has either been transfected into eukaryotic cells (see Chapter 10 by Trowbridge and Newman on the transferrin receptor) or cloned in bacteria using expression vectors. Immunological dissection of receptors (i.e. positioning of epitopes) is already proving most useful in unravelling receptor subunit structure and in identifying possible phylogenetic or evolutionary relationships of different receptors and their components through the existence of shared antigens and structural homologies. Ultimately these fundamental questions will be resolved by sequence analysis; this is now becoming feasible at the gene level as illustrated by the recent sequencing of cDNA clones of the *Torpedo* acetylcholine receptor subunits, Noda *et al.*, 1983.

1.4 BIOLOGICAL ACTIVITY OF ANTIBODIES TO RECEPTORS

One of the recurring themes in this volume is the capacity of some anti-receptors to mimic the natural or physiological ligand in inducing a cellular response (see Chapter 2). This is seen, for example, with some (but not all) polyclonal antibodies against receptors for insulin (see Chapter 11 by Jacobs *et al.*), polyclonal and monoclonal antibodies to the receptor for TSH (see Chapter 9 by Kohn *et al.*), polyclonal anti-prolactin receptor antibodies (Friesen *et al.*, 1982), monoclonal antibodies against the antigen receptors on T lymphocytes (Meuer *et al.*, 1983; Kaye *et al.*, 1983), and with one particular monoclonal antibody to the EGF receptor (see Chapter 12 by Schlessinger *et al.*). The implication of such mimicry is that the natural regulator probably has no 'instructional' role on the cell surface or inside following internalization other than binding to the receptor itself. It also appears that antibodies do not have to bind close to the natural ligand-recognition site on a receptor in order to elicit, via the receptor, a cellular response. This suggests that allosteric or other ligand-induced receptor changes essential for signal transduction, in these cases at least, do not have stringent stereochemical requirements linked to the binding site of the receptor itself.

I must say that I am rather surprised that monoclonal antibodies have biological activity in some of these systems. We have come to expect that receptor aggregation induced by cross-linkage may be a normal requirement for signal generation. In contrast to polyclonal antisera, monoclonal antibodies specific for individual determinants on receptor proteins cannot effectively cross-link but can only form dimers unless, that is, the receptors are arranged as homopolymers or have repeating determinants on a single molecule. Several of the receptors discussed in this volume appear, when isolated or analysed *in situ,* to be homodimers but others are clearly heterodimers or single polypeptides. Perhaps bridging of two receptor protein molecules is indeed suffiicient to generate a membrane signal in some systems. Jacobs *et al.* (Chapter 11) draw attention to the observation that as far as insulin receptors are concerned polyclonal antisera may mimic insulin (i.e. have agonist activity) but the relatively small number of monoclonal antibodies so far tested are devoid of this biological activity.

Monoclonal antibodies raised against whole cells and selected for cell type specificity may also provide insight into hitherto undefined cell surface structures with important regulatory functions. This is best exemplified by studies on murine and human T lymphocytes and other blood cells (reviewed by Terhorst in Chapter 7, and by Springer *et al.*, 1982) which have led to the identification of a number of distinct glycoproteins involved in cell–cell interactions. In some instances, however, these structures on T cells and other hemopoietic cells, turn out to be functional receptors previously defined by other assays, e.g. antigen receptor for T cells (Meuer *et al.*, 1983), complement (C3b) receptors (see Chapter 5 by Ross, and also Beller *et al.*, 1982) and receptor for IL-2/T cell growth factor (see Chapter 3 by Leonard *et al.*). Such serendipitous availability of anti-receptor reagents is having a major impact in immunology and is also typified by the way in which monoclonal antibodies to transferrin receptors were first identified (see Chapter 10 by Trowbridge and Newman). The structures responsible for most cell–cell interaction systems (e.g. in embryogenesis, neural development) have generally proved elusive (Garrod, 1978); monoclonal antibodies may now be able to open the door (Edelman, 1983; McKay *et al.*, 1981).

Finally, studies with receptor antibodies either naturally occurring, or experimentally induced, have important implications for understanding the aetiology of some autoimmune disease, as discussed in several chapters of this volume and elsewhere (see *Ciba Foundation Symposium 90: Receptors, Antibodies and Disease,* 1982). In special circumstances there is also the intriguing possibility of therapy targeted specifically to receptors (of cancer cells, for example) via monoclonal antibodies (see Chapter 10 by Trowbridge and Newman), or directed at the immunoglobulin idiotype (of cell surface antigen receptors) of lymphocyte clones which are themselves autoanti-receptor- and disease-associated (e.g. in myasthenia gravis; see Chapter 8 by Fuchs *et al.*).

REFERENCES

Anderson, D.J. and Blobel, G. (1981), *Proc. Natl. Acad. Sci. U.S.A.*, **78,** 5598–5602.

Ballivet, M., Patrick, J., Lee, J. and Heinemann, S. (1982), *Proc. Natl. Acad. Sci. U.S.A.*, **79,** 4466–4470.

Beisiegel, U., Schneider, W.J., Goldstein, J.L., Anderson, R.G.W. and Brown, M.S. (1981), *J. Biol. Chem.*, **256,** 11923–11931.

Beller, D.I., Springer, T.A. and Schreiber, R.D. (1982), *J. Exp. Med.*, **156,** 1000–1009.

Brown, M.S., Anderson, R.G.W. and Goldstein, J.L. (1983), *Cell*, **32,** 663–667.

Cleveland, W.L., Wasserman, N.H., Sarangarajan, R., Penn, A.A. and Erlanger, B.F. (1983), *Nature (London)*, **305,** 56–57.

Cuatrecasas, P. and Roth, T. (eds) (1983), *Receptor-Mediated Endocytosis. Receptors and Recognition,* Series B, Vol. 15, Chapman and Hall, London.

Edelman, G.M. (1983), *Science*, **219,** 450–457.

Enns, C.A., Larrick, J.W., Soumalainen, H., Schroder, J. and Sussman, H.H. (1983), *J. Cell Biol.*, **97,** 579–585.

Friesen, H.G., Shiu, R.P.C., Elsholtz, H., Simpson, S. and Hughes, J. (1982), *Ciba Found. Symp.*, **90,** 263–278.

Froehner, S. (1981), *Biochemistry*, **20,** 4905–4915.

Garrod, D.R. (ed.) (1978), *Specificity of Embryological Interaction. Receptors and Recognition,* Series B, Vol. 4, Chapman and Hall, London.

Gatter, K.C., Brown, G., Trowbridge, I.S., Woolston, R.-E. and Mason, D.Y. (1983), *J. Clin. Pathol.*, **36,** 536–545.

Geuze, H.J., Slot, J.W., Strous, G.J.A.M., Lodish, H.F. and Schwartz, A.L. (1983), *Cell*, **32,** 277–287.

Giraudat, J., Devillers–Thiery, A., Auffray, C., Rougeon, F. and Changeux, J.P. (1982), *EMBO J.*, **1,** 713–717.

Goodfellow, P. and Solomon, E. (1982), in *Monoclonal Antibodies in Clinical Medicine* (J. McMichael and J.W. Fabre, eds), Academic Press, London, pp. 365–393.

Goodfellow, P.N., Banting, G., Sutherland, R., Greaves, M., Solomon, E. and Povey, S. (1982), *Somatic Cell Genet.*, **8,** 197–206.

Gooi, H.C., Thorpe, S.J., Hounsell, E.F., Rumpold, H., Kraft, D., Forster, O. and Feizi, T. (1983), *Eur. J. Immunol.*, **13,** 306–312.

Gullick, W.J. and Lindstrom, J.M. (1983), *Biochemistry*, **22,** 3312–3320.

Hakomori, S.I. and Kannagi, R. (1983), *J. Natl. Cancer Inst.*, **71,** 231–251.

Hansson, G.C., Karlsson, K.-A., Larson, G., McKibbin, J.M., Blaszczyk, M., Herlyn, M., Steplewsky, Z. and Koprowski, H. (1983), *J. Biol. Chem.*, **258,** 4091–4097.

Harford, J., Lowe, M., Tsunoo, H. and Ashwell, G. (1982), *J. Biol. Chem.*, **257,** 12 685–12 690.

Hedo, J.A., Kahn, C.R., Hayashi, M., Yamada, K.M. and Kasuga, M. (1983), *J. Biol. Chem.*, **258,** 10 020–10 026.

Hopkins, C.R. and Trowbridge, I.S. (1983), *J. Cell Biol.*, **97,** 508–521.

Huang, L.C., Brockhaus, M., Magnani, J.L., Cuttitta, F., Rosen, S., Minna, J.D. and Ginsburg, V. (1983a), *Arch. Biochem. Biophys.*, **220,** 318–320.

Huang, L.C., Civin, C.I., Magnani, J.L., Shaper, J.H. and Ginsburg, V. (1983b), *Blood*, **61,** 1020–1023.

Jacobs, S. and Cuatrecasas, P. (eds) (1981), *Membrane Receptors. Receptors and Recognition*, Series B, Vol. 11, Chapman and Hall, London.
Kaye, J., Porcelli, S., Tite, J., Jones, B. and Janeway, C.A. (1983), *J. Exp. Med.*, **158,** 836–856.
Kohler, G. and Milstein, C. (1975), *Nature (London)*, **256,** 495–497.
Korman, A.J., Knudsen, P.J., Kaufman, J.F. and Strominger, J.L. (1982), *Proc. Natl. Acad. Sci. U.S.A.*, **79,** 1844.
McKay, R., Raff, M.C. and Reichardt, L.F. (1981), *Monoclonal Antibodies Against Neural Antigens*, Cold Spring Harbor Laboratory Publication.
Merlie, J.P., Sebbane, R., Gardner, S. and Lindstrom, J. (1983), *Proc. Natl. Acad. Sci. U.S.A.*, **80,** 3845–3849.
Meuer, S.C., Hodgdon, J.C., Hussey, R.E., Protentis, J.P., Schlossman, S.F. and Reinherz, E.L. (1983), *J. Exp. Med.*, **158,** 988–993.
Noda, M., Takahashi, H., Tanabe, T., Toyosato, M., Furutani, Y., Hirose, T., Asai, M., Inayama, S., Miyata, T. and Numa, S. (1982), *Nature*, **299,** 793–797.
Noda, M., Takahashi, H., Tanabe, T., Toyosato, M., Kikyotani, S., Hirose, T., Asai, M., Takashima, H., Inayama, S., Miyata, T. and Numa, S. (1983), *Nature*, **303,** 251–255.
Sanchez-Madrid, F., Simon, P., Thompson, S. and Springer, T.A. (1983), *J. Exp. Med.*, **158,** 586–602.
Schneider, C., Kurkinen, M. and Greaves, M. (1983), *EMBO J.*, **2,** 2259–2263.
Schneider, C., Sutherland, R., Newman, R. and Greaves, M. (1982), *J. Biol. Chem.*, **257,** 8516–8522.
Schneider, W.J., Beisiegel, U., Goldstein, J.L. and Brown, M.S. (1982), *J. Biol. Chem.*, **257,** 2664–2673.
Schwartz, A.L., Marshak–Rothstein, A., Rup, D. and Lodish, H.F. (1981), *Proc. Natl. Acad. Sci. U.S.A.*, **78,** 3348–3352.
Springer, T.A., Davignon, D., Ho, M.K., Kurzinger, K., Martz, E. and Sanchez-Madrid, F. (1982), *Immunol. Rev.*, **68,** 111.
Sutherland, R., Delia, D., Schneider, C., Newman, R., Kemshead, J. and Greaves, M. (1981), *Proc. Natl. Acad. Sci. U.S.A.*, **78,** 4515–4519.
Swanson, L.W., Lindstrom, J., Tzartos, S., Schmued, L.C., O'Leary, D.D.M. and Cowan, W.M. (1983), *Proc. Natl. Acad. Sci. U.S.A.*, **80,** 4532–4536.
Tolleshaug, H., Goldstein, J.L., Schneider, W.J. and Brown, M.S. (1982), *Cell*, **30,** 715–724.
Tzartos, S.J., Rand, D.E., Einarson, B.L. and Lindstrom, J.M. (1981), *J. Biol. Chem.*, **256,** 8635–8645.
Tzartos, S.J., Seybold, M.E. and Lindstrom, J.M. (1982), *Proc. Natl. Acad. Sci. U.S.A.*, **79,** 188–192.
Waterfield, M.D., Mayes, E.L.V., Stroobant, P., Bennet, P.L.P., Young, S., Goodfellow, P.N., Banting, G.S. and Ozanne, B. (1982), *J. Cell. Biochem.*, **20,** 149–161.
Watters, D. and Maelicke, A. (1983), *Biochemistry*, **22,** 1811–1819.
Wright, S.D., Rao, P.E., Van Voorhis, W.C., Craigmyle, L.S., Iida, K., Talle, M.A., Westberg, E.F., Goldstein, G. and Silverstein, S.C. (1983), *Proc. Natl. Acad. Sci. U.S.A.*, **80,** 5699–5703.

2 Antibodies to Receptors and Idiotypes as Probes for Hormone and Neurotransmitter Receptor Structure and Function

A. DONNY STROSBERG and
ALAIN B. SCHREIBER

Editor's note on terminology

1. Idiotype: characteristic of an individual antibody. Identified as an antigenic determinant in the variable region of the antibody (immunoglobulin) molecule (i.e. in the antibody-combining site region). Anti-idiotypes are therefore anti-antibodies.
2. Anti-receptor antibody as used here and throughout this volume signifies simply an antibody reactive with a receptor and *not* an anti-(receptor) antibody (i.e. anti-idiotype).

Acknowledgements

The original research reviewed in this paper was supported by grants from the CNRS (79.7.054), INSERM (80.1017), DGRST (79.7.080), PIRMED (83.5-84), Neuromediateurs (82.7), University Paris VII ('Credits C, 1981'), Fonds de la Recherche Medicale Francaise and Association pour le Developpement de la Recherche sur le Cancer. Special recognition is given to Drs C. Delavier, O. Durieu, P.-O. Couraud, J. Hoebeke, B.-Z. Lu and A. Schmutz for their invaluable collaboration, to R. Schwartzman for photographic and artistic assistance, and to Mrs M. Kindt for careful and patient editing.

Monoclonal Antibodies to Receptors: Probes for Receptor Structure and Function
(*Receptors and Recognition*, Series B, Volume 17)
Edited by M. F. Greaves
Published in 1984 by Chapman and Hall, 11 New Fetter Lane, London EC4P 4EE

2.1 INTRODUCTION: ANTIBODIES AS TOOLS FOR THE ISOLATION OF RECEPTORS AND PROBES FOR THEIR FUNCTION

Antibodies have emerged as tools for the study of structure and function of hormone and neurotransmitter receptors. In principle, antibodies to hormone receptors may be used to isolate the receptor either by large-scale immunoprecipitation or by antibody-containing affinity chromatography gels (see Chapter 13 by Schneider). In dynamic studies, the biosynthesis, membrane insertion and up- or down-regulation of the receptors may be rigorously analyzed in the presence or absence of the hormone. Polysomes of the translated products of the receptor's mRNA may be isolated by specific immunoprecipitation, opening the way to the cloning of the genes of the hormone receptors (see Chapter 13 by Schneider). Labeled with fluorophores, radioisotopes or coupled to enzymes, the antibodies may be used to localize visually the receptor and its distribution on cells or tissues. Using the antibody as an alternative ligand or in conjunction with the hormone, the valence of the receptor may be determined *in situ,* and important mechanistic questions concerning the nature of the transmembrane signal may be addressed. Finally antigenic mapping of the receptor molecule with the antibody may allow phylogenetic, ontogenetic and tissue-related structural studies (Fig. 2.1).

Using receptor-bearing cells or membranes, or alternatively affinity-purified receptors as immunogens, antibodies have been raised either as polyclonal and polyspecific reagents or, alternatively, made monospecific by the preparation of clones of hybrid cells each producing a monospecific, homogeneous anti-receptor antibody. Anti-receptor antibodies have also been isolated from patients with rare autoimmune diseases. Invariably these latter antibodies have been found to interfere with the normal function of the receptor.

Anti-receptor antibodies may also be raised as anti-idiotypes. By this method, one first prepares anti-hormone antibodies which are used as antigens in order to obtain anti-idiotypic antibodies, directed against the variable region of the original antibodies. The anti-idiotypes either recognize similar structures on the combining site of anti-hormone antibodies and hormone receptors or are complementary to the active site for the hormone, and thus mimic the structure of the small ligand, literally constituting an 'internal image' of the hormone (Strosberg, 1983; Strosberg *et al.*, 1981).

The anti-idiotypic antibodies effectively interact with the receptor and may either block or mimic hormone action on the effector systems.

In this chapter we will discuss and compare the three types of anti-receptor antibodies and outline their various uses.

For most of the purposes cited above, the use of conventionally raised polyclonal immune sera or fortuitously uncovered autoantibodies is hampered by their natural heterogeneity and their limited amounts. The antibodies may be directed against each of the various antigenic determinants of the receptor

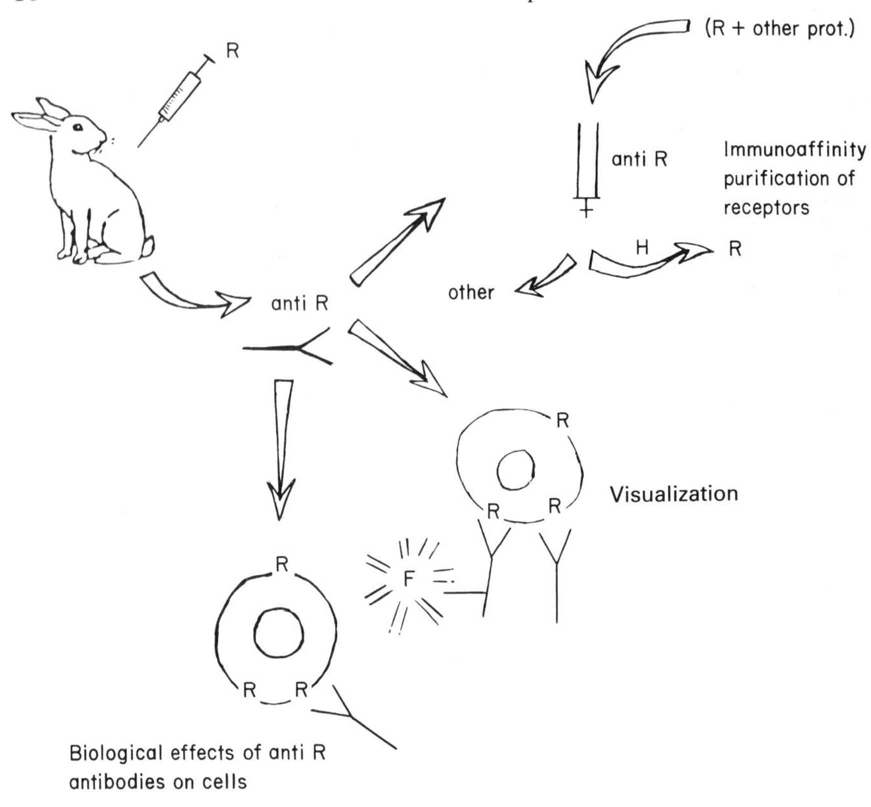

Fig. 2.1 Purified receptor (R) is injected into a rabbit and the immune response is monitored by specific screening methods. Anti R antibodies may then either be used directly on receptor tissues or cells for visualization or biological signalling, or, insolubilized, may be used for receptor purification.

or neighboring entities. Their overall effect may actually reflect the summation of a variety of biologic effects mediated by multiple antigenic sites on the receptor. If the hormone receptor cannot be isolated by independent means to purify specific antibodies, the anti-receptor antisera will be of little use in the final structural analysis of the receptor molecule.

The way to overcome these difficulties is clearly the use of somatic cell hybridization technology in which large amounts of monospecific, monoclonal antibodies are generated (see recent reviews by Goding, 1980; Galfre and Milstein, 1981) (Fig. 2.2).

Some words of caution about unexpected properties of monoclonal antibodies in general, and in our experience in particular, may be in order. Although the technicalities of hybridoma long-term culture and cloning are now routine to most laboratories, the screening and 'quality control' of monoclonal antibodies are still not trivial procedures. Monoclonal antibodies by their homogeneity may be unusually sensitive to storage, changes in pH or salt concentrations and even mild protein derivatization may sometimes result in significant antibody denaturation. Furthermore, the qualitative type of screening assay employed to select a monoclonal antibody may favor some predetermined properties: an antibody selected in a functional assay may only recognize the receptor in its native form on the cell surface and give only poor yields in immunopurification of solubilized receptor; an antibody selective for immunopurification of the receptor may be biologically totally inactive. Several experimental strategies dependent on the starting immunogen may be contemplated for the selection of monoclonal antibodies to cell surface hormone receptors.

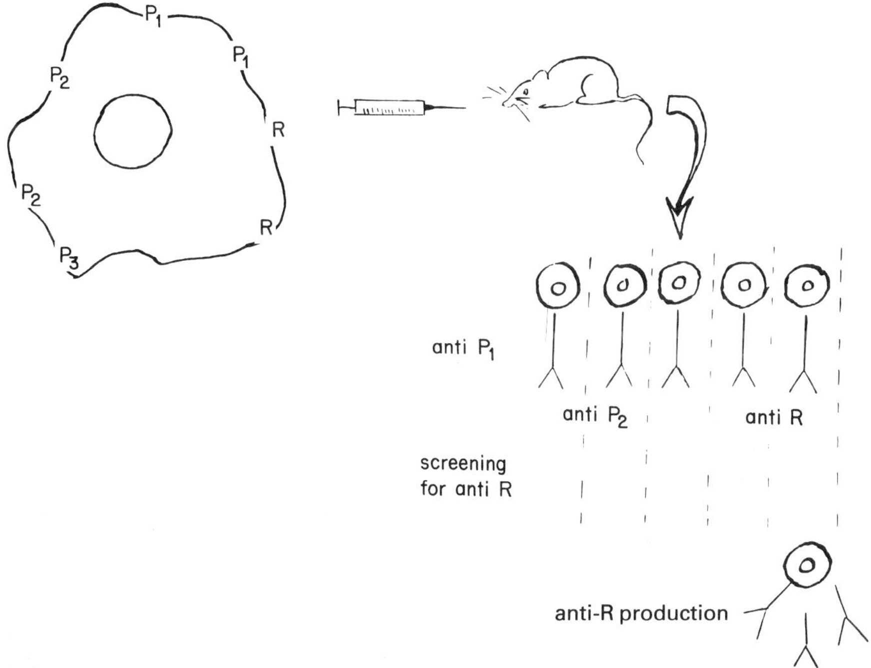

Fig. 2.2 Preparation of monoclonal anti receptor (anti R) antibodies. Whole cells are injected into mice or rats and the immune response is screened. Spleen cells are isolated and fused with myeloma cells. Clones producing anti R antibodies are selected by monitoring the activity of the cell supernatants, these cells are subsequently amplified while clones producing antibodies (anti P1, anti P2, etc.) to other cell surface proteins are discarded.

2.2 ANTIBODIES RAISED EXPERIMENTALLY AGAINST PURIFIED RECEPTORS, OR MEMBRANES OR CELLS RICH IN RECEPTORS

2.2.1 Selection of immunogen and possible specificities produced

Purified receptor can be used as an immunogen if available in sufficient amounts (e.g. as for the nicotinic acetylcholine receptor; Moshly-Rosen *et al.*,

1979; Gomez *et al.*, 1979; Tzartos and Lindstrom, 1980; James *et al.*, 1980; see also Chapter 8 by Fuchs *et al.*); this approach is to be preferred. Potential drawbacks include the possible loss of the native conformation and subcomponents of the receptor moiety. Furthermore antibodies will be generated in order of frequency according to the immunogenic potency of domains of the isolated receptor; these do not necessarily overlap with the functionally important domains *in situ*.

Purified receptor may be obtained by using various forms of affinity chromatography combined with gel filtration and ion-exchange or electrophoretic separation methods. Very small amounts of protein, often denatured by solubilization procedures involving detergents will be used to raise antibodies. These will be directed against every part of the unfolded receptor, including those buried in the membrane. These antibodies will be most useful to compare the evolution of receptors purified from various origins, to identify species-specific or organ-specific structures. They also provide invaluable tools for large-scale purification of receptor by immunoaffinity chromatography, using agarose-coupled antibodies.

Cells or cell membrane bearing a high amount of receptor as immunogen can alternatively be used to raise antibodies. This approach presents the advantage that one does not need to purify the receptor. Assuming the availability of such a cellular source, each fusion will yield a library of hybridomas secreting antibodies to a great variety of cell surface antigens. With progress in the development of new screening assays one may in principle 'go back' to this collection of monoclonals and select for antigens either functionally related to the hormone receptor or other functional domains on the receptor molecule. The obvious drawback of this approach is that several selection assays are required; these will have to include cellular binding assays on different cell types and eventually immunoprecipitation of the receptor from solubilized cells or membranes to ensure the specificity of the antibody.

Antibodies raised by injection of intact cells or membranes with receptors at their surface will mostly be directed against the parts of the receptor exposed to the solvent whether extra- or intra-cellular. The number and the nature of the antigenic determinants 'seen' by the antibodies depend on how much of the receptor protein is buried in the hydrophobic portion of the membrane. Often, the 'hidden' part is quite modest, and most of the transmembrane proteins will be immunogenic. However, a number of intrinsic proteins appear to be mostly intramembranous and will display very little to the extracellular space: cytochrome oxydase, the 'receptor' for cytochrome *b* is a good example.

Antibodies directed against the outer parts of the membrane receptors will not only be specific for the protein; the carbohydrate portion of glycoproteins appears to be quite immunogenic, and anti-carbohydrate antibodies have often been raised as anti-receptor antibodies. These reagents have, however, limited usefulness since they do not discriminate between numerous glycoproteins sharing sugar moieties.

Antibodies directed against the ligand-binding site will only constitute a minor subfraction among the various proteins directed against other parts of the receptor. The only way of identification is the ability to compete with and inhibit ligand binding.

2.2.2 Specificity of some experimentally produced receptor antibodies

Since other chapters in this volume provide detailed analyses of antibodies to several different receptors only a brief review of the specificity of some of the reported data in the field will be reviewed here (Table 2.1).

(a) Antibodies to nicotinic and muscarinic receptors for acetylcholine

Polyclonal antibodies elicited in animals against purified nicotinic acetylcholine receptor from the electric organs of electric fish inhibit the physiological activity of the receptor as well as the binding of cholinergic ligands to the receptor (Patrick *et al.*, 1973; Green *et al.*, 1975; see Chapter 8 by Fuchs *et al.*). In these reports, the inhibition of ligand binding was attributed to steric hindrance. In these respects the experimentally induced antibodies have similar properties to the autoantibodies of patients with myasthenia gravis.

Several groups (Moshly-Rosen *et al.*, 1979; Gomez *et al.*, 1979; Tzartos and Lindstrom, 1980; James *et al.*, 1980; Lennon *et al.*, 1980; see Chapter 8 by Fuchs *et al.*) have obtained a large repertoire of monoclonal antibodies from mice immunized with purified acetylcholine receptor from *Torpedo* fish. Specific antibody-producing lines were selected by radioimmunoassay with radiolabeled purified receptor. The fine specificity of the antibodies was investigated by assaying the binding to Triton-solubilized, membranous, trypsinated or denatured receptor.

The antibodies were used to purify receptor in large quantities by immunoaffinity chromatography (Lennon *et al.*, 1980) and to map the antigenic determinants on the receptor molecule (Gullik *et al.*, 1981; Conti-Tranconi *et al.*, 1981). In one study (Moshly-Rosen and Fuchs, 1981) only one out of 32 antibodies was directed against the cholinergic-binding site of the receptor; neurotoxins and cholinergic ligands competed for binding of the antibody-displaced cholinergic ligands in accordance with their respective affinities to the receptor. In other studies (Gomez *et al.*, 1979; James *et al.*, 1980) only two out of several hundred antibodies were directed to the cholinergic-binding site. These antibodies bound to the acetylcholine receptors from various species and organs, demonstrating a wide structural homology between cholinergic sites of receptors. From these studies it emerges that the cholinergic-binding site only represents a minor immunogenic specificity. Only through the monoclonal antibody technique could this minor specificity be augmented and analysed separately. It would thus appear that either antibodies which affect the turnover and/or presence of the receptor or antibodies with several defined

Table 2.1 Effects of antibodies on various receptor systems

Hormone receptor	Type of antibody	Biochemical effect	Biological effect	Disease	References
Insulin	Polyclonal from patient	Glucose uptake	Receptor disappearance	Acanthosis nigricans; insulin dependent diabetes	Grunfeld *et al.* (1980) Maron *et al.* (1983)
	Monoclonal	Inhibition kinase activation			Roth *et al.* (1982)
Thyrotropin	Polyclonal	Adenylate cyclase activation	Thyroid cell growth	Autoimmune thyroid disease (Graves,	Rees-Smith and Hall (1981)
	Monoclonal	Adenylate cylase activation	Thyroid cell growth	(Hashimoto)	Valente *et al.* (1982)
Acetylcholine (nicotinic)	Polyclonal	Cation channel opening inhibition	Receptor disappearance	Myasthenia gravis	Lindstrom *et al.* (1976)
	Monoclonal	Cation channel opening inhibition			Tzartos *et al.* (1982)
Acetylcholine (muscarinic)	Monoclonal	Guanylate cyclase activation Adenylate cylase inhibition	Contraction of myometrium		Andre *et al.*
Catecholamine (β-adrenergic)	Polyclonal, monoclonal	Adenylate cyclase activation		Asthma, Chagas (?)	Couraud *et al.* (1981) Venter *et al.* (1980) Strosberg *et al.* (1981)
Epidermal growth factor	Polyclonal, monoclonal	Protein kinase activation	Receptor redistribution and internalization		Schreiber *et al.* (1983) Waterfield *et al.* (1983)
LDL	Polyclonal, monoclonal	LDL uptake	Cholesterol metabolism	Familial hyper-cholesterolemia	Goldstein and Brown (1983)
Prolactin	Polyclonal	Casein gene activation	Growth of mammary tissue		Djiane *et al.* (1981)

specificities together may be more relevant to the pathogenic mechanisms of the disease state in myasthenia gravis.

Two different monoclonal antibodies were raised against the *muscarinic* receptor from calf forebrain and purified by affinity chromatography on dexitimide–Sepharose (Andre *et al.*, 1983, 1984). While both antibodies recognized muscarinic receptors on membranes from calf, rat and human brain, only one partially inhibited the binding of ligands such as dexitimide or atropine. This antibody immunoprecipitated ligand-binding activity present in solubilized membranes but did not identify SDS-denatured receptor in Western-blot electrophoretic analyses. In contrast, the other antibody was unable to inhibit ligand binding or precipitate active receptor, but strongly reacted with the denatured receptor in the Western blot.

Recognition of the active form of the receptor by the first antibody was correlated with an agonist-like activity: on guinea pig uterus, the antibody acted as carbamoylcholine; see Note 1, added in proof, at the end of this chapter.

(b) Antibodies directed against β-adrenergic catecholamine receptors

Only a few groups have described the preparation of antibodies directed against β-adrenergic catecholamine receptors. Couraud *et al.* (1981) have immunized mice with turkey erythrocyte β_1-adrenergic receptor purified by affinity chromatography. The resulting polyclonal antibodies were shown by immunofluorescence to stain β_1- or β_2-adrenergic receptors of a variety of cells and to immunoprecipitate a major component of the receptor. Inhibition by the antibodies of catecholamine binding to the cell-bound β-adrenergic receptors was only partial, suggesting that the antibodies and the specific ligands interact with these membrane proteins at different sites. This may explain why the anti-receptor antibodies stimulated neither basal nor catecholamine-sensitive adenylate cyclase. The synergistic effect is not observed between the antibodies and guanyl trinucleotides.

Monoclonal antibodies to β-adrenergic receptors have been raised by two groups (see Chapter 6 by Fraser). Fraser and Venter (1980) have described four monoclonal antibodies directed against the turkey erythrocyte β_1-adrenergic receptor and one antibody to the calf lung β_2-receptor. The anti-turkey erythrocyte receptor antibodies precipitated partially purified β-receptors and inhibited adrenergic ligand binding. One of these antibodies cross-reacted equally with calf liver and lung β_2-receptors as well as calf heart β_1-receptors, again suggesting the existence of some molecular homology between β-adrenergic receptors of substantially diverse pharmacological classes. The monoclonal antibodies were used in the final stage of turkey erythrocyte β_1-receptor purification.

In another study (Couraud *et al.*, 1983a), several monoclonal antibodies were raised against affinity-purified turkey erythrocyte membrane β_1-receptors. While these antibodies specifically stained and immunoprecipitated β-receptor, they were unable to inhibit ligand binding.

(c) *Antibodies to the insulin receptor (see Chapter 11 by Jacobs et al. for detailed discussion)*

Both the polyclonal (Jacobs *et al.*, 1978) and the monoclonal (Kull *et al.*, 1982) antibodies obtained by immunization with purified receptor immunoprecipitated radiolabeled insulin–receptor complexes as well as unoccupied receptor without interfering with the binding or biologic activity of insulin. Whereas the polyclonal antibodies showed little tissue or species specificity (Jacobs *et al.*, 1978), the monoclonal antibody specifically recognized the insulin receptor from human placenta and lymphocytes and to a lesser extent from human erythrocytes (Kull *et al.*, 1982). Roth *et al.* (1982) raised a monoclonal anti-insulin receptor by immunization with IM-9 human lymphocytes which have large numbers of these receptors. This monoclonal antibody blocked insulin binding by more than 90% to IM-9 cells, to human adipocytes and to placenta cells, but did not inhibit binding to rat adipocytes or liver cells, suggesting that this monoclonal antibody was also species specific. The antibody inhibited 2-deoxyglucose uptake by 60% by human adipocytes and amino acid uptake by fibroblasts to the same degree. This monoclonal antibody therefore behaves as an antagonist of insulin action. Antibodies obtained up to now thus clearly differ qualitatively from the autoantibodies which lead to a disease state of diabetes (*vide infra*). One might speculate that autoantibodies to the insulin receptor of the same type as the experimentally induced antibodies may remain unnoticed as long as they do not cause a pathologic disturbance of insulin homeostasis.

(d) *Antibodies to the EGF receptor (see Chapter 12 by Schlessinger et al. for detailed discussion of this topic)*

Epidermal growth factor (EGF) is a single-chain polypeptide hormone which can be isolated from the submaxillary glands of mice and from human urine. EGF binds to specific plasma membrane receptors on a variety of target cells and initiates and maintains a complex programme of biochemical events culminating in the stimulation of cell division both *in vitro* and *in vivo*. After binding of EGF to its receptors which initially laterally diffuse and rotate in the plasma membrane, the hormone–receptor complexes aggregate into small clusters, collect over coated pits, become internalized and eventually fuse with lysosomes where the hormone and presumably also the receptor molecules are degraded (see reviews by Carpenter and Cohen, 1981; Schlessinger *et al.*, 1982). It is not known when and how EGF generates biochemical signal(s) which correlate with this binding phenomenology. Up to now no disease state has been linked to a deficiency of either EGF homeostasis or the EGF receptor system. See Note 2, added in proof.

Polyclonal antibodies to the EGF receptor were obtained in rabbits after immunization with a membrane preparation of A431 cells (Haigler and Carpenter, 1980); the A431 cell line bears an unusually high number of EGF receptors (2×10^6) compared to other cells ($2\times10^4-2\times10^5$/cells). The heterogeneous antibodies inhibited the binding of EGF to its receptors as well as EGF-mediated early and delayed biologic effects.

At the time when studies were initiated to obtain monoclonal antibodies to the EGF receptor, purified receptor was not available in amounts sufficient for a full immunization schedule. Schreiber *et al.* (1981, 1983) and Waterfield *et al.* (1983) therefore used intact A431 cells as immunogen. Immunoglobulins secreted by the hybridomas were first screened for specific binding to A431 cells and several lines of fibroblasts bearing EGF receptors and lack of reactivity with several lymphoid cell types devoid of EGF receptors. Antibodies were then selected according to the following criteria: (1) Inhibition of binding of radiolabeled EGF to its receptors on cells and membrane preparations; (2) immunoprecipitation of EGF receptor, either from biosynthetically labeled cells or associated with a tyrosyl-specific protein kinase activity, previously shown to be an integral part of the receptor moiety (Buhrow *et al.*, 1982; Cohen *et al.*, 1982); and (3) immunoprecipitation of a complex of radiolabeled EGF cross-linked to the receptor.

Several monoclonal antibodies were generated for which the specificity was established by immunoprecipitation of the EGF receptor (Schreiber *et al.*, 1981; Waterfield *et al.*, 1983). One monoclonal antibody of the IgM class competed for the binding of the hormone to its receptors and induced the aggregation and internalization of the receptor molecules. The antibody also possessed intrinsic bioactivity and mediated all the early and delayed EGF-mediated biologic effects *in vitro* (see Chapter 12 by Schlessinger *et al.*). A monovalent Fab′ fragment of this antibody competed with EGF binding, stimulated EGF-specific biologic effects but failed to induce receptor aggregation and mitogenesis. When cell-bound Fab′ fragment of this antibody competed with EGF binding, it stimulated membrane-related EGF biologic effects but failed to induce receptor aggregation and mitogenesis. When cell-bound Fab′ fragments of the antibody were artificially cross-linked with a second layer of anti-Fab′ antibodies, both receptor clustering and mitogenic activity were restored (Schreiber *et al.*, 1983). Another monoclonal antibody to EGF receptor of the IgG class appeared to be an excellent high-affinity ($K_{app.}$ $10^9\ \mathrm{M}^{-1}$) immunoprecipitation tool but did not interfere with EGF binding, nor did it possess any intrinsic bioactivity. The cross-linking of receptors by this antibody on the cell surface leads, however, to a sufficient biologic signal to trigger mitogenesis.

In the case of EGF receptor the use of monoclonal antibodies as alternative ligands to the hormone has thus provided valuable information concerning the mode of action of the receptor system: (1) The receptor, when properly triggered, contains all the biochemical attributes necessary for the initiation of biologic effects. The hormone molecule and its internalizing and degradation do not appear necessary. (2) Receptor aggregation (and subsequent internalization), even when mediated via some domains distinct from the hormone-binding site on the receptor molecule, appears as a necessary and sufficient signal for the triggering of mitogenesis in this system.

(e) Antibodies to the prolactin receptor

Prolactin interacts with specific membrane receptors that respond both to a down- and up-regulation by the hormone.

Prolactin controls the development and activity of the mammary gland; it induces casein synthesis by activation of casein gene transcription and the enhancement of casein mRNA stability (see review by Houdebine, 1980). The mechanisms by which prolactin transfers its biologic information from the membrane receptors to the genome are unknown.

Polyclonal antibodies to the prolactin receptor were first prepared by Shiu and Friesen (1978) by immunization with partially purified receptor. This antibody preparation was shown to block both prolactin binding and its biologic activity both *in vitro* (Shiu and Friesen, 1978) and *in vivo* (Bohnet *et al.*, 1978). Another antibody preparation, obtained in response to the same immunogen, possessed an intrinsic prolactin-like activity (Djiane *et al.*, 1981). At low concentrations the anti-receptor antibodies were capable of mimicking prolactin action on casein gene expression and on the stimulation of DNA synthesis in mammary tissue; at higher concentrations the antibodies inhibited their own actions. The stimulatory effect of the antibodies was also shown to be modulated by the same agents (e.g. glucocorticoids) which affect the action of prolactin (Teyssot *et al.*, 1982). As in the case of the anti-EGF receptor antibodies which mimic the action of EGF, it would appear that also for the prolactin–receptor system, the hormone molecule is not strictly necessary for the transfer of biologic information beyond the binding step to the receptor. Several reservations have to be made for the polyclonal anti-prolactin receptor) antibodies; indeed some antibodies directed against impurities of the receptor preparations might induce the observed biological effects, while the antibodies truly binding to the receptor molecule may be biologically inactive. Moreover, the observed biologic activity might be the result of the synergistic interaction of several antibodies with different antigenic determinants on the receptor. This system will most certainly benefit from the monoclonal antibody technique for the fine analysis of hormone transmembrane signaling (Shiu and Friesen, 1978).

2.3 ANTIBODIES TO HORMONE AND NEUROTRANSMITTER RECEPTORS IN AUTOIMMUNE DISEASES

The probable causal significance of autoantibodies to receptors in various disease states is attracting increasing attention (see Harrison (1984) for recent review). Details of some of these antibodies are given in individual chapters in this book. Here we review and interpret some of these observations especially from the viewpoint of the specificity and biological effect of antibodies.

The clinical incidence of autoantibodies to hormone receptors varies considerably. The type B syndrome of insulin-resistant diabetes and acanthosis nigricans with anti-insulin receptor antibodies has been documented for less than 50 cases to date. See Note 3, added in proof. Graves' disease (antibodies to the TSH receptor) and myasthenia gravis (antibodies to the nicotinic acetylcholine receptor) are moderately prevalent, while asthma and other atopic disorders (antibodies to the β_2-adrenergic receptor?) are very common.

2.3.1 Specificity and biological effect of human autoantibodies to receptors

(a) Autoantibodies to the insulin receptor

Autoantibodies to the insulin receptor were first discovered during evaluation of diabetic patients with tenacious insulin resistance. Some of these patients had evidence of a systemic immune disease characterized by elevated serum globulins, anti-nuclear and anti-DNA antibodies (Flier *et al.*, 1975). A type A syndrome of insulin resistance was found to be caused by a genetic defect resulting in a decrease of available cell-surface insulin receptors (Kahn *et al.*, 1976). A type B syndrome of insulin resistance was shown to be associated with circulating antibodies, mostly of the IgG class, which reacted with insulin receptors from various tissues, and also across species lines. The anti-insulin receptor antibodies were shown to compete for insulin binding to cell surface receptors; moreover, insulin competed with the binding of the antibodies. The properties of the antibodies varied from patient to patient: in some sera the affinity of the receptors for insulin binding was primarily altered while other sera decreased the apparent number of available binding sites. Recently, the specificity of the autoantibodies has been unequivocally established by immunoprecipitation of biosynthetically labeled insulin receptor from various cell surfaces (Kasuga *et al.*, 1981; Van Obberghen and Kahn, 1981; Van Obberghen *et al.*, 1981). *In vitro,* the autoantibodies behave as full agonists of insulin, they mimic both membrane-related and intracellular effects mediated physiologically by insulin virtually with equimolar potency as compared to the hormone (see reviews by Harrison and Kahn, 1980; Kahn *et al.*, 1981). It is noteworthy that the bioactivity of the autoantibodies is not in a simple relationship to their inhibitory potency for insulin binding. DEAE fractionation of the autoimmune sera will yield very different antibody populations in terms of binding and agonistic properties, reflecting the heterogeneity of antigenic determinants recognized by the antibodies.

Kahn *et al.* (1978) have shown that divalence is essential to the intrinsic bioactivity of the autoantibodies. Monovalent Fab′ and divalent F(ab′)$_2$ fragments of the autoantibodies to the insulin receptor are fully active in inhibiting the binding of insulin to its receptors. Divalent F(ab′)$_2$ also retains the insulin-like potency. In contrast, the monovalent Fab′ is a competitive antagonist of insulin. Cross-linking the Fab′–receptor complexes at the cell surface with a second layer of anti-Fab′ antibodies results in the restoration of the insulin-like biologic activity.

Autoantibodies to the insulin receptor induce the same topographical rearrangement of the receptor molecules as is observed after insulin binding (Schlessinger *et al.*, 1978; see Chapter 12): the receptors aggregate, collect in coated pits, and become internalized. This phenomenology may explain the apparent paradox of agonistic antibodies and a clinical syndrome of qualitative insulin deficiency. Indeed Grunfeld *et al.* (1980) showed that long-term incubation of the antibodies with cells caused desensitization towards insulin on subsequent incubation. See Note 4, added in proof.

(b) Autoantibodies to the TSH receptor

Autoimmune thyroid disease consists of a spectrum of clinical disorders extending from thyrotoxicosis through Hashimoto's thyroiditis to atrophic hypothyroidism. In Graves' disease, characterized by hyperthyroidism with diffuse goiter, an opthalmopathy and a dermopathy, the presence of serum antibodies which disrupt TSH homeostasis was long recognized (Adams and Kennedy, 1967; Drexhage *et al.*, 1981). It was further demonstrated that these autoantibodies reacted with a thyroid component identical, or closely associated, with the TSH receptor (Rees-Smith and Hall, 1974; Claque *et al.*, 1976). The autoantibodies, were shown to inhibit TSH binding to its receptors (see Chapter 9 by Kohn *et al.*) with interpatient variability with respect to decrease of TSH-binding sites and/or change in affinity and species specificity (Endo *et al.*, 1982). Solubilized TSH receptors may be immunoprecipitated with the autoantibodies. Ternary complexes of labeled TSH, TSH receptor and antibody cannot be formed, indicating that the antibodies bind to a domain of the receptor similar to or identical with the binding site for the hormone (Rees-Smith and Hall, 1981). In most cases the autoantibodies will stimulate the TSH-specific adenylate cyclase-mediated effect *in vitro* (Zakarija *et al.*, 1980). Some antibodies, however, are not intrinsically stimulatory while in a few cases the antibodies appeared to trigger TSH-like anabolic effects rather than cyclic AMP (cAMP) linked biologic effects.

Many features of a diffuse toxic goiter can be reproduced in the thyroid of animals after passive transfer of the autoantibodies from patients with Graves' disease. However, titers of autoantibodies in patients' sera do not correlate either with the presence or absence of thyrotoxicosis nor with its degree of severity. Moreover, significant titers of autoantibodies with demonstrable stimulatory activity *in vitro* may be associated with a clinically normal thyroid function. The association of Graves' disease with other autoimmune disorders as for example systemic lupus erythematosus, Sjogren's syndrome and also Hashimoto's disease has led to the general belief that the autoantibodies to the TSH receptor represent an epiphenomenon amplifying a causal factor with genetic susceptibility that perturbs both humoral and cell-mediated immunity (Strakosch *et al.*, 1978).

(c) Autoantibodies to the nicotinic cholinergic receptors

Autoantibodies to the nicotinic cholinergic receptors are found in the sera of most patients with myasthenia gravis (Almon *et al.*, 1974; Abramsky *et al.*, 1975; see also Chapter 8 by Fuchs *et al.*). The disease is characterized by a functional abnormality of the neuromuscular junction which leads to muscle weakness and enhanced fatiguability (Farmbrough *et al.*, 1973). Autoantibodies to the acetylcholine receptor block the binding of ligands to the receptor molecule; this competition is however not mutual as cholinergic ligands will not interfere with the binding of the antibodies (Lindstrom, 1976). It was suggested that the antibodies interfere with ligand binding by steric hindrance, rather than by occupying the cholinergic-binding site on the receptor (Aharonov *et al.*, 1977). The inhibition of ligand binding by the antibodies also results in the blocking of cholinergic-induced stimulation of ion fluxes. The specificity of the autoantibodies was confirmed both by immunoprecipitation of the acetylcholine receptor from various sources and direct binding assays to purified receptor (Appel *et al.*, 1977). The population of autoantibodies is heterogeneous and polyclonal; antibody binding appears to be restricted to certain regions of the receptor molecule, but the pattern of reactivity differs between patients.

Experimental autoimmune myasthenia gravis can be induced by immunization of animals with purified acetylcholine receptor (Patrick and Lindstrom, 1973; Patrick *et al.*, 1973) or by passive transfer of anti-receptor antibodies (Toyka *et al.*, 1977). (See review by Fuchs, 1980.) The mechanism by which the autoantibodies may lead to the pathogenesis of myasthenia gravis is not clear. Inhibition of ligand binding and blocking of function do not appear to be sufficient since there is no good correlation between the clinical state of patients and antibody titer or antibody activity *in vitro*. Increased receptor internalization and lysosomal destruction were demonstrated ultrastructurally, but appeared to be balanced by a concomitant increase in receptor biosynthesis (Kao and Drachman, 1977; Drachman *et al.*, 1978). Complement-dependent lysis (Rash *et al.*, 1976) as well as activation of cell-mediated immunity have also been envisaged (Fuchs *et al.*, 1980). Recently Drachman *et al.* (1982) reported that the quantitative extent of both increased rate of receptor degradation and blockade of ligand binding correspond best to the clinical severity of the disease.

(d) Autoantibodies to the catecholamine β-adrenergic receptor

In asthma a reduction in lung tissue β_2-adrenergic receptors as measured by agonist binding or β_2 physiological responses has long been established. Clinically, asthma patients are characterized by a hyporesponsiveness to β_2-adrenergic drugs. Most recently autoantibodies to the β_2-adrenergic receptor were detected in the serum of a small number of patients with asthma or other atopic disorders (Fraser and Venter, 1980). The autoantibodies were identified by inhibition of β_2 agonists and immunoprecipitation of solubilized receptors. In patients, antibody titers appear to correlate with clinical β_2-adrenergic hyporesponsiveness. As β_2-adrenergic receptors modulate IgE-mediated release, a contributory role of the autoantibodies to the etiopathogeny of asthma seems an attractive rationale. The clinical relevance of the autoantibodies may at this point however be questioned: too few patients have been investigated and equal numbers of apparently healthy individuals also possess circulating antibodies to the β_2-adrenergic receptor.

In Chagas' disease a large proportion of patients develop cardiovascular deficiencies which have been explained by the appearance of autoantibodies against the cardiac tissue. These antibodies have been extensively studied and some have indeed been shown to inhibit the binding of catecholamine antagonists or agonists to the cardiac β-adrenergic receptors (Khoury *et al.*, 1978). More studies are needed, however, to determine the origin of these autoanti-catecholamine receptor antibodies.

It is worth mentioning that in both diseases, asthma and Chagas, the identification of the autoantibodies to the β_2-adrenergic receptor was the result of a deliberate search to fit clinical and pharmacological observations together, combined with the application of advances in receptor biochemistry techniques.

2.3.2 Human or human/mouse hybridomas producing autoantibodies to receptors

Immortalization of lymphocytes of patients with autoimmune disease holds great potential for the elucidation of the role of antibodies with defined specificity in the pathogenesis of autoimmune diseases and presents the advantage that the receptor as immunogen needs not to be purified. Valente *et al.* (1982) have obtained human monoclonal antibodies by fusing mouse myeloma cells with peripheral lymphocytes from patients with active Graves' disease (see Chapter 9 by Kohn *et al.*). The problems encountered with human inter- and intra-species lymphocyte fusion have probably delayed the application of this approach. The advent of appropriate cell lines as fusion partners (Croce *et al.*, 1980; Chiorazzi *et al.*, 1982) and alternative immortalization procedures such as Epstein Barr virus infection (Steinitz and Klein, 1981) will probably soon allow successful generation of patients' monoclonal anti-receptor antibodies.

2.4 ANTI-IDIOTYPIC ANTIBODIES WHICH RECOGNIZE RECEPTORS

2.4.1 Antibodies to hormones as immunogens for the induction of anti-idiotypic antibodies

Antibodies to hormones which share with hormone receptors binding properties for the same ligands may also serve as immunogens to raise antibodies. If the animals used for immunization are genetically similar to those from which the anti-hormone antibodies were taken, it is then likely that the anti-antibodies will be exclusively directed against the variable part of the anti-hormone immunoglobulins. Such anti-variable region antibodies have been defined as 'anti-idiotypic' (Fig. 2.3).

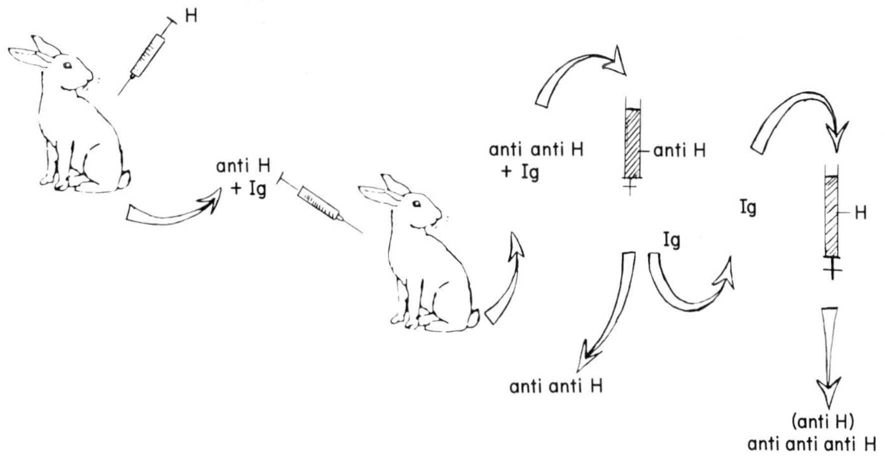

Fig. 2.3 Diagram describing the production and purification of antibodies directed against hormone and against anti-hormone antibodies. H represents the hormone coupled on a carrier protein; anti H is the antibody against the hormone; Ig represents immunoglobulin of other specificities; anti anti H is the anti-idiotypic antibody which is retained on an anti-hormone immunoadsorbent which allows the separation from other Ig. If this Ig is then applied to a hormone immunoabsorbent column, anti-hormone antibody and anti anti anti hormone antibody can be purified.

The anti-idiotypic antibodies interact with the anti-hormone antibodies through their respective variable regions, and in some cases, exclusively through the combining sites. This interaction may resemble the binding of the antibody to the hormone. Thus, Jerne, has suggested in his immune network theory that some anti-idiotypic antibodies bind to the anti-hormone immunoglobulins by mimicking the hormone, thus behaving as an 'internal image' of the initial antigen (Jerne, 1973) (Fig. 2.4) or 'homobodies' (Lindenmann, 1973).

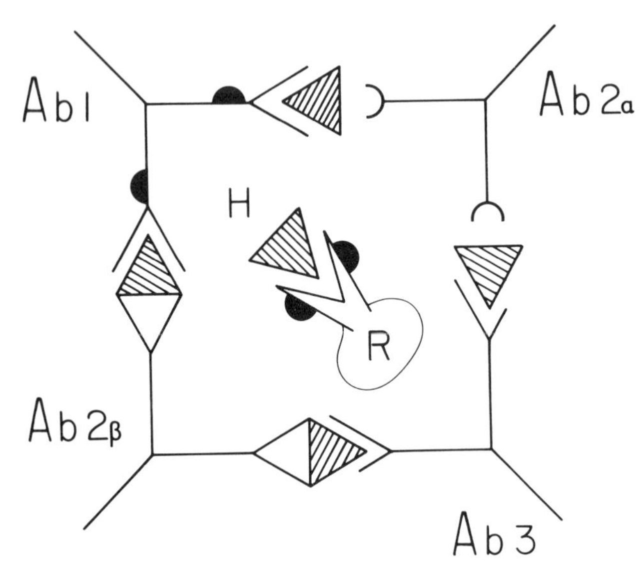

Fig. 2.4 Interacting components of an idiotype–anti-idiotype network involving hormones (H), receptors (R), and antibodies (Ab1, Ab2, Ab3). In this representation, Ab1 antibodies specific for hormone H display binding sites which are similar to those found on the hormone receptor R. Anti-idiotypic Ab2 antibodies recognize determinants on Ab1 and on R which are either adjacent to the hormone combining site (Ab2α) or are complementary to this site (Ab2β). Antibodies directed against the second kind of Ab2 molecules will be able to bind hormone. (From Strosberg, 1983.)

Such antibodies could possibly interact with other kinds of hormone-binding macromolecules, such as the hormone receptors and a number of examples have indeed been described (reviewed in Strosberg *et al.*, 1981; Strosberg, 1983). Alternatively, one could imagine that hormone-binding macromolecules share structural features in their combining sites which will be recognized by antibodies raised against any of these proteins. This hypothesis has certainly been verified in the case of antibodies raised in different species to the same ligand. We have represented these two alternatives in Fig. 2.4, where Ab1 represents the anti-hormone antibodies and Ab2α and Ab2β stand for the cross-reactive or the internal-image anti-idiotypes. Ab3 represents the molecules raised against the Ab2β antibodies.

Most situations described below involve the experimental induction of such anti-idiotypic antibodies. In at least two cases, the spontaneous occurrence of such molecules has been studied. Furthermore, the experimental myasthenia gravis-like symptoms following the induction of anti-idiotypic antibodies against the nicotinic acetylcholine receptor will be discussed in the light of possible pathological developments which hitherto remain unexplained.

2.4.2 Some examples of anti-idiotypes which act as antibodies to receptors

(a) Anti-idiotypic anti-insulin receptor antibodies

Sege and Peterson (1978a,b) were the first to show that the injection into rabbits of anti-insulin antibodies resulted in the synthesis of anti-idiotypic antibodies which mimicked the action of the hormone by binding to the insulin receptors on rat epididymal fat cells and by stimulating the uptake by rat thymocytes of aminoisobutyric acid. These observations were recently confirmed and extended by Shechter *et al.* (1982), who showed that the simple injection of insulin provoked the production of not only anti-insulin antibodies but also of autologous anti-idiotypic antibodies, which competed with insulin both for binding to the receptor and for stimulation of glucose uptake by insulin-sensitive cells. The binding of the anti-idiotypes to the receptor could be blocked by anti-insulin antibodies, presumably through the idiotype–anti-idiotype interaction.

(b) Anti-idiotypic anti-β-adrenergic receptor antibodies

Antibodies raised in rabbits against alprenolol, a potent β-adrenergic catecholamine antagonist, were shown to bind other antagonists as well as agonists. Anti-idiotypic antibodies were produced in other rabbits by immunization with the anti-alprenolol antibodies. The anti-idiotypic antibodies inhibited the binding of alprenolol to both its specific antibodies and to the β-adrenergic receptors of different types of cells. Direct binding of the anti-idiotypic antibodies to the receptor was demonstrated by the use of immunofluorescent or radiolabeled anti-rabbit immunoglobulin antibodies. The number of cell-bound molecules was compatible with the number of receptors calculated from hormone-binding studies (reviewed in Lu *et al.*, 1983).

Binding of the anti-idiotypic antibodies resulted in inactivation of basal and catecholamine-sensitive adenylate cyclase (Schreiber *et al.*, 1980; Homcy *et al.*, 1982). The synergistic effect between the hormone and the anti-idiotypic antibodies confirmed that binding to the receptor, while producing similar results, did not need to be competitive, and thus did not necessarily occur through exactly the same interactions.

The synthesis of anti-idiotypic anti-receptor antibodies was transient and occurred during short periods which differed from rabbit to rabbit. Decrease of receptor-binding activity could be correlated with increase of hormone-binding responses, due to the emergence of a third type of antibody with an anti-anti-idiotypic activity (Couraud *et al.*, 1983b). The successive antibody fractions induced by the anti-hormone immunization may reflect different stages in the network of idiotypic–anti-idiotypic interactions proposed by Jerne (1974), as depicted in Fig. 2.5.

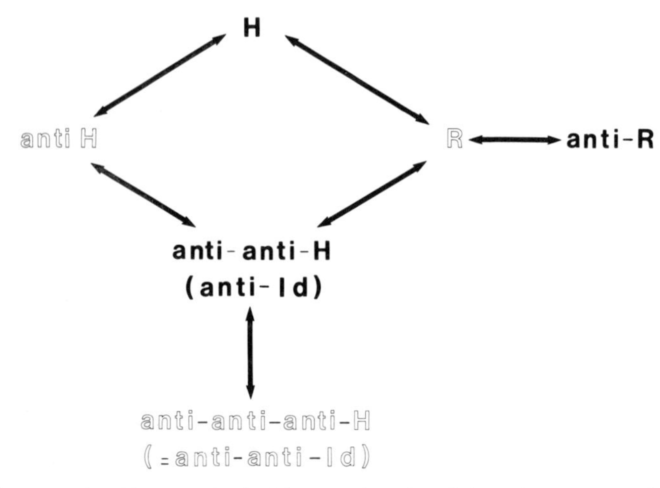

Fig. 2.5 A network of immunological interactions involving a hormone and its specific receptor and antibodies. The symmetrical positions of the anti-hormone antibodies (anti H) and the receptor (R) around the hormone (H) are mirrored by those of the anti-R and the anti-anti-H antibodies around R. The network of interactions is extended by the emergence of anti-anti-anti-H antibodies which are anti-idiotypic (anti-Id) towards the anti-anti-H immunoglobulins. (From Lu *et al.*, 1983.)

According to this model, Ab1 (the idiotype) would induce the production of Ab2 (anti-idiotype) which in turn would stimulate the synthesis of Ab3 (anti-anti-idiotype). Each of these pairs of antigen–antibodies would interact by mutual complementarity resulting in neutralization. Subsets of the Ab2 and the Ab3 antibodies may sufficiently resemble the original antigen against which either Ab1 or Ab2 was directed. This molecular mimicry by immunological internal images would result in specific interactions which can be accurately measured in a system in which radiolabeled ligands and sensitive enzymic assays are available.

The neutralization of Ab1 by Ab2 and of Ab2 by Ab3 could indeed be demonstrated after separating the various antibody subpopulations on the appropriate affinity columns using alprenolol–agarose for Ab1 or Ab3, and idiotype (Ab1)–agarose for Ab2. The purified Ab2 fraction contained no catecholamine-binding antibodies, whereas the purified Ab3 had little or no β-adrenergic-receptor-binding activity. Both Ab1 and Ab3 bound agonists and antagonists but with different affinities. Ab3 initially bound the ligands with a higher affinity than Ab1, but in successive weeks, the affinity of Ab3 decreased progressively to a level below the affinity of Ab1, as if Ab3 were drifting away from the initial model (Couraud *et al.*, 1983b). As expected, quantitative studies showed that not all anti-idiotypic antibodies bound receptor, and that anti-hormone antibodies constituted only a subset of the anti-anti-idiotypic antibodies.

(c) Induction of experimental myasthenia gravis by anti-idiotypic antibodies

Wasserman *et al.* (1982) used a derivative of Bis Q (trans-3,3′-bis-a-(trimethylammonio)methylazobenzene bromide), a potent agonist of the acetylcholine receptor, to prepare rabbit antibodies which mimicked the binding characteristics of the acetylcholine receptor with respect to the order of binding of decamethonium (an agonist) over hexamethonium (an antagonist). Immunization of rabbits with the purified anti-Bis Q antibodies yielded anti-idiotypic sera which recognized rat, *Torpedo* or eel acetylcholine receptor by complement fixation and enzyme immunoassay. The binding was inhibited by the free-ligand Bis Q. Two of the three rabbits showed signs of muscle weakness similar to that seen after immunization with the purified receptor. One of the rabbits was injected intramuscularly with neostigmine and showed temporary improvement. Another showed post-tetanic exhaustion of hindlimb muscles after stimulation of the sciatic nerve. The third rabbit, which had a significant titer of anti-idiotypic antibodies, showed no signs of muscle weakness.

As was the case for the anti-idiotypic response to anti-(β-adrenergic) ligands, the respose to anti-(Bis Q) was transient in one animal, both with respect to anti-receptor titer and to signs of experimental myasthenia gravis. Maximal titer and muscle weakness occurred after the first boost and remained high until the second boost, after which the titer dropped three- to four-fold and signs of weakness disappeared. Subsequent boosting apparently caused tolerance: the anti-receptor titer dropped to the range found in non-immunized animals.

(d) Anti-idiotypic anti-TSH receptor antibodies

Rabbits immunized with rat anti-human thyroid-stimulating hormone antibodies produced immunoglobulins which did not bind the hormone but inhibited the binding of bovine TSH to the porcine thyroid receptor in a dose-dependent manner up to 50% of the total binding. Direct interaction of the anti-idiotypic antibodies to porcine thyroid cell membranes was saturable and could be inhibited up to 64% by increasing concentrations of the free bovine hormone. In the presence of Gpp[NH]p (guanosine 5′-[βγ-imido]triphosphate) the anti-idiotypic antibodies increased thyroid membrane adenylate cyclase activity 40% over the enzymic activity induced by non-specific immunoglobulins. The rate of incorporation of [^{131}I]Na into cultured thyrocytes was also increased in a dose-dependent manner by the anti-idiotypic antibodies which induced the organization of these cells into follicles between 5 and 7 days of culture (Farid *et al.*, 1982).

(e) Anti-idiotypic anti-chemotactic peptide receptor
Marasco and Becker (1982) raised anti-idiotypic antibodies in mice, guinea pigs and goats against rabbit antibodies to the chemo-attractant peptide formylmethionylleucylphenylalanine (fMet-Leu-Phe). The goat antibodies bound to rabbit polymorphonuclear leukocytes and the corresponding $F(ab')_2$ fragments partially inhibited the binding of the fMet-Leu-Phe peptide to the same cells. Both the anti-(polymorphonuclear receptor) and the anti-idiotype activities were lost after passage over an affinity gel containing antibody against the fMet-Leu-Phe peptide. The authors could not demonstrate that the anti-idiotypic antibodies mimicked the biological activity of this peptide, but this was explained by the fact that the preimmune control IgG preparations themselves induced locomotion and, in the presence or absence of cytochalasin B, granule enzyme release. These 'non-specific' effects thus could mask a hypothetical specific effect by the anti-idiotypic 'internal image'.

2.4.3 Antibodies to receptors arising as autoanti-idiotypes: possible role in initiating disease

The origin of anti-receptor autoantibodies is generally thought to be the accidental release into the blood system of receptor, possibly owing to tissue injury. However, the experiments by Shechter *et al.* (1982) in the insulin system and by Wasserman *et al.* (1982) in the acetylcholine system suggest that pathological developments need not be induced by the receptor itself or a cross-reactive antigen, but may also be caused by an anti-idiotypic response against anti-ligand antibodies. Such anti-hormone antibodies are more likely to be set free in the blood system than the membrane-bound receptor. Autologous anti-insulin antibodies have been described as early as 1957. The presence in the sera of myasthenia gravis patients of anti-idiotypic antibodies against anti-acetylcholine antibodies has been reported recently (Dwyer *et al.*, 1983).

2.5 ANTIBODIES TO RECEPTORS AS TRIGGERS OF RECEPTOR–MODULATED BIOLOGICAL FUNCTIONS

2.5.1 Pharmacodynamic properties of antibodies to receptors

Whether found naturally or experimentally induced, the properties of antibodies to receptors cannot easily be predicted with respect to pharmacological activity. In addition to antigenic specificity, other intrinsic parameters of the antibodies must be taken into account for a comparison of their effects with those of the natural ligand. Under physiological circumstances the di- or multi-valency of the antibodies may lead to an artificial cross-linking of the membrane receptors, sometimes followed by internalization and processing of the antibody–receptor complexes. The dynamics of binding of an antibody to cell surface receptors are obviously intertwined with both its antigenic specificity and multivalency. Antibodies – originally devised to 'catch' circulating antigens – are usually characterized by a low dissociation constant, while their equilibrium binding constants may span a broad range. *In vivo,* the pharmacodynamic characteristics of anti-receptor antibodies are very different from those of hormones; moreover, Fc-mediated processes (complement activation, antibody-dependent cellular cytotoxicity) may occur and amplify or change the biological effect triggered by the binding of antibody.

2.5.2 Inhibitory anti-receptor antibodies

Anti-receptor antibodies may inhibit the binding of the natural and related ligands to the receptor. The inhibition may be of a competitive nature that primarily alters the affinity of the hormone binding. In this case, the antibody is presumed to recognize the combining site for the hormone on the receptor molecule, although allosteric mechanisms cannot be excluded. Alternatively the inhibition of binding may be non-competitive resulting in the masking of binding sites. Antibodies which inhibit hormone binding may act as antagonists and only block hormone-mediated effects. Some polyclonal anti-β-adrenergic receptor and anti-prolactin receptor antibodies and the reported monoclonal anti-TSH receptor antibodies (see Kohn *et al.*; Chapter 9) possess these properties. We cited several examples of anti-receptor antibodies which inhibit hormone binding and behave as partial or full pharmacologic agonists; the anti-idiotypic antibodies which recognize hormone receptors, the autoantibodies to the insulin and the TSH receptor, some anti-prolactin receptor) and some monoclonal anti-(EGF receptor) antibodies may be classified in this category. In all these cases the intrinsic biologic activity of the antibodies demonstrates that the information transmitted through the hormone–receptor system resides in the receptor rather than in the hormone. The hormone (or its fragments, after internalization and degradation) does not appear to be strictly necessary for biologic signalling beyond the binding step.

2.5.3 Antibodies without effector activity

There are several examples of anti-receptor antibodies which do not interfere with hormone binding and are biologically inert (e.g. the reported monoclonal anti-insulin receptor and a number of monoclonal anti-EGF receptor and anti-acetylcholine receptor antibodies). One may foresee an increase in the incidence of antibodies with these properties as their selection is favored when immune recognition of receptor–hormone complexes is chosen as easy screening of hybridoma secretory products. Although of little interest to the investigation of the mode of action of a hormone–receptor system, these antibodies may prove to be outstanding tools for receptor purification and useful markers to follow receptor dynamics in cell traffic, in the presence or absence of the hormone.

2.5.4 Agonist-like antibodies

Unexpectedly, anti-receptor antibodies may possess agonistic properties even though they do not interfere with hormone binding. This was observed in the case of the polyvalent antibodies directed against the β-adrenergic receptors whose presence considerably enhanced the stimulation of the epinephrine-activated adenylate cyclase (Couraud *et al.*, 1981; Schreiber *et al.*, 1980). Similar results were obtained for the TSH-stimulated receptor–adenylate cyclase complex, for anti-insulin receptor antibodies and for cross-linked monoclonal anti-EGF receptor antibodies. These antibodies induce a qualitatively distinct type of receptor regulation which recently emerged in several systems (see review by Hollenberg, 1981) and was termed heterotypic modulation (homotypic being the modulation of a receptor system by ligands via the combining site for the hormone). The anti-receptor antibodies which trigger a hormone–receptor system via domains distinct from the combining site for the hormone provide a strong argument for the view that the receptor moiety contains all the biochemical attributes for biologic activity. Should the antigenic determinants recognized by the antibodies be identified, they would represent ideal targets for pharmacologic agents to regulate receptor function without disruption of physiologic hormonal homeostasis.

2.5.5 Role of antibodies in receptor redistribution

The lateral mobility of membrane receptors and their clustering (aggregation) upon binding of ligands was suggested to play an essential role in the process of transmembrane signalling (Edelman, 1976; Schreiner and Unanue, 1976; Segal *et al.*, 1977). Hormonal induction of membrane receptor clustering, followed by internalization (receptor-mediated endocytosis) and processing of the hormone–receptor complexes has emerged in the last few years as a common pathway in several systems (see reviews by Goldstein and Brown, 1977; Schlessinger *et al.*, 1983a). To date, the mechanism by which the presumably monovalent hormones (e.g. EGF, insulin, TSH) induce receptor aggregation is not known. One possibility is that hormone binding induces a conformational change in the receptor molecule, which in turn could lead to the appearance of an 'aggregation site' on the receptor molecule which would facilitate receptor–receptor or receptor–coated pit interactions. According to this model one can view the hormone receptor as an 'allosteric' receptor (Changeux, 1981; Schlessinger *et al.*, 1983b) where the hormone would act as one of the possible allosteric regulators. With known diffusion coefficients for the lateral motion of membrane receptors, it clearly appears that receptor clustering is not the rate-determining step in hormone action (see review by Schlessinger, 1979). Rather, receptor clustering may provide a rapid, efficient means for producing a molecular species with appropriate conformation, orientation and activation energy to interact with other membranal, submembranal and intracellular components resulting in the biologic signal. The intrinsic bioactivity of some antibodies to hormone receptors may be related to their ability to induce receptor aggregation by their built-in multivalency. This may actually mimic the vertical physiologic signals normally mediated by the hormone; therefore, the comparison (rate, amplitude, dose–responses) of the biologic activity induced by the hormone and by anti-receptor antibodies should be treated with caution. Experiments in which monovalent Fab fragments of anti-receptor antibodies fail to induce biologic activity which may be restored by external cross-linking with a second layer of antibodies would indicate the importance of receptor aggregation *per se* for biologic signalling. From careful binding studies it appears, however, that the cross-linking of monovalent Fab fragments bound to a cell surface receptor – besides causing receptor aggregation – also dramatically reduces the rate of dissociation of the Fab fragments. This, in turn, leads to an increase in the apparent affinity of the Fab fragments towards the receptors, i.e. in thermodynamic terms a stabilization of the interaction perhaps in itself essential to biologic signalling. Hence, receptor clustering induced by ligands and the nature of the interaction of these ligands with the surface receptors are mutually dependent parameters and it is not straightforward to conclude which component contributes to the bioactivity of the anti-receptor antibodies (Schreiber *et al.*, 1983).

In the *in vivo* situations with heterogeneous antibody populations to hormone receptors, all the mechanisms discussed above may co-exist in an intricate interplay. This may considerably hamper the understanding of etiopathogenesis.

2.6 CONCLUSION

In this chapter we have discussed how anti-receptor antibodies raised by immunization with cells, membranes or affinity-purified receptors provide unique tools for probing receptor structure and functions. The advent of hybridoma technology has generated many studies with homogeneous monoclonal antibodies whose specificities have widely diverged from pure agonists or antagonists of receptor functions to specific, although inert, markers of receptor molecules. The preparation of anti-idiotypic antibodies which bind to receptor has opened new avenues for the immunological analysis of hormonally stimulated systems, and may yet provide new insight into the physiological and possibly pathological interactions between the endocrine and the immune networks.

REFERENCES

Abramsky, O., Aharonov, A., Teitelbaum, O. and Fuchs, S. (1975), *Arch. Neurol.*, **32**, 684–687.

Adams, D.D. and Kennedy, T.J. (1967), *J. Clin. Endocrinol. Metab.*, **27**, 173–177.

Aharonov, A., Tarrab-Hazdai, R., Silman, I. and Fuchs, S. (1977), *Immunochemistry*, **14**, 129–137.

Almon, R.R., Andrew, C.G. and Appel, J.H. (1974), *Science*, **186**, 55–57.

Andre, C., Guillet, J.G., De Backer, J.P., Hoebeke, J. and Strosberg, A.D. (1983), *EMBO J.*, **2**, 499–504.

Andre, C., Guillet, J.G., De Backer, J.-P., Vanderheyden, P., Hoebeke, J. and Strosberg, A.D. (1984), *EMBO J*, **3**, 17–21.

Appel, J.H., Anwyl, R., McAdams, M.W. and Elias, S. (1977), *Proc. Natl. Acad. Sci. U.S.A.*, **74**, 2130–2134.

Bohnet, H.G., Shiv, R.P.C., Grinwich, D. and Frieson, H.G. (1978), *Endocrinology*, **102**, 1657–1661.

Buhrow, S.A., Cohen, S. and Staros, J.V. (1982), *J. Biol. Chem.*, **257**, 4019–4022.

Carpenter, G. and Cohen, S. (1981), in *Receptors and Recognition* (Lefkowitz, R.L., ed.), Series B, Vol. 13, Chapman and Hall, London, p. 41.

Changeux, J.-P. (1981), *Harvey Lectures* (1979–1980), Series 75, pp. 85–254.

Chiorazzi, N., Wasserman, R.L. and Kunkel, H.G. (1982), *J. Exp. Med.*, **156**, 930–935.

Claque, R., Mukthar, E.D., Pyle, G.A., Nutt, J., Clark, F., Scott, M., Evered, D., Rees-Smith, B. and Hall, R. (1976), *J. Clin. Endocrinol. Metab.*, **43**, 550–556.

Cohen, S., Ushiro, H., Stoscheck, C. and Chinkers, M. (1982), *J. Biol. Chem.*, **257**, 1523–1531.

Conti-Tranconi, B., Tzartos, S. and Lindstrom, J. (1981), *Biochemistry*, **20**, 2181–2190.

Couraud, P.-O., Delavier-Klutchko, C., Durieu-Trautmann, O. and Strosberg, A.D. (1981), *Biochem. Biophys. Res. Commun.*, **99**, 1295–1302.

Couraud, P.O., Lu, B.-Z., Schmutz, A., Durieu-Trautmann, O., Klutchko-Delavier, C., Hoebeke, J. and Strosberg, A.D. (1983a), *J. Cell, Biochem.*, **3**, 187–193.

Couraud, P.-O., Lu, B.-Z. and Strosberg, A.D. (1983b), *J. Exp. Med.*, **157**, 1369–1378.

Croce, C.M., Linnenbach, A., Hall, W., Steplewski, A. and Koprowski, H. (1980), *Nature (London)*, **288**, 488–489.

Djiane, J., Houdebine, L.-M. and Kelly, P.A. (1981), *Proc. Natl. Acad. Sci. U.S.A.*, **78**, 7445–7448.

Downward, J., Yarden, Y., Mayer, E., Scrace, G., Totty, N., Stockwell, P., Ullrich, A., Schlessinger, J. and Waterfield, M. (1984), *Nature*, **307**, 521–527.

Drachman, D.B., Angus, C.W., Adams, R.N., Michelson, J.D. and Hoffman, G.J. (1978), *N. Engl. J. Med.*, **198**, 136–142.

Drachman, D.B., Adams, R.N., Josefek, L.F. and Self, S.G. (1982), *N. Engl. J. Med.*, **307**, 769–775.
Drexhage, H.A., Bottozzo, G.F., Bitensky, L., Chayen, J. and Doniach, D. (1981), *Nature (London)*, **289**, 594–596.
Dwyer, D.S., Bradley, R.J., Urquhart, C.K. and Kearney, J.F. (1983), *Nature (London)*, **301**, 611–614.
Edelman, G.M. (1976), *Science*, **192**, 218–226.
Endo, K., Borges, M., Amir, S. and Ingbar, S.H. (1982), *J. Clin. Endocrinol. Metab.*, **55**, 566–576.
Farid, N.R., Pepper, B., Urbina-Briones, R. and Islam, N.R. (1982), *J. Cell. Biochem.*, **19**, 305.
Farmbrough, D.M., Drachman, D.B. and Satyamurti, S. (1973), *Science*, **82**, 293–295.
Flier, J.S., Kahn, C.R., Jarrett, D.B. and Roth, J. (1975), *Science*, **190**, 63–65.
Fraser, C.M. and Venter, J.C. (1980), *Proc. Natl. Acad. Sci. U.S.A.*, **77**, 7034–7038.
Fuchs, S. (1980), *Trends. Biochem. Sci.*, **5**, 259–262.
Fuchs, S., Schmidt-Hopfeld, H., Tridenta, G. and Tarrab-Hazdai, R. (1980), *Nature (London)*, **287**, 162–164.
Galfre, G. and Milstein, C. (1981), *Methods Enzymol.*, **73**, 3–46.
Goding, J. (1980), *J. Immunol. Methods*, **29**, 285–308.
Goldstein, J.L. and Brown, M.J. (1977), *Annu. Rev. Biochem.*, **46**, 669–722.
Gomez, C.M., Richman, D.P., Berman, P.W., Burres, S.A., Arnason, B.G.W. and Fitch, R.W. (1979), *Biochem. Biophys. Res. Commun.*, **88**, 575–582.
Green, D.P.L., Miledi, R. and Vincent, A. (1975), *Proc. R. Soc. London, Ser. B*, **189**, 57–68.
Grunfeld, C., Van Obberghen, E., Karlsson, F.A. and Kahn, C.R. (1980), *J. Clin. Invest.*, **66**, 1124–1134.
Gullik, W.J., Tzartos, S. and Lindstrom, J. (1981), *Biochemistry*, **20**, 2173–2180.
Haigler, H.T. and Carpenter, G. (1980), *Biochim. Biophys. Acta*, **598**, 314-325.
Harrison, L.C. and Kahn, C.R. (1980), *Prog. Clin. Immunol.*, **4**, 207–226.
Harrison, L.C. (1984) in *Receptor Purification Procedures*, Alan R. Liss, New York, pp. 125–137.
Hollenberg, M.D. (1981), *Trends Pharmacol. Sci.*, **2**, 320–322.
Homcy, C.J., Rockson, S.G. and Haber, E. (1982), *J. Clin Invest.*, **69**, 1147–1155.
Houdebine, L.-M. (1980), in *Hormones and Cell Regulation* (Dumont, S. and Nunez, J., eds), Vol. 4, Elsevier, Cambridge, UK, pp. 175–196.
Jacobs, S., Chang, K.-J. and Cuatrecasas, P. (1978), *Science*, **200**, 1283–1284.
James, R.W., Kato, A.C., Rey, M.J. and Fulpius, B.W. (1980), *FEBS Lett.*, **120**, 145–148.
Jerne, N.K. (1973), *Sci. Am.*, **229**, 52–60.
Jerne, N.K. (1974), *Ann. Immunol. (Inst. Pasteur)*, **125C**, 373–389.
Kahn, C.R., Flier, J.S., Bar, R.S., Archer, J.A., Gorden, P., Martin, N.M. and Roth, J. (1976), *N. Engl. J. Med.*, **294**, 739–745.
Kahn, C.R., Baird, K.L., Jarrett, D.B. and Flier, J.J. (1978), *Proc. Natl. Acad. Sci. U.S.A.*, **75**, 4209–4213.
Kahn, C.R., Baird, K.L., Flier, J.S., Grunfeld, C., Harmon, J.T., Harrison, L.C., Karlsson, F.H., Kajuga, M., King, G.L., Lang, U.C., Poskalny, J.M. and Van Obberghen, E. (1981), *Recent Prog. Horm. Res.*, **37**, 477–533.
Kao, I. and Drachman, D.B. (1977), *Science*, **196**, 527–529.

Kasuga, M., Kahn, R., Hedo, J.A., Van Obberghen, R. and Yamada, K.M. (1981), *Proc. Natl. Acad. Sci. U.S.A.*, **78,** 6917–6921.
Khoury, E.L., Cossio, P.M., Szarfman, A., Marcos, J.C., Garcia-Morteo, O. and Arana, R.M. (1978), *Am. J. Clin. Pathol.*, **69,** 62–65.
Kull, F.C., Jr., Jacobs, S., Su, Y.-F. and Cuatrecasas, P. (1982), *Biochem. Biophys. Res. Commun.*, **106,** 1019–1026.
Lennon, V.A., Thompson, M. and Chen, J. (1980), *J. Biol. Chem.*, **255,** 4395–4398.
Leiber, D., Harbon, S., Guillet, J.G., Andre, C. and Strosberg, A.D. (1984), *Proc. Natl. Acad. Sci. U.S.A.*, (in press).
Libermann, T.A., Schreiber, A.B., Yarden, Y., Lax, I., Eshhar, Z. and Schlessinger, J. (1983), *J. Cell Biochem.*, (in press).
Lindenmann, G. (1973), *Ann. Immunol. (Inst. Pasteur)*, **124C,** 171–184.
Lindstrom, J. (1976), *J. Supramol. Struct.*, **4,** 389–403.
Lindstrom, J., Lennon, V.A., Seybold, M.E. and Wittingham, S. (1976), *Ann. N.Y. Acad. Sci.*, **274,** 283–299.
Lü, B.-Z., Couraud, P.O., Schmutz, A. and Strosberg, A.D. (1983), *Ann. N.Y. Acad. Sci.*, (in press).
Marasco, W.A. and Becker, E.L. (1982), *J. Immunol.*, **128,** 963–968.
Maron, R., Elias, D., de Jongh, B.M., Bruinings, G.J., van Rood, J.J., Shechter, Y. and Cohen, I.R. (1983), *Nature*, **303,** 817–818.
Moshly-Rosen, D. and Fuchs, S. (1981), *Biochemistry*, **20,** 5920–5924.
Moshly-Rosen, D., Fuchs, S. and Eshhar, Z. (1979), *FEBS Lett.*, **106,** 389–392.
Patrick, J. and Lindstrom, J. (1973), *Science*, **180,** 871–872.
Patrick, J., Lindstrom, J., Culp, B. and McMillan, J. (1973), *Proc. Natl. Acad. Sci. U.S.A.*, **70,** 3334–3338.
Rash, J.E., Albuquerque, E.X., Hudson, C.S., Mayer, R.F. and Salterfield, J.R. (1976), *Proc. Natl. Acad. Sci. U.S.A.*, **73,** 4584–4588.
Rees-Smith, B. and Hall, R. (1974), *Lancet*, **ii,** 427–430.
Rees-Smith, B. and Hall, R. (1981), *Methods Enzymol.*, **74C,** 405–420.
Roth, R.A., Cassell, D.J., Wong, K.Y., Maddux, B.A. and Goldfine, I.D. (1982), *Proc. Natl. Acad. Sci. U.S.A.*, **79,** 7312–7316.
Schlessinger, J. (1979), in *Physical Chemical Aspects of Cell Surface Events in Cellular Regulation* (De Lisi, C. and Blumenthal, R., eds), Elsevier, New York, pp. 89–111.
Schlessinger, J., Schechter, Y., Willingham, M.C. and Pastan, I. (1978), *Proc. Natl. Acad. Sci. U.S.A.*, **75,** 2659–2663.
Schlessinger, J., Schreiber, A.B., Libermann, T.A., Lax, I., Avivi, A. and Yarden, Y. (1983a) in *Cell Membranes: Methods and Reviews* (Elson, E.L., Frazier, W. and Glaxer, L., eds), Plenum, New York, (in press).
Schlessinger, J., Schreiber, A.B., Levi, H., Lax, I., Libermann, T.A. and Yarden, Y. (1983b), *CRC Biochem.*, (in press).
Schreiber, A.B., Couraud, P.-O., Andre, C., Vray, B. and Strosberg, A.D. (1980), *Proc. Natl. Acad. Sci. U.S.A.*, **77,** 7385–7389.
Schreiber A.B., Lax, I., Yarden, Y., Eshhar, Z. and Schlessinger, J. (1981), *Proc. Natl. Acad. Sci. U.S.A.*, **78,** 7535–7539.
Schreiber, A.B., Libermann, T.A., Lax, I., Yarden, Y. and Schlessinger, J. (1983), *J. Biol. Chem.*, **258,** 846–853.
Schreiner, G.F. and Unanue, E.R. (1976), *Adv. Immunol.*, **24,** 38–168.

Segal, D., Taurog, S. and Metzger, H. (1977), *Proc. Natl. Acad. Sci. U.S.A.*, **74,** 2993–2997.
Sege, K. and Peterson, P.A. (1978a), *Nature (London)*, **271,** 167–168.
Sege, K. and Peterson, P.A. (1978b), *Proc. Natl. Acad. Sci. U.S.A.*, **75,** 2443–2447.
Schechter, Y., Maron, R., Elias, D. and Cohen, E.P. (1982), *Science*, **216,** 542–545.
Shiu, R.P.C. and Friesen, H.G. (1978), *Science*, **192,** 259–261.
Smith, B.R., Pyle, G.A., Peterson, V.B. and Hall, R. (1977), *J. Endocrinol.*, **00,** 00–00.
Steinitz, M. and Klein, G. (1981), *Immunol. Today*, **2,** 38–39.
Strakosch, C.R., Joyner, D. and Wall, J.R. (1978), *J. Clin. Endocrinol. Metab.*, **46,** 345–348.
Strosberg, A.D. (1983), *Springer Semin. Immunopathol.*, **6,** 67–78.
Strosberg, A.D., Couraud, P.-O. and Schreiber, A.B. (1981), *Immunol. Today*, **2,** 75–79.
Teyssot, B., Djiane, J., Kelly, P.A. and Houdebine, L.-M. (1982), *Biol. Cell.*, **43,** 81–88.
Toyka, K.V., Drachman, D.B., Griffin, D.E., Pestronk, A., Winkkelstein, J.A., Fischbeck, K.H., Jr. and Kao, I. (1977), *N. Engl. J. Med.*, **296,** 125–131.
Tzartos, S.J. and Lindstrom, J. (1980), *Proc. Natl. Acad. Sci. U.S.A.*, **77,** 755–759.
Tzartos, S.J., Seybold, M.E. and Lindstrom, J.M. (1982), *Proc. Natl. Acad. Sci. U.S.A.*, **79,** 188–192.
Valente, W.A., Vitti, P., Yavin, Z., Yavin, E., Rotella, C.M., Grollman, E.F., Toccafondi, R.S. and Kohn, L.D. (1982), *Proc. Natl. Acad. Sci. U.S.A.*, **79,** 6680–6684.
Van Obberghen, E. and Kahn, C.R. (1981), *Mol. Cell Endocrinol.*, **22,** 277–294.
Van Obberghen, E., Kasuga, M., LeCam, A., Hedo, J.A., Hin, A. and Harrison, L.C. (1981), *Proc. Natl. Acad. Sci. U.S.A.*, **78,** 1052–1056.
Venter, C., Fraser, C.M. and Harrison, L.C. (1980), *Science*, **207,** 1361–1363.
Waterfield, M.D., Mayes, E.L.V., Stroobant, P., Bennett, P.L.P., Young, S., Goodfellow, P.N., Banting, G.S. and Ozanne, B. (1983), *J. Cell. Biochem.*, (in press).
Wasserman, N.H., Penn, H.S., Freimuth, P.I., Treptow, N., Wentzel, S., Cleveland, W.L. and Erlanger, B.F. (1982), *Proc. Natl. Acad. Sci. U.S.A.*, **79,** 4810–4814.
Zakarija, M., McKenzie, J.M. and Banovac, K. (1980), *Ann. Intern. Med.*, **93,** 28–32.

Notes added in proof

1. In addition, the guanylate cyclase activation and adenylate cyclase inhibition were blocked by the muscarinic antagonist atropine. Both anti-receptor antibodies triggered contractions as well (Leiber *et al.*, 1984).
2. However it was recently suggested that transformation of cells by avian erythroblastosis virus may result, in part, from the inappropriate acquisition of a truncated EGF receptor. This was based on the finding that the EGF receptor domain responsible for cell proliferation contains amino acid sequences homologous to a portion of the v-erb-B transforming protein of the virus (Downward *et al.*, 1984).

3. However, Maron *et al.* (1983) reported the occurrence of such antibodies in 10 out of 22 patients with insulin-dependent diabetes mellitus, a much more prevalent disease.
4. More recently, anti-insulin receptor antibodies were also shown in a large number of patients with insulin-dependent diabetes mellitus (Maron *et al.*, 1983). Unlike those described above, these antibodies were of the IgM class and were also found in patients who had not received exogenous insulin and were thus expected to provide negative control sera. Besides, the anti-receptor antibodies were not checked for activity towards idiotypes of insulin antibodies.

3 A Monoclonal Antibody to the Human Receptor for T Cell Growth Factor

WARREN J. LEONARD, JOEL M. DEPPER, THOMAS A. WALDMANN and WARNER C. GREENE

Glossary of terms and abbreviations

TCGF or IL-2	T cell growth factor or interleukin 2
T cells	Thymus-derived lymphocytes
IL-1	Interleukin 1, a lymphocyte-activating factor derived from macrophages ('monokine')
Anti-Tac	Monoclonal antibody specific for T lymphocytes activated by lectin or antigen. Identifies the receptor for IL-2
SDS-PAGE	Polyacrylamide gel electrophoresis in sodium dodecyl sulfate
HTLV	Human T cell leukemia virus
HUT-102B2	An HTLV-infected T leukemia cell line
'Western' blotting	Transfer of electrophoresed protein to nitrocellulose paper for probing with labeled antibody or other ligands
Ia antigens	HLA type II glycoproteins on human lymphocytes. Referred to as Ia by analogy to similar (and partially homologous) type II proteins coded by the *I*mmune response *a*ssociated region of murine histocompatibility complex, H-2.

Acknowledgement

J.M.D. is the recipient of a National Arthritis Foundation fellowship award.

Monoclonal Antibodies to Receptors: Probes for Receptor Structure and Function
(*Receptors and Recognition*, Series B, Volume 17)
Edited by M. F. Greaves
Published in 1984 by Chapman and Hall, 11 New Fetter Lane, London EC4P 4EE

3.1 INTRODUCTION: TCGF AND TCGF RECEPTORS

T cell growth factor (TCGF or interleukin-2) is a 14800-dalton glycoprotein hormone which is essential for a normal immune response. It has been extensively reviewed previously (Smith, 1980; Ruscetti and Gallo, 1981). Morgan *et al.* (1976) were the first to report that conditioned media from lectin-stimulated mononuclear cells contained a factor capable of supporting the exponential growth of lectin-activated human T cells. That factor became known as TCGF, and is now known to be essential for the expansion and continued proliferation of cytotoxic (Gillis *et al.*, 1978a), suppressor (Coutinho *et al.*, 1979), and some helper T cells (Watson, 1979). TCGF has permitted the long-term maintenance of T cell lines *in vitro* (Morgan *et al.*, 1976; Ruscetti *et al.*, 1977). It is important to recognize that the specificity of the immune response is determined by the antigen, while TCGF provides the mitogenic stimulus in an antigen non-specific and non-restricted manner (Smith *et al.*, 1979). Utilizing a sensitive bioassay (Gillis *et al.*, 1978b), this lymphokine was biochemically characterized (Mier and Gallo, 1980; Gillis *et al.*, 1980), purified to homogeneity (Robb, 1982; Robb *et al.*, 1983), and its *N*-terminal protein sequence determined (Robb *et al.*, 1983). Recently, the gene encoding TCGF has been cloned and expressed using a eukaryotic vector (Taniguchi *et al.*, 1983; Cheroutre *et al.*, 1983; Clark *et al.*, 1983; Lin *et al.*, 1983).

Robb *et al.* (1981) purified biosynthetically radiolabeled TCGF from JURKAT leukemic cells and demonstrated high-affinity specific membrane receptors for TCGF on phytohemagglutinin (PHA)-activated normal lymphocytes and certain T cell lines. T cell proliferation in response to antigen or mitogen is known to require both the *de novo* synthesis and secretion of TCGF and the expression of these specific TCGF receptors. This system is therefore an interesting example of inducible gene expression, wherein neither the growth factor nor receptor are produced by resting cells, but rather both are elaborated during T cell activation. A model of T cell activation is shown in Fig. 3.1. In the presence of antigen or lectin, macrophages are induced to synthesize and secrete the monokine, interleukin-1 (IL-1, also known as lymphocyte activating factor). In the presence of antigen or lectin and IL-1, inducer/helper T lymphocytes are activated to synthesize and secrete TCGF. Effector T lymphocytes, such as cytotoxic or suppressor T cells, are simultaneously induced to express TCGF receptors. It is unclear whether expression of TCGF receptors also requires IL-1, although Koretzky and co-workers (1983) have suggested that a factor contained in macrophage supernatants is required. In normal physiology, it is unknown whether the same cells ever synthesize both TCGF and TCGF receptors (a situation that occurs in some leukemic cells, such as HUT-102B2 cells). In the presence of TCGF, the cells with receptors are then capable of proliferation and continued growth.

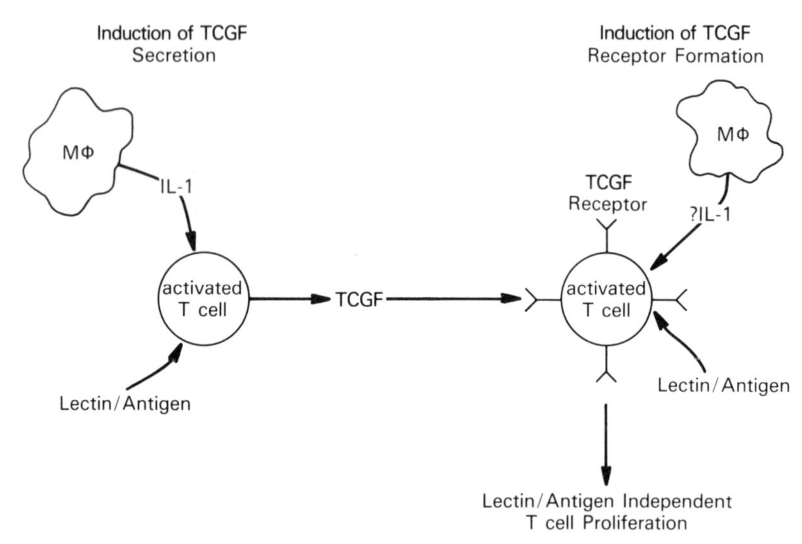

Fig. 3.1 A model for activation and growth of human T cells. Mϕ = monocytes or macrophages.

3.2 SPECIFICITY OF ANTI-TAC MONOCLONAL ANTIBODY

We have employed an anti-human TCGF receptor monoclonal antibody termed anti-Tac to characterize the TCGF receptor and to help elucidate the role of TCGF in a number of TCGF-dependent aspects of lymphocyte activation. Anti-Tac was prepared by Uchiyama and co-workers (1981a,b) using a TCGF-dependent continuous T cell line as the immunogen. The antibody was selected on the basis of its ability to bind to activated T cells (hence the name anti-'Tac', for T activated). Anti-Tac was noted to bind to lectin and antigen-activated T cells, to all TCGF-dependent T cell lines tested, but not to TCGF-independent lines (MOLT-4, 8402, CEM, HSB-2), various B cell lines (8392, PA-3), pre-B cell lines (HPB-null, NALM-1, NALM-6), nor five Epstein-Barr-transformed B cell lines. This pattern of cellular reactivity was essentially identical with the distribution of TCGF receptors reported by Robb *et al.* (1981). We therefore hypothesized that anti-Tac recognized the human receptor for TCGF.

Other data in support of this hypothesis are as follows:

(1) Anti-Tac blocks 80% of TCGF-induced DNA synthesis of TCGF-dependent continuous T cell lines, but does not inhibit DNA synthesis of TCGF-independent T cell lines (Leonard *et al.*, 1982).

(2) Anti-Tac, but not control anti-Ia monoclonal antibodies, blocks over 95% of the binding of [^{3}H]TCGF to HUT-102B2 cells and PHA-activated lymphoblasts (Leonard *et al.*, 1982; Robb and Greene, 1983).
(3) TCGF at high concentrations blocks the binding of [^{3}H]anti-Tac but not [^{125}I]anti-OKT11 (a T-cell-specific monoclonal antibody) to PHA-activated lymphoblasts (Leonard *et al.*, 1983a).
(4) After TCGF is covalently cross-linked to HUT-102B2 cells with disuccinimidyl suberate, both anti-Tac and anti-TCGF can immunoprecipitate a band that is 12 000–14 000 daltons larger than the receptor immunoprecipitated by anti-Tac in the absence of cross-linking. We believe this additional band represents TCGF cross-linked to its receptor (Leonard *et al.*, 1983a). Initially, it may seem that the ability to perform a successful cross-linking experiment is inconsistent with the fact that TCGF is able to block the binding of anti-Tac, as indicated above in (3). However, it is possible that either the receptor exists as a dimer or higher-polymeric form and that one monomer is cross-linked to TCGF and anti-Tac binds another, immunoprecipitating both uncross-linked and cross-linked components together. Alternatively, anti-Tac and TCGF may recognize different epitopes on the receptor, and the binding of TCGF may induce an allosteric change in the receptor that weakens but does not eliminate all binding of the antibody. In support of anti-Tac and TCGF recognizing different epitopes, we note that anti-Tac is a blocking antibody, and does not have TCGF-agonist-like activity.
(5) TCGF immobilized on affigel beads and used as an affinity column will purify a receptor band from extracts of PHA-activated normal lymphocytes that is identical with that identified by anti-Tac conjugated to Sepharose. Further, if the cellular extract was cleared with TCGF immobilized on affigel, it was no longer possible to purify the band identified by anti-Tac conjugated to Sepharose. Similarly, extracts cleared with anti-Tac coupled to Sepharose no longer contained material that bound to TCGF immobilized on affigel (Robb and Greene, 1983).
(6) After lectin activation of peripheral blood lymphocytes, receptors identified by anti-Tac appear within 8 hours, identical with the time course of appearance of TCGF-binding sites (unpublished observations).
(7) Miyawaki and co-workers have demonstrated that activated T lymphocytes have the ability to absorb TCGF from culture supernatants, and that addition of anti-Tac blocks such absorption (Miyawaki *et al.*, 1982).

The data against the hypothesis are essentially limited to discrepancies in receptor number that we have obtained performing Scatchard analyses with labeled anti-Tac (Leonard *et al.*, 1982, and unpublished observations) compared with those of Robb *et al.* (1981) with labeled TCGF. These analyses are performed under slightly different conditions in different laboratories, but we have no complete explanation so far to account for this disparity in which anti-Tac usually predicts approximately 5–10-fold more receptors than TCGF. It may be that not all 'TCGF receptors' recognized by anti-Tac are functional and capable of binding TCGF, or that perhaps a subset of these receptors have diminished affinity for TCGF and are not as readily identified by low concentrations of [^{3}H]TCGF. This could explain the need for large amounts of TCGF to block anti-Tac binding. In any case, we feel that the weight of evidence, as summarized above, supports the contention that anti-Tac recognizes the TCGF receptor.

3.3 PHYLOGENETIC REACTIVITY OF TCGF AND ANTI-TAC

Various mammalian TCGFs have been evaluated according to their ability to stimulate cells from another species. It has been observed that this ability acts down the phylogenetic scale (Ruscetti and Gallo, 1981; Stadler *et al.*, 1982) i.e. human TCGF stimulates proliferation of activated human, rat, and mouse T cells; rat TCGF stimulates rat and mouse but not human T cells; and murine TCGF stimulates only mouse T cells. We have therefore evaluated the ability of anti-Tac to bind to TCGF-dependent mouse cells and find that it does not. This phylogenetic restriction property was exploited by Miyawaki and co-workers (1982) and by us (Depper *et al.*, 1983) in order to assay for TCGF activity using murine HT-2 cells in the presence of anti-Tac. Spliter and co-workers have shown the anti-Tac does not react with bovine T cells (personal communication).

3.4 BIOCHEMICAL CHARACTERIZATION OF THE TCGF RECEPTOR

Having demonstrated the specificity of the antibody, we next proceeded to characterize the receptor. As shown in Fig. 3.2, anti-Tac immunoprecipitations of [^{125}I]surface-labeled PHA-activated normal lymphocytes were analyzed on 8.75% SDS-polyacrylamide gels under reducing conditions. A diffuse band of 55 000 daltons (p55) is readily identified as the putative TCGF receptor (see also Leonard *et al.*, 1983b), suggesting that this protein contains tyrosine residues on the outer cell membrane. In Fig. 3.3, when cells were biosynthetically labeled with [^{35}S]methionine, and immunoprecipitates analyzed by 7.5% SDS, a similar band (band B) is identified. As will be discussed below, this band is slightly smaller on non-reducing gels than on reducing gels. In addition, a second band of 113 000 daltons (band A) is seen. This band is routinely identified when we label with [^{35}S]methionine, but not when we label

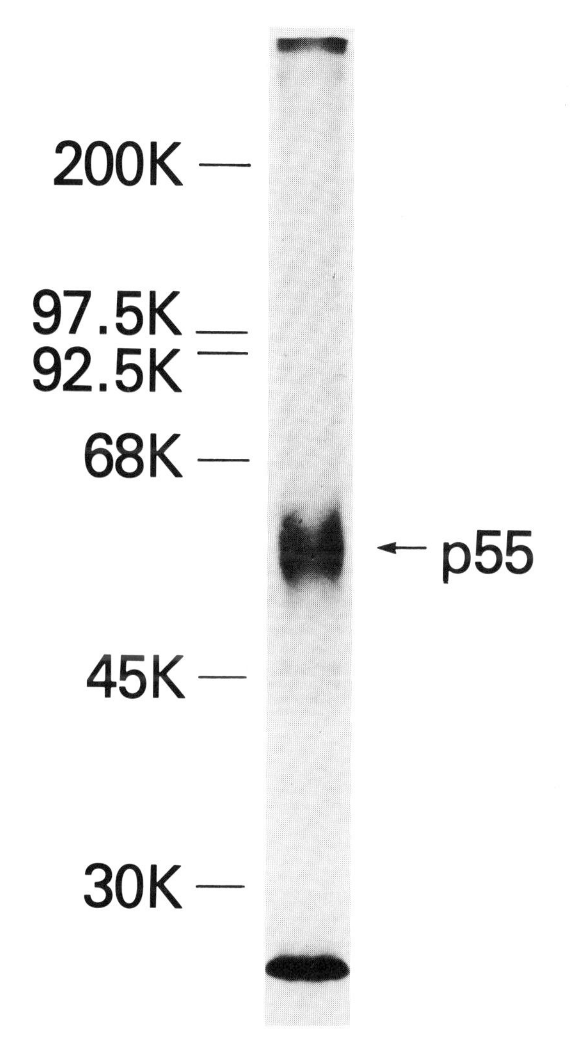

Fig. 3.2 Anti-Tac immunoprecipitation of ^{125}I-surface-labeled PHA-activated lymphoblasts analyzed on an 8.75% SDS/polyacrylamide gel electrophoresed under reducing conditions. Migration of TCGF receptor (p55) is indicated.

cell surface proteins extrinsically with [^{125}I] using lactoperoxidase. Thus, it does not appear to be on the external membrane, but could theoretically represent another subunit of the receptor. Alternatively, it may simply be fortuitously co-immunoprecipitated (Leonard *et al.*, 1982). We have performed experiments using CHAPS, a detergent that was successful in dissociating subunits of the prolactin receptor (Liscia *et al.*, 1982). However,

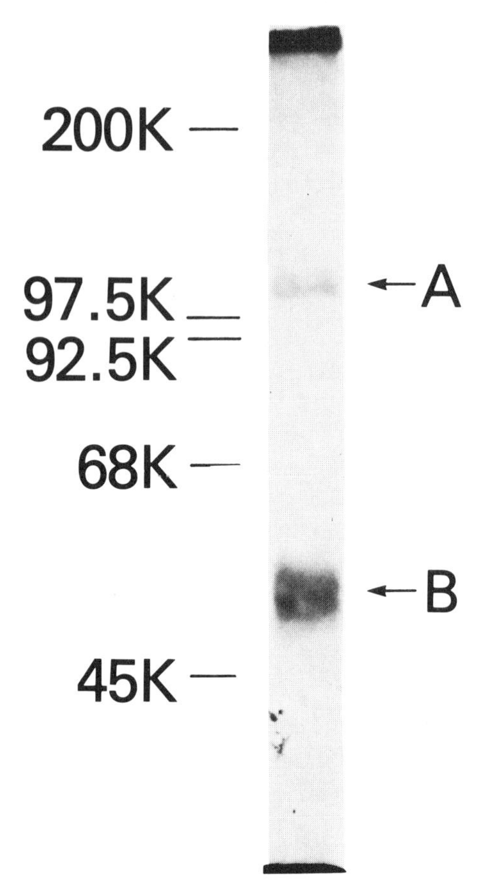

Fig. 3.3 Anti-Tac immunoprecipitation of [^{35}S]methionine-labeled PHA-activated lymphoblasts, analyzed on a 7.5% SDS/polyacrylamide gel electrophoresed under non-reducing conditions. Migration of TCGF receptor (band B) and p113 (band A) are indicated

solubilizing with CHAPS had no effect on the presence of p113 in our immunoprecipitations, making it neither more nor less likely that p113 is a subunit of the TCGF receptor. Similar bands have been noted for HUT-102B2 cells (Leonard *et al.*, 1982). We had previously identified another protein of 180 000 daltons in immunoprecipitations from [^{35}S]methionine-labeled PHA-activated normal lymphocytes and HUT-102B cells (Leonard *et al.*, 1982); however, this band is only variably present and we do not believe it to be structurally related to the TCGF receptor.

As noted above, there is a difference (approximately 5000 daltons) in the migration of p55 on reducing and non-reducing gels. This is clearly demonstrated in Fig. 3.4. In these experiments, PHA-activated normal lymphocytes

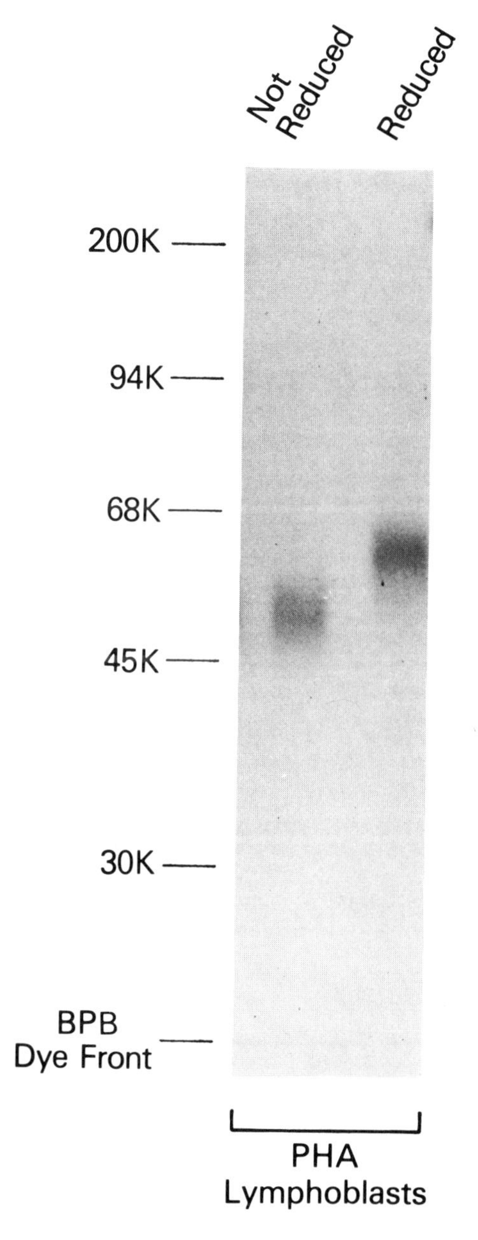

Fig 3.4 Anti-Tac immunoprecipitations of [^{3}H]D-glucosamine-labeled PHA-activated lymphoblasts, electrophoresed on 8.75% SDS/polyacrylamide gels under non-reducing (left lane) or reducing (right lane) conditions.

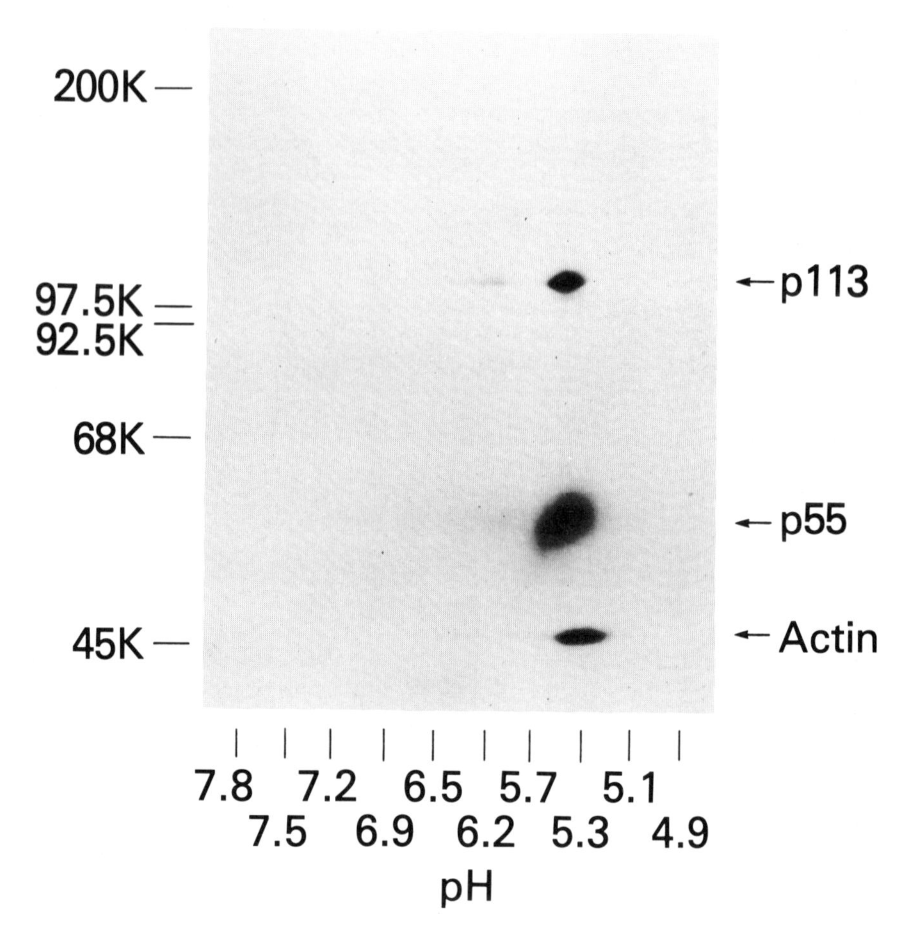

Fig. 3.5 Two-dimensional gel of anti-Tac immunoprecipitation from [^{35}S]methionine-labeled PHA-activated lymphoblasts. Migration of p113, p55, and actin are indicated.

were biosynthetically labeled with [^{3}H]D-glucosamine, solubilized with Triton X-100, immunoprecipitated with anti-Tac, and analyzed on 8.75% SDS-PAGE. Lanes A and B represent the migration of the receptor electrophoresed under non-reducing and reducing conditions, respectively. The increase in apparent receptor size following treatment with dithiothreitol suggests the presence of intrachain disulfide bond(s). Since these cells were labeled with [^{3}H]D-glucosamine, p55 is a glycoprotein. These data confirm our reports that the TCGF receptors on HUT-102B2 cells and a variety of other human T cell leukemia/lymphoma virus (HTLV)*-infected T cell lines are

*In this chapter HTLV is used to refer to the original HTLV, now also denoted HTLV-I.

glycoproteins that contain intrachain disulfide bonds (Leonard *et al.*, 1983b).

Western blotting has been performed using anti-Tac and HUT-102B2 cells to confirm the specificity of anti-Tac (Leonard *et al.*, 1983a). Such immunoblotting was only successful when gels were electrophoresed under non-reducing conditions, suggesting that the receptor's disulfide bonds must be intact for it to be recognized by anti-Tac.

We have also determined the isoelectric point for p55. [^{35}S]Methionine-labeled PHA-activated normal lymphocytes were immunoprecipitated with anti-Tac, and electrophoresed on a two-dimensional gel (Fig. 3.5), according to the method of O'Farrell (1975). p55 migrated as a diffuse spot with a pI of 5.4–5.7, slightly more acidic than that previously shown for the receptor present on HUT-102B2 cells (pI 5.5–6.0) (Leonard *et al.*, 1983b). The diffuseness of the spot is consistent with this protein being a glycoprotein of some intrinsic heterogeneity. The receptor presumably is not composed of diverse peptide chains since sharp methionine and leucine peaks were obtained when the TCGF receptor on HUT-102B2 cells was labeled with [^{35}S]methionine and [^{3}H]leucine and subjected to sequential Edman degradation on a Beckman sequenator (W.J. Leonard, S. Rudikoff and W.C. Greene, unpublished observations). Also seen on this gel is p113. In addition, since the gel was run under reducing conditions, actin is also identified (in Fig. 3.1b, performed under non-reducing conditions, actin is largely polymerized with myosin as actomyosin and does not enter the gel). The actin seen on the two-dimensional gel migrates with a pI just slightly more acidic than that which has been reported in the literature for other eukaryotic actins (Horovitch *et al.*, 1979).

In summary, we have shown so far that the TCGF receptor on PHA-activated normal lymphocytes is a 55000-dalton membrane glycoprotein that contains intrachain disulfide bonds and has a pI of 5.4–5.7. This receptor is similar in size to many other human T cell membrane glycoproteins. The surface antigens T1 (van Agthoven *et al.*, 1981), T4 (Terhorst *et al.*, 1980), T5 (Terhorst *et al.*, 1980), and T11 (Reinherz *et al.*, 1982) all have apparent M_rs of 55000 to 76000. In addition, a 60000-dalton transforming growth factor receptor has been described (Massague *et al.*, 1982). In contrast, many of the classical growth factor receptors are much larger. For example, the insulin (Massague *et al.*, 1981a) and insulin-like growth factor I (IGF-I) (Kasuga *et al.*, 1981) receptors under non-reducing conditions migrate with apparent M_rs of over 300000, and following reduction, the binding subunit is 130000 daltons (see Chapter 1 by S. Jacobs). The IGF-II receptor is 260000 daltons (Massague *et al.*, 1981a), the nerve growth factor receptor is 143000 daltons (Massague *et al.*, 1981b), the platelet-derived growth factor receptor is 164000 daltons (Glenn *et al.*, 1982), and the epidermal growth factor receptor (see Chapter 12 by Schlessinger *et al.*) is 175000 daltons (Wrann and Fox, 1979). Some of these other growth factor receptors are composed of multiple subunits. As noted above, we do not exclude the possibility that the TCGF receptor either exists as

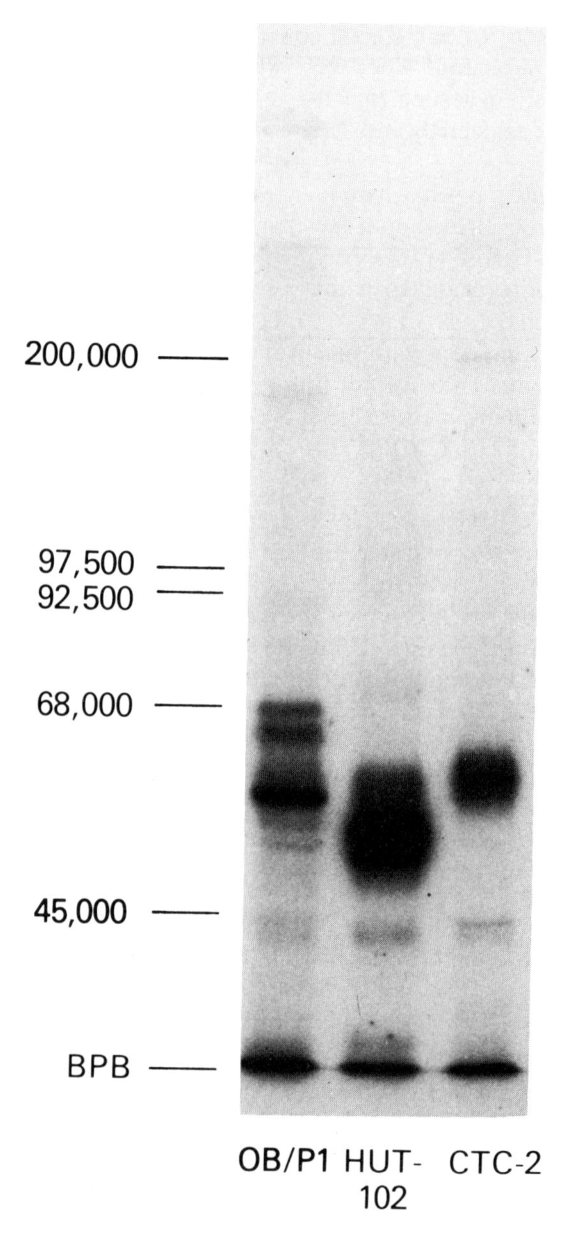

Fig. 3.6 Anti-Tac immunoprecipitations of ^{125}I-surface-labeled OB/P1, HUT-102B2 and CTC-2 cells, electrophoresed on a 7.5% SDS/polyacrylamide gel. The moderate size heterogeneity of these receptors is evident.

a dimer or higher-polymeric form of p55, or that another protein, such as p113 may also be a receptor subunit. Alternatively, the total TCGF receptor on PHA-activated T cells may simply be a single monomeric protein of 55000 daltons.

3.5 HETEROGENEITY OF TCGF RECEPTORS IN LEUKEMIC T CELL LINES INFECTED WITH HUMAN T CELL LEUKEMIA LYMPHOMA VIRUS (HTLV); POST-TRANSLATIONAL PROCESSING OF THE RECEPTOR

By screening a variety of T cell lines, we found that most T cell tumor lines do not express TCGF receptors, but that all HTLV-infected T cell lines expressed receptors (Waldmann *et al.*, 1984). HTLV is a type C retrovirus, identified by Gallo and co-workers, that is etiologically associated with adult T cell leukemia (Poeisz *et al.*, 1980, 1981). We were interested in evaluating whether the receptor on HTLV-infected cells differed in any way from that on PHA-activated normal lymphocytes, since at least some of the HTLV-infected lines have become autonomous of exogenous TCGF for growth. Such autonomy could be due to (1) the production of sufficient TCGF by these cells to support their own growth via their own TCGF receptors. Indeed, Gootenberg and co-workers (1981) have reported that some HTLV lines studied produce their own TCGF; (2) HTLV transformation which may cause cells to grow by a mechanism independent of TCGF and TCGF receptors, even though both the growth factor and receptor may be present; or (3) the cells have altered receptors that are constantly 'activated' and do not necessarily require ligand binding to produce proliferation. Consistent with this last possibility, we show that at least some HTLV-infected cell lines have aberrantly sized receptors.

Fig. 3.6 demonstrates the degree of heterogeneity that exists among three different HTLV-infected cell lines, OB/P1, HUT-102B2, and CTC-2. In this experiment, cells from each line were surface-iodinated with ^{125}I, immunoprecipitated and analyzed on SDS gels. For reference purposes, from data not shown, CTC-2 has a receptor identical in size with that of PHA-activated normal lymphocytes (unpublished observations). It is therefore of interest that although CTC-2 and HUT-102B2 are derived from the same patient, CTC-2 cells are dependent on TCGF for growth, and their growth is inhibited by anti-Tac (Table 3.1), whereas growth of HUT-102B2 cells is independent of exogenous TCGF and is not inhibited by anti-Tac. It must be emphasized, however, that these growth differences have not been shown to be due to the aberrancy of the HUT-102B2 receptor. A larger variety of HTLV-infected cells has been studied, showing additional lines with normally and slightly abnormally sized receptors (unpublished observations).

Table 3.1 Percentage suppression of TCGF-induced proliferation

	Percentage suppression				
Dilution of Anti-Tac*	10^{-6}	10^{-5}	10^{-4}	10^{-3}	10^{-2}
CTC-2 cells	0	0	68	78	85
HUT-102B2 cells	1	1	1	3	4

*Ascites fluid.

Depicted in Table 3.1 are data showing that TCGF-dependent proliferation of CTC-2 cells is inhibited by anti-Tac; in contrast, the growth of HUT-102B2 cells is neither stimulated by TCGF nor inhibited by anti-Tac. This is consistent with all three of the possible explanations of autonomous growth listed above.

We next attempted to determine the molecular basis of these differences. For convenience, we compared HUT-102B2 cells, which have the most aberrantly sized receptor, to PHA-activated normal lymphocytes. Shown in Fig. 3.7 are analyses of anti-Tac immunoprecipitations of [^{3}H]D-glucosamine-labeled PHA-activated lymphoblasts (lanes A and B) and HUT-102B2 cells (lanes C and D) digested with neuraminidase (lanes A and C) or not digested (lanes B and D). Comparing lanes B and D, the size difference of the uncleaved receptors is again evident. Both are cleaved by neuraminidase, demonstrating that both contain sialic acid; however, the PHA-activated lymphoblast receptor remains larger. Thus, the difference in receptor size is not due solely to differences of extent of sialic acid addition. It is of interest that the PHA-activated lymphoblast receptor appears to contain slightly more sialic acid on the basis of apparent molecular weights (it declines by 7000 as compared to 5000 for the HUT-102B2 receptor), which is consistent with its having a slightly more acidic pI (Leonard *et al.*, 1983b).

Given that both receptors are glycoproteins, we next evaluated the post-translational processing of the protein by employing a combination of pulse–chase, tunicamycin and endoglycosidase F experiments. First, as shown in Fig. 3.8, HUT-102B2 cells were labeled with [^{35}S]methionine for 15 minutes and then chased with a large excess of unlabeled methionine for 0, 15, 30, 60, 120 or 240 minutes. A precursor doublet of 35 000 and 37 000 daltons is identified early, and by 60 minutes of chase, the mature form of the receptor is evident. Exactly analogous findings are seen with PHA-activated normal lymphocytes, including the time course of appearance of the mature receptor (Leonard *et al.*, 1983b). The one difference, of course, is in the size of the mature receptor (55 000 daltons for PHA-activated lymphoblasts compared to 50 000 daltons for HUT-102B2 cells). If cells are labeled in the presence of tunicamycin (Tkacz and Lampen, 1975; Takatsuki *et al.*, 1975), which prevents addition of *N*-linked

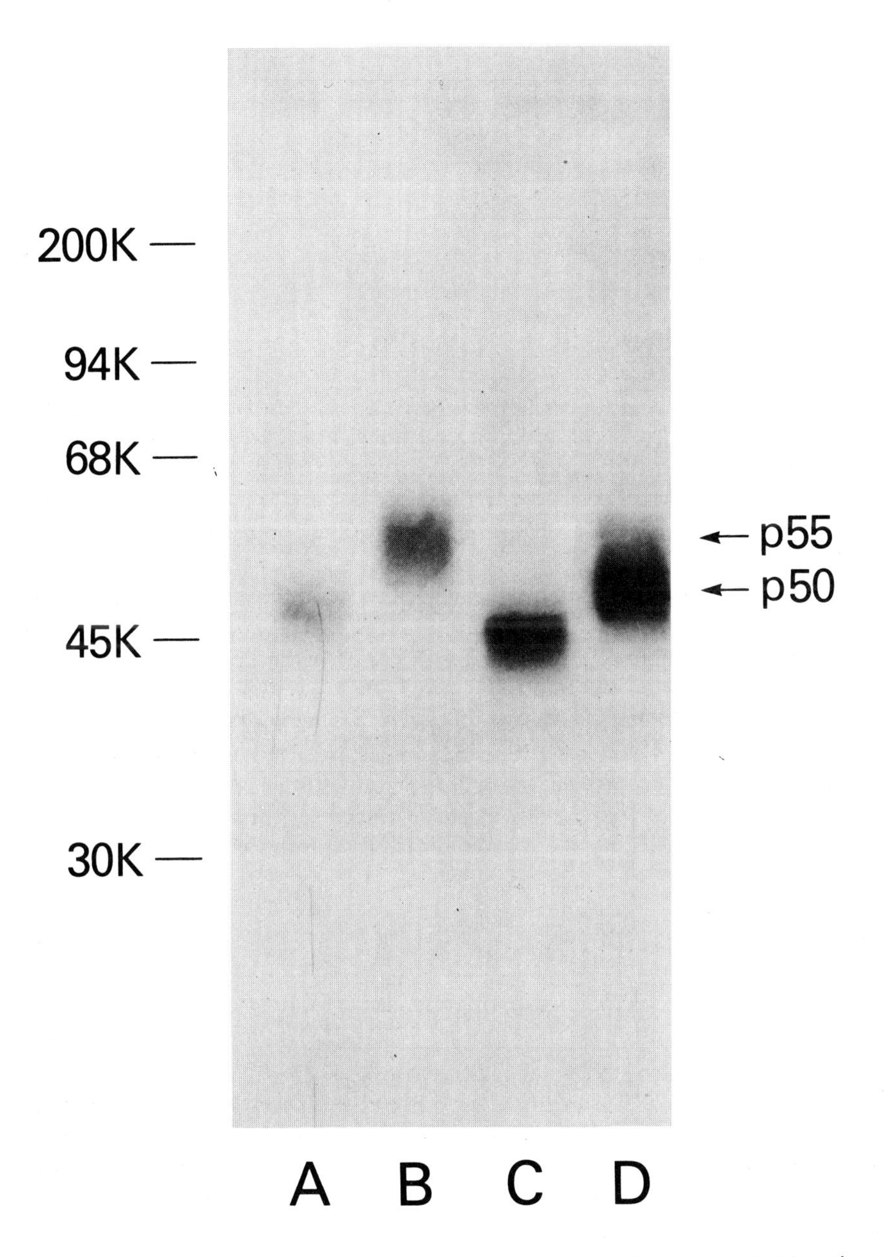

Fig. 3.7 Anti-Tac immunoprecipitations of [^{3}H]D-glucosamine-labeled PHA-activated lymphoblasts (lanes A and B) and HUT-102B2 cells (lanes C and D) digested with neuraminidase (lanes A and C) or not digested (lanes B and D).

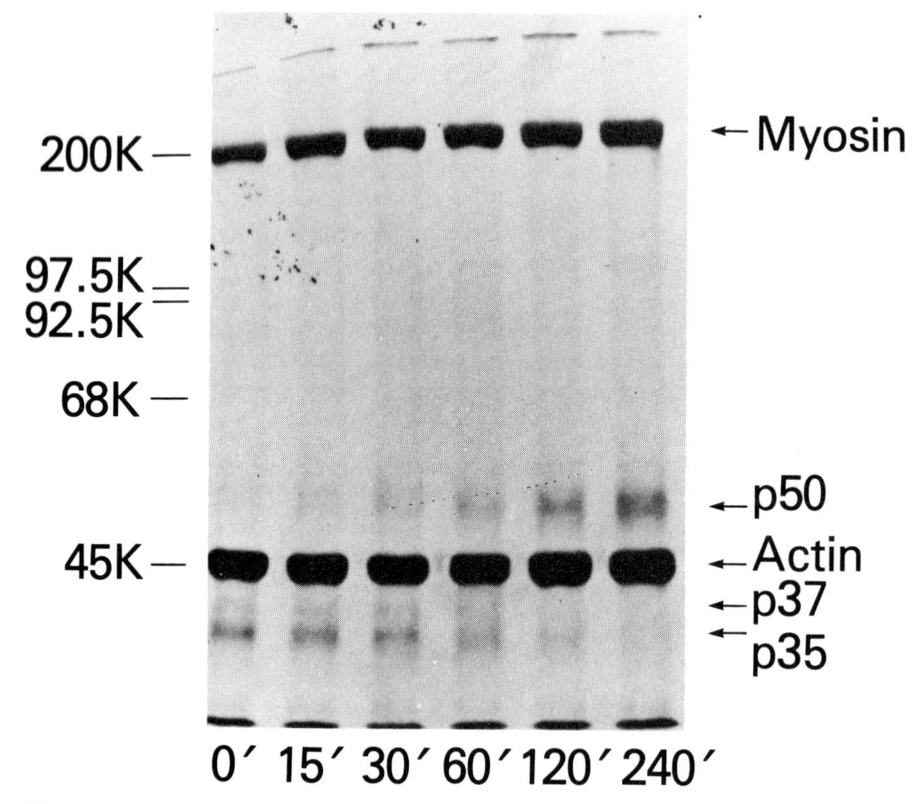

Fig. 3.8 Anti-Tac immunoprecipitations of pulse–chase labeled HUT-102B2 cells, electrophoresed on a 7.5% SDS/polyacrylamide gel under reducing conditions. Locations of the HUT-102B2 mature receptor (p50) and precursor bands p35 and p37 are indicated.

sugars, the sizes of both the precursor and mature forms of the PHA-activated lymphoblast TCGF receptor decrease by 2000–4000 daltons (Leonard *et al.*, 1983b). The p35/p37 doublet is then seen as a 33000 dalton band (p33) (unpublished data), identical with what is seen for HUT-102B2 cells (Leonard *et al.*, 1983), and the 55000-dalton band is seen as approximately 52000 daltons, analogous to the same-sized decrease manifested in the HUT-102B2 TCGF receptor (Leonard *et al.*, 1983a). It is interesting that we have not detected the 33000 dalton band in pulse–chase studies except in the presence of tunicamycin. This presumably reflects the very rapid addition of *N*-linked sugars. p33 is not seen even if only a 5-minute labeling period is used. We thus conclude that the receptor on both cell types begins as a 33000-dalton peptide that is first rapidly glycosylated by an *N*-linked mechanism to a p35/p37 doublet and then further processed to the mature receptor size (Leonard *et al.*, 1983 and unpublished data).

These findings were confirmed in experiments using endoglycosidase F cleavage of the [^{35}S]methionine-pulse-labeled HUT-102B2 and PHA-activated lymphoblast TCGF receptors. Endoglycosidase F cleaves *N*-linked carbohydrate (Elder and Alexander, 1982), and, in both cases, cleaves both the mature receptor and the p35/p37 doublet down to the sizes already demonstrated when cells were cultured in the presence of tunicamycin (Leonard *et al.*, 1983a and unpublished data). Neuraminidase digestion of the HUT-102B2 and PHA-activated lymphoblast precursor forms (p35/p37) indicate that they do not contain any sialic acid, and suggest that addition of sialic acid is a late event (Leonard *et al.*, 1983b).

We postulated that at least part of the large saltatory change between p35/p37 and the mature receptor sizes was due to *O*-linked glycosylation. Thus, we first sought to demonstrate the presence of *O*-linked carbohydrate. We labeled the HUT-102B2 TCGF receptor with [^{3}H]D-glucosamine and cleaved with either endoglycosidase F, neuraminidase, or both endoglycosidase F and neuraminidase. After endoglycosidase F digestion of [^{3}H]D-glucosamine-labeled p50, a clear band was still seen suggesting that the cleaved p50 must have *O*-linked carbohydrate. Further, digestion with both neuraminidase and endoglycosidase F yielded a smaller band than digestion with endoglycosidase F alone, suggesting the presence of sialic acid on *O*-linked carbohydrate structures. Thus, there are two forms of evidence that suggest the presence of *O*-linked carbohydrate (Leonard *et al.*, 1983a).

Whether the entire change from p35/p37 to the mature form is due to the introduction of *O*-linked carbohydrate is unknown. Such a large amount of *O*-linked carbohydrate has been observed for the low-density-lipoprotein receptor (Brown *et al.*, 1983). We cannot completely exclude the covalent association of another small peptide to the precursor protein via transglutamination. Further, post-translational modifications involving phosphorylation and sulfation may also contribute to the increase in receptor size.

Thus, the HUT-102B2 TCGF receptor has a peptide backbone identical in size with that on PHA-activated lymphoblasts. Both are processed to p35/p37 forms by *N*-linked glycosylation, and then to the mature receptor size by a process including *O*-linked glycosylation. Although some other HTLV-infected cell lines also have slightly different sized receptors, we have not yet investigated if these are similarly processed, although we do know that all are glycoproteins with intrachain disulfide bonds (unpublished data). Whether the aberrancy of these TCGF receptors is related to the malignant growth of these cells is at present unresolved.

3.6 PREVALENCE OF TCGF RECEPTORS ON HUT-102B2 CELLS

We attempted to obtain a crude estimate of the prevalence of the mRNA encoding the TCGF receptor by determining the percentage of *de novo* protein synthesis represented by the receptor compared with total cellular protein synthesis. HUT-102B2 cells routinely express more receptors than do PHA-activated lymphoblasts, and we therefore chose to perform the estimate on these cells. Cells were labeled overnight with [^{35}S]methionine or [^{3}H]leucine, solubilized in buffer containing Triton X-100, and total trichloroacetic acid (TCA)-precipitable counts in the cellular extract determined. The extract was then immunoprecipitated with anti-Tac, electrophoresed on a tube gel, and the gel was sectioned and slices solubilized in protosol/liquifluor. The number of counts specific for the receptor were then determined. Using either label, we found that the receptor accounted for 0.05% of total *de novo* protein synthesis in HUT-102B2 cells, suggesting that it is encoded for by a mRNA of low to moderate abundance (Leonard *et al.*, 1983a).

3.7 PRESENCE OF TCGF RECEPTORS ON VARIOUS LEUKEMIC CELLS

We have surveyed a variety of other leukemias for the presence of TCGF receptors. Anti-Tac is a useful diagnostic tool for distinguishing the HTLV-positive adult T cell leukemias (ATL) from the HTLV-negative Sezary Syndrome. Both of these diseases are malignancies of T cells in which the skin may be extensively involved. Histopathologically, they can be difficult to distinguish, although clinically, ATL has a more aggressive course. Waldmann and co-workers (1984) have demonstrated that all HTLV-positive leukemic cells express TCGF receptors, whereas the uninfected Sezary cells do not. In a few cases as previously noted by Uchiyama (unpublished observations), cells from ATL patients express comparatively few receptors if examined immediately after phlebotomy; however, after 24 hours of culture, these cells express far more receptors. Waldmann and co-workers have recently initiated a therapeutic trial to evaluate if anti-Tac might be clinically useful in treating patients with ATL, which, as noted, has an extremely aggressive course and has generally proven to be refractory to conventional therapy. Anti-Tac could theoretically work, at least in some patients, by (1) blocking proliferation by interfering with the binding of TCGF and thus inhibiting TCGF-induced proliferation; (2) by complement-mediated cell lysis; (3) by modulating TCGF receptors on the cell surface (although Tsudo and co-workers (1983) have reported that anti-Tac does not modulate receptors on

ATL cells); or (4) by arming for antibody-dependent cell-mediated cytotoxicity (ADCC). It is certainly possible that, should native anti-Tac not prove useful, then either anti-Tac conjugated to a toxin such as ricin or to an alpha emitting radioisotope might still prove useful. The clear advantage that anti-Tac could have therapeutically is that it would be relatively specific for the malignant cells.

Anti-Tac has also been shown to be reactive with Hairy Cell Leukemia cells. Korsmeyer and co-workers (1983) have demonstrated that Hairy Cell Leukemia is a malignant proliferation of B cells, as indicated by (1) effectively rearranged heavy- and light-chain immunoglobulin genes, (2) the presence of mRNA encoding immunoglobulin demonstrated by Northern blots, and (3) variable expression of surface immunoglobulin of appropriate heavy- and light-chain isotypes. These cells were also shown to express Tac antigen on their surface in eight out of eight cases, as evaluated by fluorescence-activated cell sorting (FACS), and by binding studies with radiolabeled anti-Tac. Two color fluorescence-activated cell-sorting experiments confirmed that anti-Tac bound to the same cells that expressed surface immunoglobulin. Anti-Tac immunoprecipitation of [^{35}S]methionine-labeled cells revealed that the TCGF receptors on Hairy Cell Leukemic cells are the same in size as those found on PHA-activated normal lymphocytes. The significance of apparent TCGF receptors on these cells is unclear. It is conceivable that B cells at some point in their maturation normally express TCGF receptors. Alternatively this may merely be a non-physiologic expression of the receptors induced by a malignant transformation. We have also now identified anti-Tac reactivity with cells of some patients with Burkitt's lymphoma, which is a malignancy of Epstein-Barr-virus-transformed B cells (unpublished observations). The degree of anti-Tac binding to these cells is very low compared with a positive T cell line, but nevertheless is specific. This again raises the possibility of a role for TCGF receptors in B cell biology.

We have also identified TCGF receptors on phorbol ester (PMA)-activated JURKAT and HSB-2 leukemic acute lymphocytic leukemia cell lines (Greene *et al.*, 1983). Both of these cell lines have been known to produce TCGF when stimulated with PHA and PMA. We have now shown that they display TCGF receptors in response to PMA or PHA plus PMA. It is of interest that the JURKAT leukemic cells express a receptor slightly aberrant in size; preliminary data suggest identically sized precursors to the receptor on PHA-activated normal lymphocytes. It is also interesting that PMA induction of JURKAT cells results in expression of TCGF receptors but minimal secretion of TCGF; thus PMA allows one largely to separate receptor expression from the production of TCGF. Dr Richard J. Robb has recently shown that these cells are capable of binding TCGF, but in amounts markedly lower than what we would predict on the basis of the extent of anti-Tac binding.

3.8 FUNCTIONAL EFFECTS OF ANTI-TAC ON LYMPHOCYTE ACTIVATION, AND EARLY EVENTS RELATED TO ACTIVATION

We have also investigated functional effects of anti-Tac on human lymphocyte activation (Depper *et al.*, 1983; see also Chapter 7 by Terhorst for effects of monoclonal antibodies on T cell function). Anti-Tac blocks T cell proliferation occurring after stimulation with soluble antigens, autologous antigens and alloantigens. It partially inhibits T cell proliferation induced by mitogenic lectins and the degree of inhibition correlates inversely with the potency of the mitogenic stimulus. Anti-Tac blocks the generation of cytotoxic T lymphocytes in allogenic cell co-cultures but does not inhibit the cytotoxic effector function by these cells once generated. Anti-Tac also inhibits T-cell-dependent pokeweed mitogen induced B cell immunoglobulin production. These effects were shown to not be mediated by a decrease in TCGF concentrations, or by production of a soluble suppressor factor, but rather appeared to be due to blockade of the TCGF receptor. Indeed, excess purified TCGF was able to overcome anti-Tac inhibition of antigen-induced proliferation.

Using anti-Tac, Neckers and Cossman (1983) have demonstrated that TCGF receptors appear in advance of transferrin receptors (see Chapter 10 by Trowbridge and Newman), and that TCGF receptors are obligatory prerequisites for the appearance of transferrin receptors. Further, anti-Tac, if added into cultures early, prevents the appearance of transferrin receptors.

Tsudo and co-workers (1982) have shown similar results for Ia antigens. Anti-Tac, when added to allo-activated T cells, blocks the appearance but not the maintenance on the cell surface of Ia antigens. They also report that anti-Tac modulates its own receptor off the cell surface of activated normal T cells, but that receptors returned if anti-Tac was washed away. In contrast, Tsudo and co-workers (1983) have also shown that anti-Tac does not modulate its receptor off ATL cells.

Maizel and co-workers (1981) have proposed that TCGF provides a signal necessary for activated lymphocytes to enter S phase. The data of Neckers and Cossman (1983) support the need for TCGF, TCGF receptors and transferrin receptors in order for cells to enter S phase. Thus, it may serve as a necessary but not sufficient signal for progression into S phase.

Cotner and co-workers (1983) have studied a variety of monoclonal antibodies to T cell activation antigens, including the transferrin receptor, and the antigens recognized by anti-Tac, 49.9, and 4F2. All of these were found on T cells by the time that cells entered the S phase of the cell cycle. The 49.9 monoclonal antibody appears similar to anti-Tac in terms of its cellular reactivity. We have shown (unpublished observations) that 49.9 and anti-Tac cross-compete for binding sites on HUT-102B2 cells. In collaboration with Dr

Richard J. Robb, we have demonstrated that, like anti-Tac, 49.9 blocks the binding of radiolabeled TCGF to PHA-activated lymphoblasts. We have confirmed Cotner's original observation (personal communication) that these two monoclonal antibodies immunoprecipitate identical diffuse bands from HUT-102B2 cells, suggesting that 49.9 also recognizes the human receptor for T cell growth factor.

3.9 ANTIBODIES TO TCGF RECEPTORS IN OTHER SPECIES

Osawa and Diamantstein (1983) have described a monoclonal antibody ART18 that appears to recognize the rat TCGF receptor. This monoclonal antibody reacts with activated rat T lymphocytes but not with thymocytes or splenocytes. It inhibits the ability of rat T lymphoblasts to proliferate in response to TCGF and to absorb TCGF activity, and identifies a protein of approximately 55 000 daltons (personal communication).

Malek and co-workers (1983) have identified a rat monoclonal antibody 7D4 that appears to react with the murine TCGF receptor. 7D4 is expressed at high level on HT2 cells and concanavalin A (Con A) ionduced murine T lymphoblasts. It is present, although at much lower levels, on lipopolysaccharide-induced B cell blasts, and not detected on $>95\%$ of thymocytes and resting T and B cells. 7D4 specifically immunoprecipitated human [^{3}H]TCGF from detergent extracts of HT2 cells, suggesting that it immunoprecipitated a TCGF–TCGF receptor complex. Initial characterization of the cell surface antigen recognized by 7D4 reveals broad diffuse bands in the 48–62 000 dalton range.

Thus, both of these other monoclonal antibodies, ART18 and 7D4, which appear to recognize the rat and murine TCGF receptors, immunoprecipitate proteins that are similar in size to that of the human receptor.

3.10 CONCLUSIONS

Since its discovery in 1976 by Morgan, Ruscetti and Gallo, TCGF has been a focus of intensive research. This lymphokine has been of tremendous value in maintaining activated T cell lines *in vitro*. The gene encoding TCGF has now been cloned, and purified recombinant TCGF should be widely available in the not too distant future. Studies of the receptor are now progressing rapidly, largely due to the availability of antibodies to the receptor in three species. At present, the human receptor is the most completely characterized. It is a 55 000-dalton glycoprotein that contains intrachain disulfide bonds and has a pI of 5.3–5.7. This protein consists of only 33 000 daltons of peptide that is post-translationally processed to the mature form. It is interesting that a number of leukemic cells have aberrantly sized receptors, apparently due to differences in post-translational processing. Some of the HTLV-infected cells with aberrantly sized receptors have become autonomous, raising the possibility, as of now unproven, that such altered receptors could relate directly to the autonomous growth of these cells.

Resting cells neither display TCGF receptors nor synthesize TCGF. Both are induced as T cells are activated by antigen. This is therefore an interesting example of inducible gene expression within the immune response, and the regulatory controls of this system at a molecular level are fertile areas for future research.

REFERENCES

Brown, M.S., Anderson, R.G.W. and Goldstein, J.L. (1983), *Cell,* **32,** 663–667.

Cheroutre, H., Devos, R., Plaetinck, G. and Fiers, W. (1983), *J. Cell. Biochem. Suppl.,* **7A,** no. 0439, p. 165.

Clark, S., Arya, S.K., Wong-Staal, F., Matsumoto-Kohayashi, M., Kay, R., Brown, G., Kaufman, R., Copeland, T., Oroszlan, S., Smith, K., Sarngadharan, M., Lindner, S. and Gallo, R. (1983), *Proc. Natl. Acad. Sci. U.S.A.,* (in press).

Cotner, T., Williams, J.M., Christenson, L., Shapiro, H.M., Strom, T.B. and Strominger, J. (1983), *J. Exp. Med.,* **157,** 461–472.

Coutinho, A., Larsson, E-L., Gronvik, K-O. and Andersson, J. (1979), *Eur. J. Immunol.,* **9,** 587.

Depper, J.M., Leonard, W.J., Robb, R.J., Waldmann, T.A. and Greene, W.C. (1983), *J. Immunol.,* **131,** 690–696.

Elder, J.H. and Alexander, S. (1982), *Proc. Natl. Acad. Sci. U.S.A.,* **79,** 4540–4544.

Gillis, S., Baker, P.E., Ruscetti, F.W. and Smith, K.A. (1978a), *J. Exp. Med.,* **148,** 1093.

Gillis, S., Ferm, M.M., Ou, W. and Smith, K.A. (1978b), *J. Immunol.,* **120,** 2027–2032.

Gillis, S., Smith, K.A. and Watson, J.D. (1980), *J. Immunol.,* **124,** 1954–1962.

Glenn, K., Bowen-Pope, D.F. and Ross, R. (1982), *J. Biol. Chem.,* **257,** 5172–5176.

Gootenberg, J.E., Ruscetti, F.W., Mier, J.W., Gazdar, A. and Gallo, R.C. (1981), *J. Exp. Med.,* **154,** 1403–1418.

Greene, W.C., Wong-Staal, F.Y., Depper, J.M., Leonard, W.J., Gallo, R.C. and Waldmann, T.A. (1983), *Clin. Res.,* **31,** 344A.

Horovitch, S.J., Storti, R.V., Rich, A. and Pardue, M.L. (1979), *J. Cell. Biol.,* **82,** 82–92.

Kasuga, M., Van Obberghen, K., Nissley, S.P. and Rechler, M.M. (1981), *J. Biol. Chem.,* **256,** 5305–5308.

Koretzky, G.A., Daniele, R.P., Greene, W.C. and Nowell, P.C. (1983), *Proc. Natl. Acad. Sci. U.S.A.,* **80,** 3444–3447.

Korsmeyer, S.J., Greene, W.C., Cossman, J., Hsu, S-M., Neckers, L.M., Marshall, S.L., Jensen, J.P., Bakhshi, A., Leonard, W.J., Jaffe, E.S. and Waldmann, T.A. (1983), *Proc. Natl. Acad. Sci. U.S.A.,* **80,** 4522–4526.

Leonard, W.J., Depper, J.M., Uchiyama, T., Smith, K.A., Waldmann, T.A. and Greene, W.C. (1982), *Nature (London)*, **300**, 267–269.

Leonard, W.J., Depper, J.M., Robb, R.J., Waldmann, T.A. and Greene, W.C. (1983a), *Proc. Natl. Acad. Sci. U.S.A.* **80**, 6957–6961.

Leonard, W.J., Depper, J.M., Waldmann, T.A. and Greene, W.C. (1983b), *Clin. Research*, **31**, 348a.

Lin, Y., Livak, K.J., Arentzen, R., Petteway, S.R. and Robb, R.J. (1983), in *Thymic Hormones and Lymphokines '83*, George Washington University, June, 1983, Abstr.

Liscia, D.S., Alhadi, T. and Vonderhaar, B.K. (1982), *J. Biol. Chem.*, **257**, 9401–9405.

Maizel, A., Mehta, S.R., Hauft, S., Franzini, D., Lachman, L.B. and Ford, R.J. (1981), *J. Immunol.*, **27**, 1058–1064.

Malek, T.R., Robb, R.J. and Shevach, E.M. (1983), *Proc. Natl. Acad. Sci. U.S.A.*, **80**, 5694–5698.

Massague, J., Pilch, P.F. and Czech, M.P. (1981a), *J. Biol. Chem.*, **256**, 3182–3190.

Massague, J., Guilette, B.J., Czech, M.P., Morgan, C.J. and Bradshaw, R.A. (1981b), *J. Biol. Chem.*, **256**, 9419–9424.

Massague, J., Czech, M.P., Iwata, K., De Larco, J. and Todaro, G.J. (1982), *Proc. Natl. Acad. Sci. U.S.A.*, **79**, 6822–6826.

Mier, J.W. and Gallo, R.C. (1980), *Proc. Natl. Acad. Sci. U.S.A.*, **77**, 6134–6138.

Miyawaki, T., Yachie, A., Uwadana, N., Ohzeki, S., Nagaoki, T. and Taniguchi, N. (1982), *J. Immunol.*, **129**, 2474–2478.

Morgan, D.A., Ruscetti, F.W. and Gallo, R.C. (1976), *Science*, **193**, 1007–1008.

Neckers, L.M. and Cossman, J. (1983), *Proc. Natl. Acad. Sci. U.S.A.*, (in press).

O'Farrell, P.H. (1975), *J. Biol. Chem.*, **250**, 4007–4021.

Osawa, H. and Diamantstein, T. (1983), *J. Immunol.*, **30**, 51–55.

Poeisz, B.J., Ruscetti, F.W., Gazdar, A.F., Bunn, P.A., Minna, J.D. and Gallo, R.C. (1980), *Proc. Natl. Acad. Sci. U.S.A.*, **77**, 7415–7419.

Poeisz, B.J., Ruscetti, F.W., Reitz, M.S., Kalyanaraman, V.S. and Gallo, R.C. (1981), *Nature (London)*, **294**, 268–271.

Reinherz., E.L., Meuer, S., Fitzgerald, K.A., Hussey, R.E., Levine, H. and Schlossman, S.F. (1982), *Cell*, **30**, 735–743.

Robb, R.J. (1982), *Immunobiology*, **865**, 21–50.

Robb, R.J. and Greene, W.C. (1983), *J. Exp. Med.*, **158**, 1332–1337.

Robb, R.J., Kutny, R.M. and Chowdhry, V. (1983), *Proc. Natl. Acad. Sci. U.S.A.*, **80**, 5990–5994.

Robb, R.J., Munck, A. and Smith, K.A. (1981), *J. Exp. Med.*, **154**, 1455–1474.

Ruscetti, F.W. and Gallo, R.C. (1981), *Blood*, **57**, 379–394.

Ruscetti, F.W., Morgan, D.A. and Gallo, R.C. (1977), *J. Immunol.*, **119**, 131–138.

Smith, K.A. (1980), *Immunol. Rev.*, **51**, 337–357.

Smith, K.A., Gillis, S. and Baker, P.E. (1979), in *The Molecular Basis of Immune Cell Function*, (Kaplan, J.G., ed.), Elsevier/North-Holland Press, Amsterdam, p. 223.

Stadler, B.M., Berenstein, E.H., Siraganian, R.P. and Oppenheim, J.J. (1982), *J. Immunol.*, **128**, 1620–1624.

Takatsuki, A., Kohno, K. and Tomura, G. (1975), *Agric. Biol. Chem.*, **39**, 2089–2091.

Taniguchi, T., Matsui, H., Fujita, T., Takaoka, C., Kashima, N., Yoshimoto, R. and Hamuro, J. (1983), *Nature (London)*, **302**, 305–310.

Terhorst, C., van Agthoven, A., Reinherz, E.L. and Schlossman, S.F. (1980), *Science,* **209,** 520–521.

Tkacz, J.S. and Lampen, J.O. (1975), *Biochem. Biophys. Res. Commun.,* **65,** 248–257.

Tsudo, M., Uchiyama, T., Takatsuki, K., Uchino, H. and Yodoi, J. (1982), *J. Immunol.,* **129,** 592–595.

Tsudo, M., Uchiyama, T., Uchino, H. and Yodoi, J. (1983), *Blood,* **61,** 1014–1016.

Uchiyama, T., Broder, S. and Waldmann, T.A. (1981a), *J. Immunol.,* **126,** 1393–1397.

Uchiyama, T., Nelson, D.L., Fleisher, T.A. and Waldmann, T.A. (1981b), *J. Immunol.,* **126,** 1398–1403.

van Agthoven, A., Terhorst, C., Reinherz, E.L. and Schlossman, S.F. (1981), *Eur. J. Immunol.,* **11,** 18–21.

Waldmann, T.A., Broder, S., Greene, W.C., Sarin, P.S., Saxinger, C., Blayney, D.W., Blattner, W.A., Goldman, C.K., Frost, K., Sharrow, S. Depper, J.M., Leonard, W.J. Uchiyama, T. and Gallo, R.C. (1984), *J. Clin. Invest.,* (in press).

Watson, J. (1979), *J. Exp. Med.,* **150,** 1510.

Wrann, M.M. and Fox, C.F. (1979), *J. Biol. Chem.,* **254,** 8083–8086.

Notes added in proof

1. Arya *et al.* (1984), Science, **223,** 1086–1087, have recently shown that several HTLV-infected, TCGF-independent cell lines do not express TCGF mRNA, ruling out a general autocrine growth mechanism for those cell lines.
2. Preliminary data suggest that TCGF receptors are both phosphorylated and sulfated.

4 Monoclonal Antibodies to Steroid Receptors

BRUNO MONCHARMONT and INDU PARIKH

Monoclonal Antibodies to Receptors: Probes for Receptor Structure and Function
(*Receptors and Recognition*, Series B, Volume 17)
Edited by M. F. Greaves
Published in 1984 by Chapman and Hall, 11 New Fetter Lane, London EC4P 4EE

4.1 INTRODUCTION

The original idea of the interaction of toxins and hormones with specific recognition sites on the cell was proposed at the beginning of this century by Paul Ehrlich (1900). Furthermore, he proposed that such sites were responsible for the physiological sensitivity and selectivity of the target cell. It was only two decades ago, however, that the existence of a specific site of interaction in the cells of target tissue for a steroid hormone, estradiol, was discovered (Glascock and Hoekstra, 1959; Jensen and Jacobson, 1962). In the following years this receptor was characterized (Toft and Gorski, 1966; Puca and Bresciani, 1969) and receptors for other steroid hormones were discovered. A receptor for androgen was observed in the rat ventral prostate (Baulieu and Jung, 1970; Fang *et al.*, 1969; Mainwaring, 1969) and the presence of progesterone receptor established in guinea pig uterus (Milgrom *et al.*, 1970) and chick oviduct (O'Malley *et al.*, 1970). A receptor for the glucocorticoid hormones was described in thymus cells (Munck and Wira, 1971) and hepatoma cells (Baxter and Tomkins, 1971) and for the mineralocorticoids in the toad bladder (Sharp *et al.*, 1966). Specific receptors for 1,25-dihydroxyvitamin D (Brumbaugh *et al.*, 1975) and for the insect hormone ecdysone (Yund *et al.*, 1978) were also discovered.

In the 1970s, the efforts of several groups led to the purification to homogeneity of some of the steroid receptors including the estrogen receptor (Sica *et al.*, 1973; Molinari *et al.*, 1977), the progesterone receptor (Schrader *et al.*, 1977; Coty *et al.*, 1979) and the glucocorticoid receptor (Govindan and Sekeris, 1978; Wrange *et al.*, 1979). This led to the production of polyclonal and later monoclonal antibodies to the receptor proteins, which will be discussed in the following sections.

4.2 GENERAL REMARKS ON STEROID RECEPTORS

The unifying hypothesis of the mechanism of action of steroid hormones appears to stand on seven 'pillars' which may be summarized as follows: (1) the steroid molecule enters the cell by passive diffusion through the plasma membrane; (2) the selectivity and specificity may be determined by the presence in the cells of a protein able to bind the hormone with high affinity and, generally, low capacity; (3) such a protein is present in the cytoplasmic compartment of the cell; (4) binding of the hormone to its receptor protein triggers a modification in the latter promoting (5) an elective association of the receptor with nuclear structures; (6) such association is able to modify, in a specific way, the gene expression of the cell producing new mRNA; (7) after the completion of its action, the receptor is destined to be recycled or processed. The reader can find an extensive description of our current understanding of steroid hormone action in some of the recent reviews (Jensen and DeSombre, 1972, 1973; O'Malley and Means, 1974; O'Malley and Schrader, 1976; Gorski and Gannon, 1976; Yamamoto and Alberts, 1976; Jensen, 1978; Muldoon, 1980; Jensen *et al.*, 1982; Grody *et al.*, 1982).

Although a great amount of work has been dedicated to elucidating and validating this basic hypothesis, our overall knowledge about the action of steroid hormones is still limited. Furthermore, experimental results have been produced, which are difficult to reconcile with the hypothesis or, in some instances, are in direct contradiction to it. Recent reports indicating the presence of a possible receptor or transport protein (Munck and Brinck-Johnsen, 1968; Baulieu *et al.*, 1978; Parikh *et al.*, 1980) on the cell surface for selective uptake of steroids into the target cells questions the passive diffusion mechanism of steroid uptake. Among other findings which do not fit with the unified hypothesis mention should be made of the direct effect of steroid hormones on protein synthesis and the interaction of the estradiol receptor with ribonucleoprotein particles (Liang and Liao, 1974; 1975), the presence of estrogen- and androgen-binding sites on the nuclear matrix (Barrack and Coffey, 1980), in the microsomes (Parikh *et al.*, 1980) or on the plasma membrane (Pietras and Szego, 1977) and the binding of the estrogen receptor to cytoskeletal structures (Puca *et al.*, 1981).

The investigations performed during the past few years have added a large amount of detail to the unified model of steroid action. It is clear, however, that the receptor plays the key role in eliciting the hormonal response. The presence of the receptors in the cells of the target tissue will determine the response to the hormonal stimulus. The concentration of steroid receptors in target cells is of the order of 10^4 copies per cell.

All the receptors for steroid hormones (estrogens, androgens, progestins, glucocorticoids, mineralocorticoids, vitamin D) share certain common characteristics. The molecular weight is between 70000 and 110000 with high affinity for the respective steroids ($K_d \leq 1$ nM). They also have a marked tendency to aggregate in their native state, and the steroid-binding site is relatively heat-labile. Furthermore, they have a domain in the molecule that is able to interact with DNA and/or with the nuclear chromatin. The physicochemical characteristics of the receptor for a given steroid hormone are usually similar, regardless of the target tissues or the animal species. More recently, it has been demonstrated that the progesterone receptor of chick oviduct (Weigel *et al.*, 1981a; Dougherty *et al.*, 1982), the glucocorticoid receptor from mouse fibroblast (Housley and Pratt, 1983) and the estrogen receptor of the calf uterus (Migliaccio *et al.*, 1982) are phosphoproteins. In the latter, phosphorylation modulates the activity of the steroid-binding site. The existence of a similar phenomenon has also been postulated for the glucocorticoid receptor (Sando *et al.*, 1979). The majority of these studies used the steroid-binding site of the receptor protein for its own recognition. Such recognition is undoubtedly highly specific and useful for practical purposes in the light of the high affinity and relative specificity of the steroid-binding sites and of the low dissociation rate of the hormone from such sites.

The extensive purification of the steroid receptors has made it possible to produce antibodies against these proteins which provides scientists with a new tool for the recognition of the receptors independent from the steroid-binding site (Table 4.1).

4.3 POLYCLONAL ANTIBODIES

The importance of the availability of antibodies to the receptors for the study of the mechanism of action of the related hormones was recognized a few years after the discovery of steroid receptors and, indeed, much before the receptors were sufficiently purified. The presence of only small amounts of receptor protein in the cell (between 10^4 and 5×10^4 copies/cell) combined with its relative lability had frustrated earlier attempts at purification. With the advent of affinity chromatography (Cuatrecasas *et al.*, 1968), the first steroid hormone receptor, namely the 4 S form of estrogen receptor (Sica *et al.*, 1973), was extensively purified.

The first report of the production of a rabbit antiserum to estrogen receptor dated back to 1968 (Soloff and Szego, 1969). This antiserum was produced with a preparation of rat uterus estrogen receptor, and the presence of antibodies was determined by the shift of the estrogen receptor peak on a sucrose gradient in the presence of such antibodies. However, no further characterization of this antiserum was reported. In 1976 another group published the production of rabbit antiserum to calf uterus estrogen receptor (Fox *et al.*, 1976). Although a poorly purified receptor was used as antigen in the latter report, the authors exploited the natural tendency of the receptor protein to form large aggregates in order to increase its antigenicity. Antibodies to the receptor were present in the serum, although at very low titer, and their interaction with the antigen was established by gel filtration chromatography and immunoprecipitation, using the specific, high-affinity binding of radiolabeled estradiol as marker for the antigen. The purification of the estrogen receptor from calf uterus, exploiting an *in vitro* nuclear translocation of the receptor, led to the production of another rabbit antiserum (Greene *et al.*, 1977). In addition to the demonstration of the presence of antibodies to the estrogen receptor by immunoprecipitation, by interaction of the receptor with insoluble immunoabsorbants and by sucrose density gradient analysis, Greene *et al.* (1977) also reported that the antiserum reacted with both the nuclear and cytosolic receptor. This observation suggested a similar or identical nature for the two receptors and supported the 'two-step' theory for the mechanism of action of steroid hormones (Jensen *et al.*, 1968). Furthermore, they showed that other steroid

Table 4.1 Antibodies to steroid receptors

Receptor	References	Antigen injected: Origin	Antigen injected: Purity	Antigen injected: Amount of 1st inject./booster	Antibodies raised in	Affinity K_d	Cross-reactivity with receptor from other species: Mammalian	Cross-reactivity with receptor from other species: Non-mammalian	Effect on steroid binding
Glucocorticoid receptor	Soloff and Szego (1969)	Rat uterus	n.r.*	n.r.	Rabbit	n.r.	n.r.	n.r.	n.r.
	Fox *et al.* (1976)	Calf uterus	< 0.1%	20 μg/60 μg	Rabbit	n.r.	n.r.	n.r.	n.r.
	Greene *et al.* (1977)	Calf uterus	10–30%	20 μg/20–50 μg	Rabbit	n.r.	Yes	n.r.	No
	Greene *et al.* (1979)	Calf uterus	20–40%	100 μg/150 μg	Goat	n.r.	Yes	n.r.	Yes
	Radanyi *et al.* (1979)	Calf uterus	5–20%	170 μg/80 μg	Rabbit	~ 1 nM	n.r.	Yes	n.r.
	Al-Nuaimi *et al.* (1979)	Rat mammary tumors	n.r.	n.r.	n.r.	n.r.	Yes	n.r.	n.r.
	Coffer *et al.* (1980, 1981)	Human myometrium	50%	20 μg	Sheep	n.r.	Yes	n.r.	n.r.
	Raam *et al.* (1981)	Human breast cancer	n.r.	80 ng	Rabbit	n.r.	Yes	n.r.	No
	Greene *et al.* (1980a)	Calf uterus	12–20%	36 μg	Rat–mouse myeloma	0.1 nM	No	No	No
	Greene *et al.* (1980b)	MCF-7 cell line	14%	1 nmol	Rat–mouse myeloma	n.r.	Yes	Yes†	No
	Moncharmont *et al.* (1982)	Calf uterus	20%	10 μg/10 μg	Mouse–mouse myeloma	0.06 nM	Yes	Yes	No
Progesterone receptor	Logeat *et al.* (1981)	Rabbit uterus	20%	70–80 μg	Goat	n.r.	Yes	No	n.r.
	Renoir *et al.* (1982)	Chick oviduct	80–100%	50–90 μg	Goat	n.r.	Yes	Yes	n.r.
	Feil (1983)	Rabbit uterus	n.r.	n.r.	Guinea pig	n.r.	Yes	n.r.	n.r.
	Radanyi *et al.* (1983)	Chick oviduct	5–10% 50–100%	30 μg/15–80 μg	Rat–mouse myeloma	~ 1 nM	No	Yes	No
	Logeat *et al.* (1983)	Rabbit uterus	6%	375/375 p mol	Mouse/mouse	0.1–4 nM	Yes	No	Yes
Estrogen receptor	Govindan and Sekeris (1978)	Rat liver	n.r.	2.5 μg/5 μg	Rabbit	n.r.	n.r.	n.r.	n.r.
	Eisen (1980)	Rat liver	10–30%	8 μg	Rabbit	n.r.	Yes	n.r.	n.r.
	Okret *et al.* (1981)	Rat liver	50%	100 μg/25–50 μg	Rabbit	n.r.	Yes	n.r.	n.r.
	Tsawdaroglou *et al.* (1982)	Rat thymus	n.r.	2.5 μg/5 μg	Rabbit	n.r.	n.r.	n.r.	n.r.
	Grandics *et al.* (1982)	Rat liver	n.r.	50 μg/50–100 μg	Mouse–mouse myeloma	n.r.	n.r.	n.r.	Yes
	Westphal *et al.* (1982)	Rat liver	30%	23 μg/21 μg	Mouse–mouse myeloma	0.5–77 nM	Yes	Yes	No
Vitamin D_3 receptor	Pike *et al.* (1982, 1983)	Chicken intestine	13%	30–50 μg/ 30–50 μg	Rat/mouse	n.r.	Yes	Yes	No

*n.r. = not reported.
†The rat serum showed immunoreactivity with the estrogen receptor from hen oviduct, but the monoclonal antibodies derived from this rat did not.

receptors were unreactive with the same antiserum. These studies were performed using the estradiol-binding activity of the receptor as a specific marker for the antigen; binding of antibodies did not affect hormone binding. With the same procedure, a goat antiserum was also produced (Greene *et al.*, 1979). This antiserum had similar characteristics to the previously described rabbit antiserum but, in contrast to the latter, decreased the affinity of the receptor for steroid.

A cytosolic receptor, purified by affinity chromatography from calf uterus, was used to raise a rabbit antiserum (Radanyi *et al.*, 1979). The presence of a high titer of antibodies with high affinity for the receptor (K_d = 1 nM) was demonstrated and, for the first time, a cross-reactivity with the avian estrogen receptor was reported.

A sheep antiserum was prepared to estrogen receptor from human myometrium (Coffer *et al.*, 1980; Coffer and King, 1981) which was purified by affinity chromatography and isoelectrofocusing to 50% purity. This antiserum also showed reactivity with other mammalian estrogen receptors. Other antisera to estrogen receptors from a dimethylbenzanthracene (DMBA)-induced rat mammary tumor (Al-Nuaimi *et al.*, 1979) and human breast cancer (Raam *et al.*, 1981) have also been reported.

The progesterone receptor has been the subject of more extensive studies on its structure and functions (for a review see Schrader *et al.*, 1981); however, its immunological history is much shorter. Unsuccessful attempts at producing antibodies to progesterone receptor (Schrader *et al.*, 1981) can probably be ascribed to the low antigenicity of such preparations and the presence in sheep serum of 'spontaneous' antibodies to the progesterone receptor (Weigel *et al.*, 1981b). Such antibodies have a high apparent affinity (K_d = 2 nM) for the chick progesterone receptor and are able to recognize an antigen present on both A and B subunits. A specific antiserum to the progesterone receptor from rabbit uterus has, however, been obtained by immunization of a goat with a partially purified (20% purity) receptor (Logeat *et al.*, 1981). This antiserum recognizes both nuclear and cytoplasmic receptors and cross-reacts with various mammalian progesterone receptors but not with avian receptors. It also fails to cross-react with receptors for other steroids. This latter study was performed using radiolabeled R5020 as a specific marker of the receptor protein. Antibodies raised in goats and rabbits to a highly purified and molybdate-stabilized form of chick oviduct progesterone receptor showed analogous properties in addition to the ability to cross-react with the mammalian receptor (Renoir *et al.*, 1982). An antiserum to the molybdate-stabilized progesterone receptor from rabbit uterus produced in guinea pig has also been described (Feil, 1983).

The purified glucocorticoid receptor from rat liver has been used to immunize rabbits (Govindan and Sekeris, 1978; Govindan, 1979). Two bands of purified receptor (90 K and 45 K daltons) have produced two antisera showing reciprocal cross-reactivity. Similar results have been obtained with the receptor from rat thymus (Tsawdaroglou *et al.*, 1982). Yet another preparation of an antiserum with highly purified rat liver glucocorticoid receptor was reported in an extensive study by Eisen (1980). Although successfully used for immunoaffinity chromatography, these sera had only a limited interspecies cross-reactivity. Finally, a rabbit antiserum to the rat liver glucocorticoid receptor has been produced and characterized (Okret *et al.*, 1981). This reagent has been successfully used in a competitive ELISA (enzyme-linked immunoabsorbant assay) technique for quantification of the glucocorticoid receptor independently of the ability of the receptor to bind the ligand. These antibodies were also elegantly used to pursue structural studies of the receptor protein in normal tissues (Okret *et al.*, 1983).

4.4 MONOCLONAL ANTIBODIES

The introduction of monoclonal antibodies revolutionalized the study of immunology as well as of molecular and cellular biology (Köhler and Milstein, 1975; Goding, 1980; Yelton and Schart, 1981). As discussed in other chapters of this volume, monoclonal antibodies are also proving to be of considerable value in the analysis of receptor structure and function.

4.4.1 Monoclonal antibodies to the estrogen receptor

Monoclonal antibodies to steroid receptors were obtained for the first time by Jensen and co-workers, who immunized rats with estrogen receptor protein purified from calf uterus (Greene *et al.*, 1980a). Hydridoma cell lines, produced by hybridization of Lewis rat spleen cells with a mouse mutant myeloma cell line, secreted anti-estrogen receptor immunoglobulins of the M and G_{2A} class. None of these clones showed cross-reactivity with receptors from species other than calf. The IgM antibodies reacted preferentially with the nuclear receptor, whereas the IgG antibodies had a similar affinity for both the cytoplasmic and nuclear receptor (K_d = 0.1 nM). The same group also reported the production of monoclonal antibodies to the human estrogen receptor using for immunization receptor purified by affinity chromatography from the cytosol of a human breast cancer cell line, MCF-7 (Greene *et al.*, 1980b). The production of large numbers of MCF-7 cells was necessary for this purpose. The spleen cells of the immunized Lewis rat were fused with two different mouse myeloma cell lines to yield, after cloning by limited dilution, three hybridoma cell lines. One of these produced IgM and the other two, IgG_{2A}. The antibodies produced by one of the clones were able to recognize the estrogen receptor from human and primates only, whereas the other two were able also to recognize calf and rat uterine receptors. None of the monoclonal antibodies recognized the avian receptor, although the antiserum from the immunized rat donating the spleen cells did. There appears to be some evidence that these three antibodies recognize three different epitopes on the human receptor molecule. Monoclonal antibodies to the native cytoplasmic estrogen receptor from calf uterus have also been obtained by fusion of spleen cells from an immunized mouse with a non-secreting mouse myeloma cell line (Moncharmont *et al.*, 1982). Five hybridoma lines have been isolated by this fusion, producing IgG of the subclass 1 and 2A. Antibodies from four clones show cross-reactivity with the receptor of other mammalian and avian species while one does not react with the human receptor. They exhibit a high affinity (K_d = 0.5 nM) for the estrogen receptor. The affinity for the receptor is at least an order of magnitude higher when the antibodies are immobilized on an insoluble matrix. Although the antibodies from these clones show differences in subclasses and cross-reactivity, they are not able to simultaneously bind to the antigen molecule (i.e. they compete or cross-block in binding assays).

4.4.2 Monoclonal antibodies to the progesterone receptor

Monoclonal antibodies to the chick oviduct progesterone receptor have been produced by fusion of the spleen cells of an immunized rat with a mouse myeloma cell line (Radanyi *et al.*, 1983). The antigen used for the immunization was a partially purified preparation of molybdate-stabilized progesterone receptor from chick oviduct. One clone (BF4), producing IgG_{2B}, was able to recognize the progesterone receptor. This antibody has a high affinity for the progesterone receptor of chick oviduct (K_d = 1 nM) but does not cross-react with progesterone receptors from mammalian species. One interesting and surprising feature of this monoclonal antibody is that, in contrast to all the other steroid receptor antibodies so far described, it also shows reactivity with the other steroid receptors of chick oviduct in their native state (Joab *et al.*, 1983). More recently Milgrom and co-workers (Logeat *et al.*, 1983) reported production of monoclonal (mouse/mouse hybridoma) antibodies against rabbit uterine progesterone receptor. Five of the eleven hybridomas produced were examined in detail. The equilibrium dissociation constant ranged from 0.1 to 4.0 nM. Three of the clones produced IgG_1 while two produced IgG_{2A} isotype. All five monoclonal antibodies cross-reacted with the human and other mammalian (rat, guinea pig) but not avian (chick) progesterone receptors. No cross-reactivity of these antibodies with glucocorticoid receptor from rabbit liver or corticosteroid-binding globulin from rabbit plasma was detectable. Two of the antibodies, at low ionic strength, were able to slightly increase the sedimentation of estrogen receptor; however, this interaction appears to be abolished at high ionic strength. Cross-reactivity to other steroid receptors was not tested.

4.4.3 Monoclonal antibodies to the glucocorticoid receptor

The glucocorticoid receptor from rat liver has been used as antigen for the production of monoclonal antibodies. In the two reports published to date, the receptor was purified by affinity chromatography and the antibodies were synthesized by mouse/mouse hybridomas. In one report (Grandics *et al.*, 1982) culture supernatants of the antibody-producing clones were screened for their ability to inhibit steroid-binding activity. Although these antibodies inhibited binding of the steroid to the receptor, radiolabeled steroid was used, in a chromatographic procedure, to show the presence of antibody by increase in molecular weight of the hormone–receptor complex. This report fails to show convincing evidence of the presence of antibodies to the receptor protein. Another report indicates the production of several clones which secreted antibodies to the rat liver glucocorticoid receptor (Westphal *et al.*, 1982). The presence of antibodies was confirmed by immunoprecipitation and sucrose-density-gradient analysis. While most of these antibodies are of IgM idiotypes one was shown to be IgG. They show a wide variability in their affinity for the receptor (the K_d varies by two orders of magnitude) as well as a variability in cross-reactivity with the glucocorticoid receptor of other species. One domain of the receptor molecule, containing the epitopes recognized by the IgG, is well-preserved in other species (including avian). The IgMs recognize different epitopes. These antibodies furthermore do not prevent the interaction of glucocorticoid receptor with DNA.

4.4.4 Monoclonal antibodies to the vitamin D_3 receptor

Recent reports from Haussler and co-workers (Pike *et al.*, 1982, 1983) describe production of four clones of rat/mouse hybridoma cell lines which secrete antibodies against chicken intestinal cytoplasmic 1α,25-dihydroxyvitamin D_3 receptor. The antigen used in this study was purified by a series of chromatographic procedures to about 13% homogeneity. The four clones described produced distinct subtypes of antibodies including one secreting IgM. These monoclonal antibodies react with both occupied and unoccupied chicken intestinal cytoplasmic as well as nuclear receptors. These antibodies are unreactive with estrogen and glucocorticoid receptors and with both serum- and cytosol-derived vitamin D-binding proteins. However, they display extensive cross-reactivity with $1,25(OH)_2$-vitamin D_3 receptors obtained from a variety of species including human.

4.5 APPLICATION OF ANTIBODIES IN STEROID RECEPTOR STUDIES

4.5.1 Structural studies

The first information on the steroid receptor structure, derived from immunological studies, is related to the presence of domains of the receptor molecule that are conserved through evolution. A detailed analysis of the interspecies cross-reactivity of the monoclonal and polyclonal antibodies to the estrogen receptor provides evidence that, while there is a consistent immunological similarity among the estrogen receptors of mammalian species, cross-reactivity with the antigenic domains with avian receptors is limited. Only two polyclonal antisera, one from a rabbit (Radanyi *et al.*, 1979) and one from a rat (Greene *et al.*, 1980b), raised against mammalian receptors, were able to recognize the estrogen receptor from the chick oviduct. Such cross-reactivity was however lost when spleen cells of the rat were hybridized for generation of monoclonal antibodies. Furthermore, more recently produced monoclonal antibodies (Moncharmont *et al.*, 1982) demonstrate cross-reactivity to avian receptor in four out of five clones so far investigated (unpublished data). Among the mammalian estrogen receptors there are also some differences. Thus, primates have some antigenic determinants not shared with bovine and rodent species. Some of the monoclonal antibodies to calf receptor do not recognize human (Greene *et al.*, 1980a; Moncharmont *et al.*, 1982), while among the monoclonal antibodies to the human receptor (Greene *et al.*, 1980a,b) one shows a cross-reactivity with primates only. Bovine receptors also have antigenic determinants not shared with other species. Some of the monoclonal antibodies to calf estrogen receptor do not show any cross-reactivity with any other species. No unique or exclusive antigenic determinants have been described for the estrogen receptor of rodents which share common antigenic domains with calf and human (Greene *et al.*, 1980a; Moncharmont *et al.*, 1982; Giambiagi and Pasqualini, 1982).

The progesterone receptor also exhibits different antigenic determinants in mammals and birds. An antiserum raised against the mammalian receptor (rabbit) did not cross-react with the chick receptor (Logeat *et al.*, 1981), whereas the antiserum raised against the chicken receptor did cross-react with the mammalian progesterone receptor (Renoir *et al.*, 1982). However, a monoclonal antibody (Radanyi *et al.*, 1983) to the receptor from chick oviduct did not recognize any determinants on the mammalian receptor.

An antiserum to the rat liver glucocorticoid receptor shows only a limited cross-reactivity with receptors from other rodents, indicating that a large proportion of the determinants are species-specific (Eisen, 1980). Another antiserum, however, shows cross-reactivity with the human receptor (Okret *et al.*, 1981). The cross-reactivity pattern of the monoclonal antibodies to the glucocorticoid receptor confirms the above findings (Westphal *et al.*, 1982). While the majority of the IgM class recognize a determinant present only on rat receptors, one of the antibodies recognizes a determinant which is widely conserved among various mammalian and non-mammalian (hen) species.

For all the various steroid receptors studied so far, there is no evidence for the presence of common antigenic determinants with other non-receptor steroid-binding proteins, confirming a lack of genetic relationship among these proteins. Furthermore, the various steroid receptors themselves are probably distinct proteins since antibodies to any particular steroid receptor do not usually recognize other steroid receptors in the same tissue. This is not true, however, for a remarkable monoclonal antibody to the chick oviduct progesterone receptor reported by Radanyi *et al.*, (1983). Apparently the latter antibody recognizes not only the progesterone receptor in the chicken oviduct, but also the estrogen, the glucocorticoid and the androgen receptors (Joab *et al.*, 1983) in their native form ('8 S'). None of these receptors is recognized in the dissociated ('high salt') form. This suggests that the antibody may be directed toward a protein (M_r 90000) that is a common subunit to the 'native' forms of all steroid receptors in this tissue. The validity of such a hypothesis in different tissues of different species has to be further investigated as well as its physiological significance. Finally, no differences have been detected, by immunological methods, among the individual steroid receptors in different tissues of the same species. This confirms a unitary identity of the receptor for each single steroid.

Monoclonal antibodies, by virtue of their specificity for only one antigenic determinant, can be used to investigate the subunit structure of a receptor protein. They form stoichiometric complexes with their antigens and therefore can provide information on the number of antigenic determinants present on the receptor molecule. The estrogen receptor from calf uterus has been the subject of studies using this approach (Moncharmont *et al.*, 1982; Moncharmont and Parikh, 1983). Various molecular forms of the calf uterus estrogen receptor have been investigated. The native '8 S' estrogen receptor has been found to have two antigenic determinants per molecule, whereas the 'high salt' 4 S form contains only one. This suggests the presence of a homodimer of the 4 S subunit, associated with yet another protein, in the large '8 S' form of the native receptor. The nuclear receptor, which was postulated to be a homodimer of the 4 S subunit (Notides and Nielsen, 1974; Nielsen and Notides, 1975; Notides *et al.*, 1975, 1981), has been demonstrated to have two antigenic determinants for the monoclonal antibody tested (Moncharmont *et al.*, 1984).

The glucocorticoid receptor from rat liver occurs in different molecular forms depending on the conditions used for the tissue homogenization. Frozen tissue contains a 40000-molecular-weight form whereas the fresh tissue contains a 90000-molecular-weight form (Westphal and Beato, 1980). It is assumed that the smaller form is originated from the larger one by proteolysis (Wrange and Gustafsson, 1978). Monoclonal antibodies (Westphal *et al.*, 1982), and some polyclonal antibodies (Carlsted-Duke *et al.*, 1982), recognize only the 90000-molecular-weight protein. It is likely therefore that the antigenic determinants recognized by these antibodies are contained in the portion of the molecule that is removed or lost during the isolation of the smaller form. However, another antiserum (Govindan and Sekeris, 1978) is able to recognize both forms of the glucocorticoid receptor. Similar antisera obtained against rat thymus glucocorticoid receptors (Tsawdaroglou *et al.*, 1982) show reciprocal cross-reactivity for the 90000-dalton and 45000-dalton cytoplasmic proteins but fail to recognize a 72000-dalton nuclear protein which is the only glucocorticoid-binding activity present in nuclei.

The great majority of antibodies to steroid receptors so far obtained do not have an effect on the interaction of the steroid hormone with its receptor protein. Amongst the polyclonal antibodies tested only a goat antiserum (Greene *et al.*, 1979) was able to reduce the affinity of estradiol for its receptor. A rabbit antiserum (Greene *et al.*, 1977) was reported to decrease the affinity of the estrogen receptor from calf uterus for a weak agonist, 5α-androstenediol, and an antagonist, tamoxifen (Garcia *et al.*, 1982). The reduced affinity in this case was due to a decrease in the association rate, but the affinity of 17β-estradiol and hydroxytamoxifen was unaffected. Another rabbit antiserum to the calf uterus estrogen receptor was able to dissociate the steroid from the hormone-binding site (Moncharmont, Puca and Parikh, unpublished data). In addition, there is a report of one monoclonal antibody to the glucocorticoid receptor that is able to inhibit hormone binding (Grandics *et al.*, 1982). In this case inhibition of steroid binding has been used as the initial assay for screening and selection of hybridomas (see Chapter 1). However, it would be wrong to conclude that the steroid-binding domain is necessarily a weak immunogen. In fact, all the studies of the steroid receptor–antibody interaction, as well as screening for monoclonal antibodies, have been performed using the steroid-binding activity as specific marker for the receptor proteins, thus biasing selection against any antibodies inhibiting such interaction.

The DNA-binding domain of the glucocorticoid receptor does not appear to be affected by either monoclonal (Westphal *et al.*, 1982) or polyclonal (Carlstedt-Duke *et al.*, 1982) antibodies as assessed by receptor binding to DNA–cellulose. There is also evidence in the case of the estrogen receptor that monoclonal antibodies (Moncharmont *et al.*, 1982) do not affect the site of interaction with the nuclear component, neither do they interfere with nuclear translocation *in vitro* (Moncharmont *et al.*, 1982). Furthermore, radiolabeled monoclonal antibodies are able to bind in a specific way to the estrogen receptor in the intact nuclei, when the receptor is already complexed with the acceptor structure(s) (Moncharmont and Parikh, 1983).

4.5.2 Morphological studies

The intracellular localization of the steroid receptors has always been a subject of debate. The biochemical study of steroid hormone–receptor interactions involves procedures which depend upon tissue homogenization and fractionation and which do not distinguish between individual parts of the organs that contain heterogeneous types of tissues and cells. In contrast, histochemical methods utilizing receptor-specific antibodies on tissue sections can visualize intracellular receptor sites which are not easily accessible by biochemical techniques. This may be especially important for biological specimens such as biopsies or surgical samples where biochemical techniques are often not feasible or may yield only crude results. Many morphological studies have been performed to investigate the intracellular localization of the receptor. For this purpose, either fluorescent steroids (Pertschuk *et al.*, 1980) or antibodies to the steroid molecule (Nenci *et al.*, 1976; Nenci, 1979; Walker *et al.*, 1980) have been used. However, these methods may generate artifacts (McCarty *et al.*, 1981; Chamness *et al.*, 1980; Zehr *et al.*, 1981; Underwood *et al.*, 1982). The availability of antibodies to the receptor protein certainly opened new possibilities. However, the published literature in this field has not been as abundant as might have been expected, a sign perhaps of significant major technical difficulties. In the case of steroid receptors, the investigator has to deal with a relatively low number of antigenic sites per cell. In addition such proteins have the tendency to become soluble and therefore can be washed off the tissue sections during various incubation and washing procedures.

Polyclonal antibodies to the estrogen receptor have been used for detecting estradiol receptor in the rat pituitary cells. The Sternberg immunoperoxidase method was used, and the sections were examined with the electron microscope (Morel *et al.*, 1981). In addition to a preferential distribution of the estrogen receptor in the gonadotropic, lactotropic and somatotropic cells, an estradiol-dependent redistribution of receptor to the nuclear compartment was seen upon exposure of the animal to the hormone. Immunohistochemical localization of the estrogen receptor in frozen (Raam *et al.*, 1982) or paraffin-embedded (Parikh *et al.*, unpublished data) sections of human mammary carcinoma using rabbit antisera to estrogen receptor indicates a cytoplasmic localization of the receptor (Raam *et al.*, 1982).

With monoclonal antibodies to the estrogen receptor, however, no extensive *in situ* labeling studies have been performed. Nuclear and cytoplasmic staining in the macaque endometrium (McCellan *et al.*, 1982) with a monoclonal antibody conjugated with peroxidase has been reported. Also nuclear staining in breast cancer cells with the immunoperoxidase method has been reported (King *et al.*, 1982). These latter two reports, however, were only presented at scientific meetings and no detailed reports have been published to date. Although pictures of immunostaining of breast cancer cells with monoclonal antibodies have been presented in other reports (Nadji *et al.*, 1982; Jensen *et*

al., 1982), a detailed presentation of these findings has not been published at the time of writing this review. While the monoclonal antibodies to estrogen receptor generated by Jensen and co-workers exhibit exclusively nuclear staining, the monoclonal antibodies produced in our laboratory provide immunohistochemical staining localized predominantly in cytoplasm of estrogen responsive cells (unpublished data).
have been presented in other reports (Greene and Jensen, 1981; Jensen *et al.*, 1982), a detailed presentation of these findings has not been published at the time of writing this review.

An immunocytochemical study has been performed on the chick oviduct with antibodies to progesterone receptor (Gasc *et al.*, 1982). Antibodies raised in rabbit and goat (Renoir *et al.*, 1982) and monoclonal antibodies (Radanyi *et al.*, 1983) were used in this study with the immunoperoxidase technique. Specific staining was seen in most of the cells of the luminal epithelium, but not in the glandular cells. Whereas polyclonal antibodies to progesterone receptor showed either nuclear or cytoplasmic staining, only cytoplasmic staining was evident with monoclonal antibody.

The immunocytochemical detection of glucocorticoid receptor has been performed with polyclonal antibodies in a hepatoma cell line (Govindan, 1980; Papamichail *et al.*, 1980) or in breast cancer sections (Ioannidis *et al.*, 1982), and its differential compartmentalization (cytoplasmic or nuclear) upon exposure to the hormone has been presented.

4.5.3 Quantitative aspects

The quantitative evaluation of the receptor content in tissue sample and/or extracts is a fundamental necessity in studies on the mechanism of action of steroid hormones. In addition to its basic research value, a quantitative evaluation of steroid receptor content provides clinical information as a guide to prognosis and therapy in some hormone-dependent neoplastic processes.

Steroid receptors have usually been studied using hormone-specific binding activity for their identification and quantitation. Radiolabeled ligands allow highly specific and functional identification of the receptor protein. However, binding ability is labile under certain experimental conditions. Furthermore, in some instances the steroid-binding site is occupied by non-radiolabeled hormones of endogenous or pharmacological origin. In order to circumvent this problem, many assays have been developed that use high temperatures (Anderson *et al.*, 1972) or chaotropic salt (Sica *et al.*, 1980) to cause the exchange of the receptor-bound endogenous hormone with the labeled one. Antibodies offer an alternative method of receptor recognition and quantification, independent of the occupation of the steroid-binding site.

The independent recognition of different epitopes on the human estrogen receptor molecule by monoclonal antibodies forms the basis for a new receptor assay (Greene and Jensen, 1982). This uses two different monoclonal antibodies, one linked to an insoluble support (polystyrene beads), the other bearing an enzymatic or radiolabeled marker. The binding of the labeled antibody to the insoluble bead is linearly related, over a wide range, to the concentration of the antigen in the sample. This assay can be used for the detection of estrogen receptor in breast cancer tissue samples and shows a good correlation with the more expensive and time-consuming assays that use charcoal or a sucrose gradient. Commercial development of such assays is under way.

One other possible immunological approach to the quantitation of estrogen receptor uses the finding that a monoclonal antibody is able to bind in a specific and saturable way to the nuclear receptor in the intact nuclei (Moncharmont and Parikh, 1983). Such binding is linearly dependent upon the receptor content of the nuclei. This method would also be valuable for clinical purposes, since it measures the receptor present at its site of action.

Another quantitative assay of the steroid receptor using antibodies has been described for the glucocorticoid receptor (Okret *et al.*, 1981). This is an indirect, competitive ELISA. This assay uses microtiter plates coated with a fixed amount of purified glucocorticoid receptor. After preincubation of a fixed aliquot of immune antiserum dilution with the unknown amount of antigen, the sample is incubated with the precoated plate. The amount of antibody still available for interaction with the coated plate is inversely proportional to the amount of receptor present in the unknown sample. The antibodies bound to the plate are then exposed to a second anti-rabbit antibody conjugated with peroxidase and, after the appropriate washing, the enzymatic reaction is measured.

4.6 CONCLUDING REMARKS

The availability of monoclonal antibodies has opened a new era in steroid receptor studies. More sophisticated applications of the existing antibodies will follow in the near future and, hopefully, the other steroid receptors will be more extensively purified and appropriate monoclonal antibodies to them produced. Probably the most exciting findings will come from morphological studies at the electron microscopic level, giving more insight to the ultrastructural aspects of the receptor localization. However we can be confident that when the monoclonal antibodies become available to a large number of scientists, many novel applications will be created, bringing deeper insight to our knowledge of steroid hormone action.

REFERENCES

Al-Nuaimi, N., Davies, P. and Griffiths, K. (1979), *Cancer Treat. Rep.*, **63,** 1147.
Anderson, J., Clark, J.H. and Peck, E.J. (1972), *Biochem. J.*, **126,** 561–567.

Barrack, E.R. and Coffey, D.S. (1980), *J. Biol. Chem.*, **255**, 7265–7275.
Baulieu, E.E. and Jung, I. (1970), *Biochem. Biophys. Res. Commun.*, **38**, 599–606.
Baulieu, E.E., Godeau, F., Schorderet, M. and Schorderet-Slatkine, S. (1978), *Nature (London)*, **275**, 593–598.
Baxter, J.D. and Tomkins, G.M. (1971), *Proc. Natl. Acad. Sci. U.S.A.*, **68**, 932–937.
Brumbaugh, P.F., Hughes, M.R. and Haussler, M.R. (1975), *Proc. Natl. Acad. Sci. U.S.A.*, **72**, 4871–4875.
Carlstedt-Duke, J., Okret, S., Wrange, Ö. and Gustafsson, J.-A (1982), *Proc. Natl. Acad. Sci. U.S.A.*, **79**, 4260–4264.
Chamness, G.C., Mercer, W.D. and McGuire, W.L. (1980), *J. Histochem. Cytochem.*, **28**, 792–798.
Coffer, A.I. and King, R.J.B. (1981), *J. Steroid Biochem.*, **14**, 1229–1235.
Coffer, A.I., King R.J.B. and Brockas, A.J. (1980), *Biochem. Int.*, **1**, 126–132.
Coty, W.A., Schrader, W.T. and O'Malley, B.W. (1979), *J. Steroid Biochem.*, **10**, 1–12.
Cuatrecasas, P., Wilchek, M. and Anfinsen, C.B. (1968), *Proc. Natl. Acad. Sci. U.S.A.*, **61**, 636–643.
Dougherty, J.J., Puri, R.K. and Toft, D.O. (1982), *J. Biol. Chem.*, **257**, 10831–10837.
Ehrlich, P. (1900), *Proc. R. Soc. London*, **66**, 424–448.
Eisen, H.J. (1980), *Proc. Natl. Acad. Sci. U.S.A.*, **77**, 3893–3897.
Fang, S., Anderson, K.M. and Liao, S. (1969), *J. Biol. Chem.*, **244**, 6584–6595.
Feil, P.D. (1983), *Endocrinology*, **112**, 396–398.
Fox, L.L., Redeuilh, S., Baskevitch, P., Baulieu, E.-E. and Richard-Foy, H. (1976), *FEBS Lett.*, **63**, 71–76.
Garcia, M., Greene, G.L., Rochefort, H. and Jensen, E.V. (1982), *Endocrinology*, **110**, 1355–1361.
Gasc, J.-M, Renoir, J.-M, Radanyi, C., Joab, I. and Baulieu, E.-E (1982), *C.R. Hebd-Séances Acad. Sci. Ser. D.*, **295**, 707–713.
Giambiagi, N. and Pasqualini, J.R. (1982), *Endocrinology*, **110**, 1067–1075.
Glascock, R.F. and Hoekstra, W.G. (1959), *Biochem. J.*, **72**, 673–682.
Goding, J.W. (1980), *J. Immunol. Methods*, **39**, 285–308.
Gorski, J. and Gannon, F. (1976), *Annu. Rev. Physiol.*, **38**, 425–450.
Govindan, M.V. (1979), *J. Steroid Biochem.*, **11**, 323–332.
Govindan, M.V. (1980), *Exp. Cell. Res.*, **127**, 293–297.
Govindan, M.V. and Sekeris, C.E. (1978), *Eur. J. Biochem.*, **89**, 95–104.
Grandics, P., Gasser, D.L. and Litwack, C. (1982), *Endocrinology*, **111**, 1731–1733.
Greene, G.L. and Jensen, E.V. (1982), *J. Steroid Biochem.*, **16**, 353–359.
Greene, G.L., Closs, L.E., Fleming, H., DeSombre, E.R. and Jensen, E.V. (1977), *Proc. Natl. Acad. Sci. U.S.A.*, **74**, 3681–3685.
Greene, G.L., Closs, L.E., DeSombre, E.R. and Jensen, E.V. (1979), *J. Steroid Biochem.*, **11**, 333–341.
Greene, G.L., Fitch, F.W. and Jensen, E.V. (1980a), *Proc. Natl. Acad. Sci. U.S.A.*, **77**, 157–161.
Greene, G.L., Nolan, C., Engler, J.P. and Jensen, E.V. (1980b), *Proc. Natl. Acad. Sci. U.S.A.*, **77**, 5115–5119.
Grody, W.W., Schrader, W.T. and O'Malley, B.W. (1982), *Endocrinol. Rev.*, **3**, 141–163.
Housley, P.R. and Pratt, W.B. (1983), *J. Biol. Chem.*, **258**, 4630–4635.
Ioannidis, C., Papamichail, M., Agnanti, N., Garas, J., Tsawdaroglou, N. and Sekeris, C.E. (1982), *Int. J. Cancer*, **29**, 147–152.

Jensen, E.V. (1978), *Pharmacol. Rev.*, **30**, 477–491.
Jensen, E.V. and DeSombre, E.R. (1972), *Annu. Rev. Biochem.*, **41**, 203–230.
Jensen, E.V. and DeSombre, E.R. (1973), *Science*, **182**, 126–134.
Jensen, E.V. and Jacobson, H.I. (1962), *Recent Prog. Horm. Res.*, **18**, 387–414.
Jensen, E.V., Suzuki, T., Kawashima, T., Stumpf, W.E., Jungblut, P.W. and DeSombre, E.R. (1968), *Proc. Natl. Acad. Sci. U.S.A.*, **59**, 632–638.
Jensen, E.V., Greene, G.L., Closs, L.E., DeSombre, E.R. and Nadji, M. (1982), *Recent Prog. Horm. Res.*, **38**, 1–40.
Joab, I. Radanyi, C., Ruwir, M., Mesler, J. and Baulieu, E.-E (1983), *Nature (London)*, (in press).
King, W.J., Jensen, E.V., Miller, L. and Greene, G.L. (1982), *64th Endocrine Society Annual Meeting, June 16–18*, San Francisco, Abstr. 713.
Köhler, G. and Milstein, C. (1975), *Nature (London)*, **256**, 495–497.
Liang, T. and Liao, S. (1974), *J. Biol. Chem.*, **249**, 4671–4678.
Liang, T. and Liao, S. (1975), *Proc. Natl. Acad. Sci. U.S.A.*, **72**, 706–709.
Logeat, F., Vu Hai, M.T. and Milgrom, E. (1981), *Proc. Natl. Acad. Sci. U.S.A.*, **78**, 1426–1430.
Logeat, F., Vu Hai, M.T., Fournier, A., Legrain, P., Buttin, G. and Milgrom, E. (1983), *Proc. Natl. Acad. Sci. U.S.A.*, **80**, 6456–6459.
McCarty, K.S., Jr., Rintgen, D.S., Seigler, H.F. and McCarty, K.S., Sr. (1981), *Breast Cancer Res. Treat.*, **1**, 315–325.
McClellan, M.C., West, N.B. and Brenner, R.M. (1982), *64th Endocrine Society Annual Meeting, June 16–18*, San Francisco, Abstr. 369.
Mainwaring, W.I.P. (1969), *J. Endocrinol.*, **45**, 531–541.
Migliaccio, A., Lastoria, S., Moncharmont, B., Rotondi, A. and Auricchio, F. (1982), *Biochem. Biophys. Res. Commun.*, **109**, 1002–1010.
Milgrom, E., Atger, M. and Baulieu, E.E. (1970), *Steroids*, **16**, 741–754.
Molinari, A.M., Medici, N., Moncharmont, B. and Puca, G.A. (1977), *Proc. Natl. Acad. Sci. U.S.A.*, **74**, 4886–4890.
Moncharmont, B. and Parikh, I. (1983), *Biochem. Biophys. Res. Commun.*, **114**, 107–112.
Moncharmont, B., Su J.-L. and Parikh, I. (1982), *Biochemistry*, **21**, 6916–6921.
Moncharmont, B., Anderson, W.L. and Parikh, I. (1984), *Biochemistry*,
Morel, G., Dubois, P., Benassayag, C., Nunez, E., Radanyi, C., Redeuilh, G., Richard-Foy, H. and Baulieu, E.-E. (1981), *Exp. Cell Res.*, **132**, 249–257.
Muldoon, T.G. (1980), *Endocrinol. Rev.*, **1**, 339–364.
Munck, A. and Brinck-Johnsen, T. (1968), *J. Biol. Chem.*, **243**, 5556–5565.
Munck, A. and Wira, C. (1971), *Adv. Biosci.*, **7**, 301–330.
Nadji, M., Morales, A.R., Greene, G.L. and Jensen, E.V. (1982), *Laborat. Invest.*, **46**, 60.
Nenci, I. (1979), *J. Histochem. Cytochem.*, **27**, 1053–1055.
Nenci, I., Beccati, M.D., Piffanelli, A. and Lanza, G. (1976), *J. Steroid Biochem.*, **7**, 505–510.
Nielsen, S. and Notides, A.C. (1975), *Biochim. Biophys. Acta*, **381**, 377–383.
Notides, A.C. and Nielsen, S. (1974), *J. Biol. Chem.*, **249**, 1866–1873.
Notides, A.C., Hamilton, D.E. and Auer, H.E. (1975), *J. Biol. Chem.*, **250**, 3945–3950.
Notides, A.C., Lerner, N. and Hamilton, D.E. (1981), *Proc. Natl. Acad. Sci. U.S.A.*, **78**, 4926–4930.

Okret, S., Carlstedt-Duke, J., Wrange, Ö., Kjell, C. and Gustafsson, J.-A (1981), *Biochim. Biophys. Acta,* **677,** 205–219.

Okret, S., Stevens, Y.-W., Carlstedt-Duke, J., Wrange, Ö., Gustafsson, J.-A and Stevens, J. (1983), *Cancer Res.,* **43,** 3127–3131.

O'Malley, B.W. and Means, A.R. (1974), *Science,* **183,** 610–620.

O'Malley, B.W. and Schrader, W.T. (1976), *Sci. Am.,* **234,** 32–43.

O'Malley, B.W., Toft, D.O. and Sherman, M.R. (1970), *Proc. Natl. Acad. Sci. U.S.A.,* **67,** 501–508.

Papamichail, M., Tsokos, G., Tsawdaroglou, N. and Sekeris, C.E. (1980), *Exp. Cell Res.,* **125,** 490–493.

Parikh, I., Anderson, W.L. and Neame, P. (1980), *J. Biol. Chem.,* **255,** 10266–10270.

Pertschuk, L.P., Tobin, E.H., Tanapat, P., Gaetjens, E., Carter, A.C., Bloom, N.D., Macchia, R.J. and Eisenberg, K. (1980), *J. Histochem. Cytochem.,* **28,** 799–810.

Pietras, R.J. and Szego, C.M. (1977), *Nature (London),* **265,** 69–72.

Pike, J.W., Donaldson, C.A., Marion, S.L. and Haussler, M.R. (1982), *Proc. Natl. Acad. Sci. U.S.A.,* **79,** 7719–7723.

Pike, J.W., Marion, S.L., Donaldson, C.A. and Haussler, M.R. (1983), *J. Biol. Chem.,* **258,** 1289–1296.

Puca, G.A. and Bresciani, F. (1969), *Nature (London),* **223,** 745–747.

Puca, G.A., Nola, E., Molinari, A.M., Armetta, I. and Sica, V. (1981), *J. Steroid Biochem.,* **15,** 307–312.

Raam, S., Peters, L., Rafkind, I., Putnum, E., Longcope, C. and Cohen, J.L. (1981), *Mol. Immunol.,* **18,** 143–156.

Raam, S., Nemeth, E., Tamura, H., O'Brien, D.S. and Cohen, J.L. (1982), *Eur. J. Clin. Oncol.,* **18,** 1–12.

Radanyi, C., Redeuilh, G., Eigenmann, E., Lebeau, M.-C., Massol, N., Secco, C., Baulieu, E.-E. and Richard-Foy, H. (1979), *C.R. Hebd. Séances Acad. Sci. Ser. D,* **288,** 255–258.

Radanyi, C., Joab, I., Renoir, J.-M, Richard-Foy, H. and Baulieu, E.-E (1983), *Proc. Natl. Acad. Sci. U.S.A.,* **80,** 2854–2858.

Renoir, J.-M, Radanyi, C., Yang, C.-R and Baulieu, E.-E (1982), *Eur. J. Biochem.,* **127,** 81–86.

Sando, J.J., La Forest, A.C. and Pratt, W.B. (1979), *J. Biol. Chem.,* **254,** 4772–4778.

Schrader, W.T., Kuhn, R.W. and O'Malley, B.W. (1977), *J. Biol. Chem.,* **252,** 299–307.

Schrader, W.T., Birnbaumer, M.E., Hughes, M.R., Weigel, N.L., Grody, W.W. and O'Malley, B.W. (1981), *Recent Prog. Horm. Res.,* **37,** 583–633.

Sharp, G.W.G., Komack, C.L. and Leaf, A. (1966), *J. Clin. Invest.,* **45,** 450–459.

Sica, V., Parikh, I., Nola, E., Puca, G.A. and Cuatrecasas, P. (1973), *J. Biol. Chem.,* **248,** 6543–6558.

Sica, V., Puca, G.A., Molinari, A.M., Buonaguro, F.M. and Bresciani, F. (1980), *Biochemistry,* **19,** 83–88.

Soloff, M. and Szego, C.M. (1969), *Biochem. Biophys. Res. Commun.,* **34,** 141–147.

Toft, D. and Gorski, J. (1966), *Proc. Natl. Acad. Sci. U.S.A.,* **55,** 1574–1581.

Tsawdaroglou, N., Govindan, M.V., Schmid, W. and Sekeris, C.E. (1982), *Eur. J. Biochem.,* **114,** 305–313.

Underwood, J.C.E., Sher, E., Reed, M., Eisman, J.A. and Martin, T.J. (1982), *J. Clin. Pathol.,* **35,** 401–406.

Walker, R.A., Cove, D. and Howell, A. (1980), *Lancet,* **i,** 171–173.
Wiegel, N.L., Pousette, A., Schrader, W.T. and O'Malley, B.W. (1981a), *Biochemistry,* **20,** 6798–6803.
Weigel, N.L., Tash, J.S., Means, A.R., Schrader, W.T. and O'Malley, B.W. (1981b), *Biochem. Biophys. Res. Commun.,* **102,** 513–519.
Westphal, H.M. and Beato, M. (1980), *Eur. J. Biochem.,* **119,** 101–106.
Westphal, H.M., Moldenhauer, C. and Beato, M. (1982), *EMBO J.,* **1,** 1467–1471.
Wrange, Ö. and Gustafsson, J.-A (1978), *J. Biol. Chem.,* **253,** 856–865.
Wrange, Ö., Carlstedt-Duke, J. and Gustafsson, J.-A (1979), *J. Biol. Chem.,* **254,** 9284–9290.
Yamamoto, K.R. and Alberts, B.M. (1976), *Annu. Rev. Biochem.,* **45,** 721–746.
Yelton, D.E. and Scharf, M.D. (1981), *Annu. Rev. Biochem.,* **50,** 657–680.
Yund, M.A., King, D.S. and Fristrom, J.W. (1978), *Proc. Natl. Acad. Sci. U.S.A.,* **75,** 6039–6043.
Zehr, D.R., Satyaswaroop, P.G. and Sheehan, D.M. (1981), *J. Steroid Biochem.,* **14,** 613–617.

Note added in proof

In a recent publication Gorski and co-workers (Welshons *et al., Nature,* **307,** 747–749, 1984) challenged the 'two-step' model for the mechanism of action of steroid hormones. In this controversial but elegant report, the authors propose that the putative cytoplasmic estrogen receptor is an artifact of homogenization; there is, in reality, no nuclear translocation as such of the receptor upon steroid treatment. Instead, they suggest that most of the unoccupied estrogen receptors are associated with the nuclear fraction. This is indeed a serious challenge to the currently accepted model of steroid hormone action and must be substantiated in other systems.

Another simultaneous report (King, W.J. and Greene, G.L., *Nature,* **307,** 745–747, 1984) arrives at a similar conclusion by an independent approach. These authors report the exclusive nuclear localization of estrogen receptors in target tissues by immunohistochemical methods, although they fail to consider the possibility that lack of cytoplasmic staining may be a property of their monoclonal antibodies. This may explain the apparent contradiction with results obtained with our monoclonal antibodies in which immunohistochemical staining was observed predominantly in the cytoplasm (unpublished data). The above two reports raise numerous questions which must be very carefully tackled to settle this new round of controversy.

5 Analysis of Complement Receptors with Polyclonal and Monoclonal Antibodies

GORDON D. ROSS

Glossary of terms and abbreviations

C	Complement – a series of serum proteins operating in sequence or as a cascade and serving to facilitate interactions between antibody molecules and cells (usually of the immune system)
CR (CR_1, CR_2, CR_3)	Complement receptors of cell surfaces (see Table 5.1 for details of specificity of these different receptors)
H-R	Factor H receptors (see also Table 5.1)

Note added in proof

After this chapter was written, it was subsequently demonstrated that the molecular weight of CR_2 was 140000 daltons (Iida *et al.* (1983), *J. Exp. Med.*, **158,** 1021–1033; Weis *et al.* (1984), *Proc. Natl. Acad. Sci. U.S.A.*, (in press, February) and that the 72000-dalton receptor material described in this chapter represented a proteolytic fragment of CR_2 that contained the C3d-binding site (R. Frade, B. L. Myones and G. D. Ross, unpublished observation).

Monoclonal Antibodies to Receptors: Probes for Receptor Structure and Function
(*Receptors and Recognition*, Series B, Volume 17)
Edited by M. F. Greaves
Published in 1984 by Chapman and Hall, 11 New Fetter Lane, London EC4P 4EE

5.1 THE EIGHT DIFFERENT TYPES OF MEMBRANE COMPLEMENT RECEPTORS

Various assays for binding of either individual complement (C) fragments or complexed C have demonstrated the existence of eight distinct types of membrane C receptors: CR_1 (C3b-receptor), CR_2 (C3d-receptor), CR_3 (iC3b-receptor), H-R (factor H-receptor), C1q-R, C3a-R, C5a-R and C3e-R (Ross, 1980, 1982). Specific antibodies have been prepared against only the first four types of these C receptors, and so only these four types of receptors and their investigation with anti-receptor antibodies will be discussed in this chapter. Table 5.1 summarizes studies of the structure, specificity and cell type distribution of these four C receptor types.

Table 5.1 Structure, specificity and cell type distribution of CR_1, CR_2, CR_3 and H-R

Receptor type	Specificity	Structure	Cell type distribution
CR_1	C3b, C4b, iC3b, C3i, C3c	205 K	Erythrocytes, granulocytes, B-lymphocytes, monocycte–macrophages, kidney podocytes
CR_2	iC3b, C3d,g, C3d	72 K	B-lymphocytes, granulocytes, monocytes
CR_3	iC3b	175 K 105 K	Granulocytes, monocyte–macrophages, large granular lymphocytes
H-R	Factor H	100 K 50 K	B-lymphocytes, granulocytes, monocytes

5.1.1 Complement receptor type one (CR_1)

By analysis with sodium dodecyl sulfate-polyacrylamide gel electrophoresis (SDS-PAGE), CR_1 has been shown to be a single-chain glycoprotein with a molecular weight (M_r) of 190000–260000 (Fearon, 1979, 1980; Dobson *et al.*, 1981; Wong *et al.*, 1983; Dykman *et al.*, 1983). On the cell surface, CR_1 may be a tetramer or pentamer, as non-ionic-detergent-solubilized CR_1 had an apparent M_r of 1×10^6 when analyzed by gel filtration (Fearon, 1979). CR_1 binds to C3b and C3i (native C3 with broken internal thioester bond) and with lower affinity also binds to C4b, iC3b and C3c (Nishioka and Linscott, 1963; Cooper, 1969; Ross and Polley, 1975; Berger *et al.*, 1981; Ross *et al.*, 1983). CR_1 is expressed on primate erythrocytes, monocyte–macrophages, neutrophils, eosinophils, B lymphocytes, some T lymphocytes and kidney podocytes

(Ross *et al.*, 1973, 1978; Gupta *et al.*, 1976; Fearon, 1980; Kazatchkine *et al.*, 1982; Wilson *et al.*, 1982a).

5.1.2 Complement receptor type two (CR_2)

CR_2 has been shown to be a 72000 M_r single-chain glycoprotein that is expressed primarily on B lymphocytes (Ross *et al.*, 1978; Lambris *et al.*, 1981). CR_2 is specific for a site in the d region of C3 that is exposed only in the iC3b, C3d,g and C3d fragments. CR_2 has a considerably higher affinity for C3d,g than for C3d, and analysis of CR_2 expression with C3d,g-coated sheep erythrocytes (EC3d,g) has demonstrated a very low level of CR_2 expression on monocytes and neutrophils which was previously not detected by assay with EC3d (Ross *et al.*, 1983).

5.1.3 Complement receptor type three (CR_3)

By SDS-PAGE, CR_3 has been shown to consist of two polypeptide chains of 155000–190000 and 94000–105000 M_r (Beller *et al.*, 1982; Todd and Schlossman, 1982; G.D. Ross and P.J. Lachmann, unpublished observations). Cr_3 is expressed on 90–95% of peripheral blood monocytes and neutrophils, 40–50% of large granular lymphocytes (natural killer cells), but is absent from B and T lymphocytes. The binding-site characteristics of human leukocyte CR_3 resemble bovine serum conglutinin (K) in that CR_3 has a dual specificity for both fixed iC3b and certain mannan-containing proteins contained in yeast cell walls (Ross *et al.*, 1983; Ross *et al.*, 1984a). Thus, in addition to being a C3 receptor, CR_3 is also a lectin-like receptor that is able to bind to the carbohydrates contained in certain types of yeast and perhaps also certain types of bacteria.

5.1.4 Factor H receptors (H-R)

H-R are expressed on B lymphocytes, monocytes and neutrophils. The B cell H-R has been shown to consist of 100000- and 50000-M_r components on unreduced SDS-PAGE gels, and a single 50000-M_r component on reduced gels (Lambris *et al.*, 1980; Lambris and Ross, 1982). H-R are able to bind aggregated or complexed H, and apparently also purified monomeric H. It is unknown, however, whether H-R can bind 'native' H in plasma.

5.2 PREPARATION OF ANTIBODIES TO COMPLEMENT RECEPTORS

Several different methods have been used to prepare either polyclonal or monoclonal antibodies to CR_1, CR_2, CR_3 and H-R. Specific antisera to CR_1

and CR_2 have been raised by immunization of rabbits with the purified receptor preparations. Monoclonal antibodies to CR_1 and CR_3 have also been generated successfully by immunization with intact monocytes or macrophages. When a purified receptor was used as the immunogen and the resulting antibodies could be shown to inhibit receptor activity, it was relatively easy to confirm that the antibodies were specific for C-receptors. By contrast, it was considerably more difficult to demonstrate the C-receptor specificity of non-inhibitory antibodies raised to intact cells.

5.2.1 Preparation of anti-CR_1

Polyclonal rabbit anti-CR_1 have been prepared by immunization with either purified erythrocyte CR_1 (Fearon, 1980; Dobson *et al.*, 1981) or SDS-PAGE gel slices derived from electrophoresis of solubilized neutrophil membranes (Melamed *et al.*, 1982). These rabbit antibodies completely inhibited CR_1 activity, and have been shown to detect approximately nine different CR_1 antigenic determinants (Wilson *et al.*, 1982b). Purified erythrocyte CR_1 has also been used to generate mouse monoclonal antibodies specific for at least four different CR_1 epitopes (Iida *et al.*, 1982; Gerdes *et al.*, 1982).

(a) *Isolation of erythrocyte CR_1*

Because normal erythrocytes contain an average of only 600 CR_1 per cell, large amounts of blood are required to purify enough CR_1 for rabbit immunization. We have obtained 1.6–3.2 mg of CR_1 from 50 units of outdated red blood cells. The primary difficulty in the isolation procedure is the preparation of washed membranes from such a large quantity of blood. The Millipore Pellicon filtering system with 0.5 μm pore size Durapore filter cassettes (Millipore Corp., Lexington, MA) is highly suited for preparing large quantities of washed red cell 'ghosts'. Following solubilization of membrane CR_1 with Nonidet P-40 (NP-40), CR_1 is isolated by essentially the same procedure described originally by Fearon (1979). A sensitive hemolytic assay is necessary to detect the very small amounts of CR_1 in column fractions. CR_1-containing samples are incubated with EC3b bearing only 1500 molecules of C3b per E. The EC3b are then tested for lysis by mixture with factors B and D in nickel-containing buffer (Fishelson and Müller-Eberhard, 1982) for 2 min at 37°C, followed by addition of guinea pig serum C diluted 1:25 with EDTA-buffer, and incubation with mixing for 45 min at 37°C. Any EC3b–CR_1 complexes generated by fluid-phase CR_1 in test samples are unable to bind factor B, so that the presence of CR_1 in this test system is indicated by inhibition of hemolysis. Because factor H inhibits C3-convertase formation by the same mechanism, the sensitivity of the CR_1 assay system can be analyzed by titration of purified H. CR_1 has also been measured by assay for inhibition of the classical pathway C3-convertase formed with EAC14b and C2 (Iida and

Nussenzweig, 1981), and by assay for accelerated decay dissociation of the properdin-stabilized C3-convertase, EC3b,Bb,P (Fearon, 1979). The most important purification steps are column chromatography with Bio-Rex 70 (Bio-Rad Laboratories, Richmond, CA) and C3-agarose affinity chromatography (Fearon, 1979). No apparent reduction in purity resulted from omission of a gel filtration step. Because of the sensitivity of CR_1 to proteolytic digestion, yields of active CR_1 are improved by addition of freshly dissolved phenylmethane sulfonyl fluoride (PMSF) during erythrocyte lysis and other steps of the isolation procedure.

(b) *Generation and analysis of rabbit anti-CR_1 sera*

Rabbits were immunized weekly at two intramuscular sites with 100 μg of purified CR_1 emulsified in Freund's complete adjuvant. After 4 weeks, immune sera were collected, and booster injections of 100 μg of CR_1 in Freund's incomplete adjuvant were given at 2–3-week intervals. Very small amounts of anti-CR_1 were detected by microtiter plate assay for inhibition of human E rosettes with EC3b (immune adherence). The maximum titer observed with sera obtained from two rabbits was 1:2048. For this test, one drop of anti-CR_1 dilution was incubated with one drop of human E (6×10^7/ml) for 30 min at 37 °C in U-bottom microtiter plates. Next, one drop of EC3b (4×10^7/ml) was added to the mixture, the plate was agitated for 15 min at 37 °C, and then the cells were allowed to settle for formation of immune adherence (agglutination) patterns. Analysis of anti-CR_1 by Ouchterlony immunodiffusion techniques demonstrated contaminating antibodies to C3 and certain unidentified components contained in some of the pooled and concentrated CR_1-negative fractions saved from the CR_1 isolation procedure. These other antibodies were removed by absorption with agarose coupled to both human serum (to remove anti-C3) and to the pooled CR_1-negative fractions from both the Bio-Rex 70 step and the C3–agarose-affinity step. This adsorbed anti-CR_1 inhibited CR_1 activity completely with all cell types, and specifically bound to a single erythrocyte component of approximately 200000 M_r. In addition, the anti-CR_1 did not inhibit either CR_2 or CR_3 activity. Similar characteristics have been reported for other rabbit anti-CR_1 sera prepared by Fearon (1980) and by Melamed *et al.* (1982).

(c) *Generation and identification of anti-CR_1 monoclonal antibodies*

Monoclonal anti-CR_1 have been produced by immunization with either purified CR_1 or intact monocytes. Hybridoma culture supernatants have been analyzed by radioimmune assay for reaction with purified CR_1 or tonsil lymphocytes (Iida *et al.*, 1982; Gerdes *et al.*, 1982). We have also used microtiter plate wells containing adherent monocytes and an anti-(mouse Ig)–peroxidase reagent for an ELISA assay of anti-CR_1. Positive culture supernatants are then tested for inhibition of immune adherence as described above.

Recently, a monoclonal anti-CR_1 was identified that had been produced by immunization with intact monocytes (Hogg *et al.*, 1984). Several different tests were required to confirm the CR_1 specificity of this antibody, as it did not inhibit CR_1 activity. This antibody (E11) is presumably specific for a determinant that is not part of or even close to the C3b-binding site. First, E11 was shown to react with a membrane surface antigen of approximately 200 000 M_r that was expressed on erythrocytes, monocyte–macrophages, neutrophils and kidney cells. E11 also stained by immunofluorescence the same peripheral blood lymphocytes that formed EC3b rosettes. Second, a rabbit (polyclonal) $F(ab')_2$–anti-CR_1 was shown to inhibit both the E11 fluorescence staining of neutrophils and the uptake of ^{125}I-labeled E11 by human E by 75%. Third, when lymphocytes were fluorescence-stained with E11 at 37 °C in the absence of azide, capping of the fluorescent E11 antigen was observed and EC3b rosetting (CR_1 activity) was inhibited by 88%. By contrast, capping of lymphocyte E11 antigen did not produce any detectable inhibition of EC3d,g rosetting (CR_2 activity). Similar treatment of monocytes and neutrophils induced endocytosis of the fluorescence-labeled E11 antigen and inhibited EC3b rosetting by 59% and 25% respectively, with no effect on EC3bi rosetting (CR_3 activity). Thus, even though E11 did not block the C3b-binding site of CR_1, it did selectively inhibit CR_1 activity by inducing either capping or endocytosis of CR_1. Finally, quantitation of E11 antigen on neutrophils and erythrocytes with ^{125}I-labeled E11, detected amounts of E11 antigen per cell (45 000 and 650) that were similar to the amounts of CR_1 detected previously by others with either monoclonal anti-CR_1 (Iida *et al.*, 1982) or C3b-dimers (Arnaout *et al.*, 1981; Wilson *et al.*, 1982b).

5.2.2 Preparation of anti-CR_2

Purified CR_2 has been used for immunization of rabbits to produce a polyclonal anti-CR_2 (Lambris *et al.*, 1981). This same CR_2 immunogen is now being used in attempts to produce a mouse monoclonal anti-CR_2. The procedure for CR_2 isolation is considerably easier than the CR_1 isolation procedure.

(a) *Purification of CR_2 from Raji cell culture supernatant*

Raji cells, and all other B cell lines that have been examined, shed their CR_2 into the culture media during growth, so that the spent culture media provided a convenient source of CR_2 for preparative isolation. Moreover, because the shed CR_2 was soluble, detergents were not required in the column buffers to maintain CR_2 in solution. The fluid-phase CR_2 in the culture media was detectable by three different assays. First, fluid-phase CR_2 inhibited EC3d rosettes with Raji cells. Second, media from cells grown in [^{3}H]leucine contained intrinsically labeled CR_2 that bound to EC3d but not to EC3b. Third, fluid-phase CR_2 inhibited the agglutination of EC3d by anti-C3d. The test for inhibition of anti-C3d agglutination was the most sensitive and reproducible of the three assays, so this was used to monitor fluid-phase CR_2 during the isolation procedure. Subsequently, it was found that fluid CR_2 did not inhibit the majority of other anti-C3d sera examined, and that only one of five different monoclonal anti-C3d antibodies tested were inhibited by fluid CR_2 (Schreiber, R.D., personal communication). In addition, it now appears likely that the EC3d used to assay CR_2 originally were actually EC3d,g and not EC3d. This was fortunate because EC3d have been shown to have a much lower affinity for CR_2 than do EC3d,g (Ross *et al.*, 1983), and EC3d were found to be a very poor reagent for assaying fluid CR_2 with anti-C3d. For maximum sensitivity in the fluid-phase CR_2 assay for inhibition of anti-C3d agglutination, it is important to use very small amounts of fixed C3d,g on the EC3d,g (2000 C3d,g per E). The only recommended change in the previously published CR_2 isolation procedure (Lambris *et al.*, 1981) is substitution of C3d,g-activated thiol Sepharose (Ross *et al.*, 1982) for C3d–Sepharose.

(b) *Production and analysis of rabbit anti-CR_2 sera*

Rabbits were immunized with purified CR_2 using the same procedure described above for CR_1 immunization. The majority of non-specific antibodies generated were directed to fetal calf serum proteins and were removed by adsorption with fetal calf serum coupled to agarose. The Raji cell surface CR_2 detected with the anti-CR_2 had the same 72 000 M_r on SDS-PAGE analysis as did the fluid-phase CR_2 isolated from spent culture media. Anti-CR_2 inhibited CR_2-dependent EC3bi and EC3d rosette formation with lymphocytes, but had no detectable effect on CR_1- or CR_3-dependent rosettes (Lambris *et al.*, 1981).

5.2.3 Preparation of anti-CR_3

CR_3 is probably a major antigen on the surface of monocyte–macrophages, as four different 'macrophage-specific' monoclonal antibodies produced in different laboratories by immunization with intact monocytes have been shown subsequently to be specific for CR_3. Anti-Mac-1, that is a rat anti-mouse macrophage antibody, selectively inhibits CR_3 activity on both mouse and human leukocytes (Beller *et al.*, 1982). It was unclear that anti-Mac-1 was actually specific for CR_3 (rather than some other nearby surface antigen) until anti-Mac-1 was shown to be unreactive with the leukocytes from two patients with a genetic deficiency of both CR_3 activity and a single neutrophil membrane surface glycoprotein with the same M_r as the heavy chain of Mac-1 (Ross *et al.*, 1984b). Later, two other monoclonal antibodies, OKM-1 (Breard *et al.*, 1980; sold by Ortho Diagnostics, Raritan, NJ, U.S.A.) and Mo-1 (Todd *et al.*, 1981), which were known to detect antigens with similar cell type distribution and M_r as Mac-1, were also shown to be unreactive with CR_3-deficient leukocytes. Biochemical analysis of normal cells also indicated that OKM-1 and Mo-1 probably reacted with different antigenic determinants on the same cell surface glycoprotein (Todd and Schlossman, 1982). Unlike anti-Mac-1 and Mo-1, however, OKM-1 did not inhibit CR_3 activity (Ross *et al.*, 1984a). Recently, a mouse anti-human monocyte monoclonal antibody was identified that inhibited human leukocyte CR_3 much better than did anti-Mac-1, and detected a human membrane surface antigen with two polypeptide chains of the same 155 000 and 94 000 M_r as the Mo-1 and OKM-1 antigen. Because this monoclonal antibody (MN-41) has a much higher affinity for human CR_3 than does Mac-1, it may well prove to be the best anti-CR_3 available for analyzing CR_3-dependent functions (Eddy *et al.*, 1984).

Because EC3bi binds simultaneously to both CR_1 and CR_3, anti-CR_3 is only able to inhibit phagocyte EC3bi rosette formation under conditions that either prevent or limit CR_1 binding of EC3bi (Ross *et al.*, 1983). When EC3bi are prepared with less than 1×10^4 iC3b molecules per EC3bi, Mac-1 produces almost complete inhibition of monocyte or neutrophil rosette formation. With more iC3b molecules per EC3bi, the binding of EC3bi to CR_1 becomes more significant and complete inhibition of EC3bi rosette formation by anti-Mac-1 requires prior treatment of phagocytes with anti-CR_1 (Ross *et al.*, 1983).

5.2.4 Preparation of anti-H-R

Small amounts of anti-idiotypic antibody (see Chapter 2 by Strosberg and Schreiber) specific for the binding site of both anti-H antibody and B cell H-R were produced by immunization of rabbits with affinity-purified goat $F(ab')_2$–anti-human H (Lambris and Ross, 1982). After this rabbit anti-anti-H was adsorbed with normal goat IgG–agarose, it was affinity-purified by adsorption and elution from goat anti-H–agarose. The purified anti-anti-H bound to B cells and inhibited B cell uptake of [^{3}H]-H. In addition, anti-anti-H stimulated B cells to release factor I in the same way as did the normal factor H ligand. Thus, anti-anti-H had specificity for H-R. When analyzed by SDS-PAGE, the unreduced H-R detected with this anti-H-R consisted of 100 000- and 50 000-M_r components, whereas only a single 50 000-M_r protein component was detected with reduced H-R samples. A very similar SDS-PAGE M_r profile was also obtained with a putative H-R isolated by affinity chromatography of Raji cell spent culture media on H–agarose (Lambris and Ross, 1982). Because only small amounts of anti-anti-H were obtained from the rabbit immune sera, it would be preferable to use the H–agarose affinity-purified H-R as an immunogen to produce either a polyclonal or monoclonal anti-H-R.

5.3 USE OF ANTIBODIES TO COMPLEMENT RECEPTORS TO DETECT RECEPTOR-BEARING CELLS AND TO QUANTITATE RECEPTOR DENSITY ON THE CELL SURFACE

Anti-C-receptor antibodies may be used to identify C-receptor-bearing cells by standard immunofluorescence or immunoperoxidase techniques. Use of $F(ab')_2$ antibodies, particularly for the second antibody–fluorochrome conjugate, is essential to avoid artifactual staining of Fc receptors. When the anti-C-receptor antibody is IgG, pretreatment of the antibody with soluble protein A from *Staphylococcus aureus* (SPA) may be useful for blocking IgG–Fc receptor staining (Ross, 1979). When using an IgG type that does not bind to SPA, cells may be pretreated with 1 mg of heat-aggregated normal IgG per ml to block Fc receptors. In any case, the IgG anti-C-receptor antibody should be centrifuged at 8000 ***g*** for 10 min. to remove large soluble aggregates, and the minimum amount necessary for staining should be determined. With polyclonal antibody, saturation of C receptors usually requires 10–80 μg of IgG or $F(ab')_2$ fragments per 1×10^6 cells in a total volume of 25–50 μl. Less monoclonal IgG should be used, in the range of 0.5–2.0 μg per 1×10^6 cells. If larger amounts of the first antibody are used, Fc receptor staining is frequently detectable with a proportion of monocytes and T cells, even if the second antibody is a $F(ab')_2$ fragment. Second antibody $F(ab')_2$ fragments specific for most animal species IgG are now available as either fluorescein or rhodamine conjugates from several suppliers. The author has had good experience with the $F(ab')_2$-anti-Ig fluorochrome antibodies sold by Cappel Laboratories (Cochranville, PA, U.S.A.).

The average number of C-receptors per cell in a given cell population may be quantitated by radioimmune assay for uptake of ^{125}I-labeled anti-C-reeptor antibody. Monoclonal antibodies are ideal for this purpose, provided that the particular antibody type binds to only a single (non-repeating) determinant in each receptor. Polyclonal anti-C-receptor antibody may also be used for receptor quantitation if the number of different determinants recognized per receptor is known. As with fluorescence staining, care must be taken to avoid Fc-receptor binding of the ^{125}I-labeled anti-C receptor reagent. Before each assay, the ^{125}I-labeled antibody should be centrifuged for 20 min at 8000 ***g***,and then the antibody used for the assay should be carefully sampled from the uppermost portion of the tube. This centrifugation step will pellet only insoluble IgG aggregates, and large soluble IgG aggregates that have a high affinity for Fc-receptors will only be displaced a short distance down from the top of the centrifuge tube.

5.3.1 Fluorescence assay of CR_1, CR_2 and CR_3

(a) CR_1

CR_1 has been assayed by both direct (Dobson *et al.*, 1981) and indirect (Fearon *et al.*, 1981; Tedder *et al.*, 1983) immunofluorescence using polyclonal $F(ab')_2$-anti-CR_1. Similar results have also been obtained using the E11 monoclonal anti-CR_1 in an indirect immunofluorescence assay with rabbit $F(ab')_2$–anti-mouse-IgG–fluorescein (Hogg *e al.*, 1984). On lymphocytes, a homogeneous membrane fluorescence-staining pattern was observed that could be induced to become patchy or to cap when the stained cells were warmed to 37 °C in buffer lacking sodium azide. By contrast, a patchy staining pattern was observed with both neutrophils and monocytes, even when staining was carried out at 0 °C with either monovalent Fab′ anti-CR_1 (Fearon *et al.*, 1981) or E11 monoclonal anti-CR_1 (Hogg *et al.*, 1984). When the CR_1 fluorescence-stained phagocytic cells were warmed to 37 °C, capping of the stained receptors was not observed, and the fluorescent patches rapidly diminished in size as a result of endocytosis of the CR_1 (Fearon *et al.*, 1981). Because erythrocytes contain only 400–1200 CR_1 per cell (Arnaout *et al.*, 1981; Iida *et al.*, 1982; Wilson *et al.*, 1982; Hogg *et al*, 1984), CR_1 is difficult to visualize on erythrocytes by either direct fluorescence assay with polyclonal anti-CR_1 (Dobson *et al.*, 1981) or indirect immunofluorescence assay with monoclonal anti-CR_1 (Hogg *et al.*, 1984). Erythrocyte CR_1 has been demonstrated successfully by indirect immuno-fluorescence with affinity-purified polyclonal $F(ab')_2$–anti-CR_1 (Tedder *et al.*, 1983). In addition, the same polyclonal anti-CR_1 was used successfully to visualize CR_1 by immunofluorescence on kidney podocytes (Kazatchkine *et al.*, 1982). By use of immunoperoxidase staining techniques and a monoclonal anti-CR_1, it has also been possible to examine the distribution of CR_1-bearing cells in tissue sections derived from spleen, tonsils and kidney (Gerdes *et al.*, 1982).

(b) CR_2

CR_2 has been detected only on B lymphocytes when a polyclonal $F(ab')_2$–anti-CR_2–fluorescein was used for a direct immunofluorescence assay (Lambris *et al.*, 1981). The pattern of anti-CR_2 fluorescence staining was similar to that observed with anti-CR_1, with capping induced at 37 °C in the absence of sodium azide.

(c) CR_3

CR_3 fluorescence staining has been performed with four different monoclonal antibodies (anti-Mac-1, Mo-1, OKM-1 and MN-41) with very similar results. In all four cases, a $F(ab')_2$–anti-Ig–fluorescein was required for the second developing antibody to avoid staining of Fc receptors. The problem of Fc receptor staining was particularly evident when staining cells from two children with an inherited CR_3 deficiency. For example, when 1×10^6 of the CR_3-deficient mononuclear cells were stained with 1.0 μg of anti-Mac-1 IgG followed by 25 μg of rabbit or sheep IgG–anti-rat-IgG–fluorescein, nearly all monocytes and 26% of lymphocytes were stained. By contrast, none of the CR_3-deficient mononuclear cells were stained when a goat $F(ab')_2$–anti-rat-IgG–fluorescein reagent was used as the second antibody. With normal cells, CR_3 was detected on nearly all blood monocytes and neutrophils, and on an average of 9% of lymphocytes. The lymphocytes stained with OKM-1 and Mac-1 have been shown to be non-B, non-T cells, many of which have the morphology of large granular lymphocytes and function as effectors in either antibody-dependent or natural cell-mediated cytotoxicity (Breard *et al.*, 1980; Ault and Springer, 1981; Ortaldo *et al.*, 1981). Monocyte fluorescence staining with anti-CR_3 was patchy and appeared to be endocytosed in a manner similar to anti-CR_1 staining. Neutrophil staining with anti-CR_3, however, differed markedly from staining with anti-CR_1 in that capping of CR_3-fluorescence staining was observed with the majority of cells. When neutrophils were stained with anti-CR_3 at 0 °C in the presence of sodium azide, the fluorescence was distributed into several large patches that spontaneously formed a single large polar cap after a short period of incubation at room temperature (G.D. Ross, unpublished observation). This suggests that CR_3 is a highly mobile receptor within the neutrophil membrane, and is not fixed in clusters in the same way as is CR_1 (Fearon *et al.*, 1981).

5.3.2 Quantitation of CR_1 and CR_3 with ^{125}I-labeled antibodies

(a) *CR_1*

The amount of C-receptor per cell varies with different cell types, and with neutrophils there can be as much as a 10-fold increase in the numbers of CR_1 and CR_3 if the cells are activated during the isolation procedure (Fearon and Collins, 1983; M.J. Walport, unpublished observation). Certain diseases may also deplete selectively those cells having a high receptor density, resulting in a lower average receptor density among the remaining cells. Considerable data on the number of CR_1 per erythrocyte and per neutrophil is now available from four laboratories. The CR_1 assay systems have used either monoclonal anti-CR_1 or polyclonal anti-CR_1 and C3b-dimers. Parallel tests with the polyclonal anti-CR_1 and C3b-dimers demonstrated that the rabbit $F(ab')_2$–anti-CR_1 recognized nine determinants per CR_1 molecule (Wilson *et al.*, 1982b). Tests with C3b-dimers (Arnaout *et al.*, 1981; Wilson *et al.*, 1982b) and with one monoclonal anti-CR_1 (Hogg *et al.*, 1984) were in agreement that erythrocytes from normal individuals expressed an average of 600 CR_1 per cell. A somewhat higher estimate of 1400 CR_1 per cell was measured with another monoclonal anti-CR_1 (Iida *et al.*, 1982). Neutrophils isolated and maintained at room temperature expressed more CR_1 per cell than did neutrophils isolated at 4 °C (Fearon and Collins, 1983), and were estimated to express 25 000–45 000 CR_1 per cell by all four laboratories (Arnaout *et al.*, 1981; Iida *et al.*, 1982; Fearon and Collins, 1983; Hogg *et al.*, 1984). Neutrophils isolated at 4°C expressed only 2500–4500 CR_1 per cell, and the number of CR_1 was shown to increase rapidly to more than 40 000 CR_1 per cell following incubation of the neutrophils with different chemotactic agents (Fearon and Collins, 1983).

Three laboratories were also in agreement that erythrocytes from patients with systemic lupus erythematosis (SLE) expressed significantly fewer CR_1, 150–400 per cell, than did erythrocytes from normal individuals. There was also agreement that the detection of erythrocyte CR_1 by anti-CR_1 was not inhibited by CR_1-bound immune complexes. In particular, two of the monoclonal anti-CR_1 antibodies used were specific for determinants that were probably not part of the C3b-binding site of CR_1, as they did not block the binding of C3b complexes to CR_1. Tests by one laboratory of the relatives of SLE patients suggested that the number of CR_1 per erythrocyte might be an inherited autosomal trait (Wilson *et al.*, 1982b). However, Iida *et al.* (1982) noted that the number of CR_1 per erythrocyte appeared to increase with treatment of the SLE patients, and might correlate with improvement in various SLE disease parameters. When large numbers of normal individuals and normal family members were examined by Ross *et al.* (1984c), no evidence was obtained for the inheritance of either low or high numbers of CR_1 per erythrocyte. In addition, some SLE patients with low CR_1 were found to have children with normal numbers of CR_1 per erythrocyte. Thus, the reason for low numbers of erythrocyte CR_1 in patients with SLE remains unknown.

(b) *CR_3*

With recognition that several monoclonal 'monocyte-specific' antibodies were actually anti-CR_3, an investigation with ^{125}I-labeled monoclonal anti-CR_3 (MN-41) was undertaken to quantitate CR_3 on neutrophils and monocytes. MN-41 proved to be more useful than OKM-1 for quantitation of CR_3 because it had a much higher binding affinity than did OKM-1. Neutrophils isolated with Ficoll–Hypaque (Ross and Lambris, 1982) at 4 °C expressed 3000–4500 CR_3 per cell, whereas neutrophils isolated at room temperature from the same blood sample expressed 50 000–65 000 CR_3 per cell. Furthermore, when the 4 °C-isolated neutrophils were incubated at 37 °C for 30 min, the number of CR_3 increased to more than 45 000 per cell. Parallel analysis of CR_1 with ^{125}I-labeled E11, indicated that this increased expression of CR_3 occurred simultaneously with a similar increase in expression of CR_1 (M.J. Walport, unpublished observation).

Two children from a family in Birmingham, England, have been identified who have a complete genetic deficiency of CR_3 on their neutrophils, monocytes and NK cells (Ross *et al.*, 1984b). These children have had serious recurrent bacterial infections that appear to be linked to a diminished ability of their phagocytic cells either to phagocytose or mount a respiratory burst in response to either opsonized or unopsonized zymosan. Quantitative assays of neutrophil CR_3 with ^{125}I-labeled monoclonal anti-CR_3 will thus provide a simple, rapid and precise way of identifying other children with CR_3-deficiency disease in the future. In addition, quantitative analysis of CR_3 may also help identify apparently normal individuals that are heterozygous for the CR_3-deficiency disease, as each of the parents of the CR_3-deficient children expressed half-normal amounts of CR_3 on their neutrophils.

5.4 ANALYSIS OF COMPLEMENT RECEPTOR FUNCTIONS WITH ANTIBODIES

As indicated by other chapters in this volume, antibodies to receptors have proven to be extremely valuable reagents for defining receptor functions. In most cases, when antibodies are to be used for analyzing receptor functions, it is important that they be specific for determinants that are located close enough to the active site of the receptor that they are able to block ligand binding to the receptor. All of the reported polyclonal anti-C-receptor antibodies have blocked ligand binding completely (Fearon, 1980; Dobson *et al.*, 1981; Lambris *et al.*, 1981; Lambris and Ross, 1982), whereas several monoclonal anti-C-receptor antibodies have had no demonstrable effect on receptor ligand-binding activity (Iida *et al.*, 1982; Hogg *et al.*, 1984). Polyclonal antibodies may also cross-link and aggregate membrane surface receptors, whereas single types of monoclonal antibodies lack this ability because each receptor can bind only one antibody molecule. To avoid triggering Fc-receptor-mediated functions, $F(ab')_2$ and Fab′ antibody fragments are preferred to whole IgG antibody for leukocyte functional assays. Polyclonal $F(ab')_2$ fragments may be able to cross-link receptors and trigger cell functions in the same way as might a polyvalent ligand, whereas Fab′ fragments of polyclonal antibody or $F(ab')_2$ fragments of monoclonal antibody may be able to block normal ligand interactions with the receptor and thereby inhibit normal receptor-mediated functions. It must always be kept in mind that the normal receptor ligand may have secondary interactions with non-receptor membrane components, and that accordingly anti-receptor antibody may not be able to mimic exactly the normal ligand in triggering cell functions.

5.4.1 Functions of CR_1

(a) ***Adherence to C3- or C4-bearing complexes or particles***

One of the major functions of CR_1 is to adhere C-coated bacteria or immune complexes on to cells of the reticuloendothelial system. After CR_1-mediated adherence has occurred, then killing and/or ingestion may be triggered by either CR_1 or some other type of membrane receptor. CR_1 facilitates these secondary membrane interactions by generating this initial attachment phase of clearance (Ehlenberger and Nussenzweig, 1977; Newman and Johnston, 1979). Although the CR_1 on lymphocytes can also be shown to mediate adherence *in vitro*, the role of this adherence in lymphocyte immune functions is unknown.

Both polyclonal and monoclonal anti-CR_1 antibodies have been used to inhibit selectively CR_1-mediated adherence reactions (Fearon, 1980; Dobson *et al.*, 1981; Iida *et al.*, 1982; Gerdes *et al.*, 1982). Anti-CR_1 was also invaluable for demonstrating that CR_1 bound C4b (Dobson *et al.*, 1981) and iC3b (Ross *et al.*, 1983) in addition to C3b. Furthermore, because iC3b binds simultaneously to phagocyte CR_1 and CR_3, the characteristics of the CR_3 interaction with iC3b could only be defined with anti-Cr_1-treated cells (Ross *et al.*, 1983; Ross and Lachmann, 1983).

(b) *Phagocytosis of C3- or C4-coated particles*

Only inflammatory macrophages (Bianco *et al.*, 1975) or monocyte-derived macrophages (Newman *et al.*, 1980) are able to phagocytose C3-coated (or C4-coated) erythrocytes that lack IgG and Fc-receptor-mediated triggering activity. Even though monocyte-derived macrophages phagocytosed both C3b- and iC3b-coated sheep erythrocytes (EC3b and EC3bi), it was unclear that both CR_1 and CR_3 triggered ingestion independently, as macrophages were known to synthesize and secrete factor H and I (Whaley, 1980) which could potentially convert CR_1-bound EC3b into EC3bi prior to CR_3-mediated ingestion. Anti-CR_1 completely inhibited EC3b binding and ingestion, but had no effect on macrophage uptake of EC3bi (Newman *et al.*, 1981). However, this finding did not exclude the possibility that CR_1 was required only for mediating adherence of EC3b prior to CR_3-mediated ingestion. Additional studies are required to determine whether anti-CR_3 inhibits ingestion of EC3b as well as EC3bi.

(c) *C-inhibitor and I-cofactor*

Probably because of its ability to bind with high affinity to C3b and C4b, isolated fluid-phase CR_1 is able to dissociate Bb and C2b from EC3b,Bb,P (Fearon, 1979) and EAC14b2b (Iida and Nussenzweig, 1981) respectively, thereby inhibiting both the C3- and C5-convertases of the classical and alternative pathways. Isolated fluid-phase CR_1 can also function as a factor I-cofactor for cleavage of fluid-phase C3b or C4b, and thus can substitute for both factor H and C4-binding protein (Fearon, 1979; Iida and Nussenzweig, 1981). In addition, erythrocyte CR_1 binds to fixed iC3b and serves as the essential I-cofactor for the normal physiologic breakdown of iC3b into C3c and

C3d,g (Ross *et al.*, 1982; Medof *et al.*, 1982). Studies with intact erythrocytes demonstrated that membrane-bound CR_1 was an efficient I-cofactor for I cleavage of complex-bound iC3b, and that this resulted in the release of fluid-phase C3c from the complexes. This function of erythrocyte membrane-bound CR_1 was confirmed by demonstration that erythrocyte I-cofactor activity was inhibited by treatment of the erythrocytes with either monoclonal (Medof *et al.*, 1982) or polyclonal (Medicus *et al.*, 1983) anti-CR_1. Recently, studies with monoclonal anti-CR_1 have demonstrated that lymphocyte membrane-bound CR_1 could function as a cell surface inhibitor of C activation (Iida and Nussenzweig, 1983). This observation suggests that membrane surface CR_1 may be able to protect host bystander cells during C activation. Such a function might be especially useful to cells that are exposed to C-activating immune complexes, such as kidney cells.

(d) ***Adsorptive endocytosis or pinocytosis***

The ability of CR_1 to induce the endocytosis of soluble C3b-complexes (Fearon *et al.*, 1981), or even small bacteria (Schreiber *et al.*, 1982), is well documented. Polyclonal rabbit $F(ab')_2$–anti-CR_1 was bound to neutrophils and rapidly internalized, whereas Fab′–anti-CR_1 bound to neutrophil surface CR_1 without being ingested. Because addition of $F(ab')_2$–anti-rabbit-IgG to Fab′–anti-CR_1-treated neutrophils resulted in endocytosis of the surface-bound Fab′–anti-CR_1, receptor cross-linkage with a bivalent antibody was required to induce CR_1 endocytosis. Similarly, small soluble C3b-complexes were bound to neutrophil CR_1 but were not ingested until cross-linked with $F(ab')_2$–anti-C3 (Fearon *et al.*, 1981). Monoclonal anti-CR_1 (E11) resembled polyclonal Fab′–anti-CR_1, and was only endocytosed by neutrophils when cross-linked with rabbit $F(ab')_2$–anti-mouse-IgG (Hogg *et al.*, 1984). Thus, the cross-linkage of several different CR_1 was apparently required for triggering CR_1 endocytosis, and monoclonal anti-CR_1, which generated at most only pairs of two cross-linked CR_1, was insufficient to trigger endocytosis. The relative importance of CR_1-mediated endocytosis versus CR_3-mediated phagocytosis in the clearance of C3- and C4-bearing complexes remains to be established. One of the patients with CR_3 deficiency and normal amounts of CR_1 has developed an apparent immune complex disease syndrome, suggesting that CR_1 alone may not be sufficient for efficient clearance of immune complexes (G.D. Ross, unpublished observation).

5.4.2 Functions of CR_2

CR_2 is expressed primarily on B lymphocytes (Ross *et al.*, 1978), and the very small amounts of CR_2 on monocytes and neutrophils do not efficiently bind C3d,g- or C3d-complexes (Newman *et al.*, 1981; Ross *et al.*, 1983). $F(ab')_2$–anti-CR_2, as well as complexes formed with C3d and $F(ab')_2$–anti-C3d, were not mitogenic with lymphocytes, and instead strongly inhibited blastogenesis triggered by either pokeweed mitogen or a mixed lymphocyte reaction (Lambris *et al.*, 1982). Because CR_2 has been detected only on B cells, it is unclear how the anti-CR_2 and C3d-complexes affected the mixed lymphocyte reaction, which is known to involve only blastogenesis of T cells. The same CR_2 ligands did not inhibit T cell blastogenesis triggered by either PHA or Con A. Because there is always some uncertainty about undetected anti-lymphocyte antibodies contained in polyclonal antibody preparations, it will be necessary to repeat these studies with either a monoclonal or affinity-purified polyclonal anti-CR_2 reagent.

5.4.3 Functions of CR_3

Experiments with iC3b-complexes and CR_3-deficient leukocytes have suggested several different functions for CR_3. The great importance of CR_3 in host defence was indicated by the increased susceptibility of CR_3-deficient patients to bacterial infections (Ross *et al.*, 1984b). Because anti-CR_3 reagents have only recently become available, very few tests of CR_3 function with these antibodies have been performed.

(a) ***Adherence of iC3b-complexes to monocyte–macrophages, neutrophils, and natural killer (NK) cells***

Because iC3b binds to all three C3-receptor types, blockade of CR_1 and/or CR_2 with anti-CR_1 and/or anti-CR_2 was required to demonstrate CR_3-specific adherence of iC3b-complexes. Only anti-Mac-1 (Beller *et al.*, 1982; Ross *et al.*, 1983) and Mo-1 and MN-41 (Eddy *et al.*, 1984; G.D. Ross, unpublished observation) have been found to inhibit CR_3 activity. OKM-1, even in combination with rabbit $F(ab')_2$–anti-mouse-Ig, did not produce any detectable inhibition of neutrophil CR_3 (G.D. Ross, unpublished observation). CR_3 is able to mediate adherence of iC3b-complexes independently from CR_1, because anti-CR_1 produced little inhibition of monocyte and neutrophil rosettes with EC3bi (Fearon, 1980; Ross *et al.*, 1983). Likewise, 40–50% of isolated large granular lymphocytes, which contain the NK cells, both formed rosettes with EC3bi (Ross and Lachmann, 1983), and were fluorescence-stained with anti-Mac-1 (Ault and Springer, 1981), OKM-1 (Ortaldo *et al.*, 1981) and Mo-1 (Ross and Lachmann, 1983).

(b) ***Phagocytosis by monocyte-derived macrophages***

The CR_3 expressed on monocyte-derived macrophages can mediate the binding and ingestion of iC3b-coated particles independently of other receptor types (S.L. Newman, unpublished observation). Even though EC3bi were bound to both CR_1 and CR_3 (Ross *et al.*, 1983), anti-CR_1 produced little or no inhibition of EC3bi phagocytosis, whereas anti-CR_3 prevented EC3bi phagocytosis completely. In addition to its ability to bind to fixed iC3b, CR_3 resembles bovine serum conglutinin in that it is able to bind directly to certain mannan-containing components of yeast cell walls. Tests with anti-CR_3 have confirmed that CR_3 can mediate the phagocytosis of unopsonized yeast (Ross *et al.*, 1984). Even though neutrophil C-receptors were not believed to be able to trigger particle ingestion independently of IgG and Fc-receptors (Newman and Johnston, 1979), the apparent defect in yeast phagocytosis of CR_3-deficient neutrophils (Ross *et al.*, 1984b) suggested that CR_3 on normal neutrophils might be able to trigger the ingestion of bound yeast particles. Experiments with normal neutrophils demonstrated that monoclonal anti-CR_3 (Mac-1, OKM-1 and MN-41) inhibited neutrophil binding and ingestion of unopsonized yeast by 50–81% (Ross *et al.*, 1984a). The finding that OKM-1 inhibited yeast ingestion but did not inhibit CR_3-dependent EC3bi rosettes suggests that the yeast-binding site in CR_3 is separate from the iC3b-binding site, and that the OKM-1 determinant is located closer to the yeast-binding site than to the iC3b-binding site.

(c) *Neutrophil respiratory burst and release of elastase*

The finding that normal, but not CR_3-deficient, neutrophils gave a respiratory burst (Schreiber *et al.*, 1982) and released elastase (Ross and Lambris, 1982) in response to iC3b-coated particles (Ross *et al.*, 1984b) suggests that neutrophil CR_3 may be able to trigger these functions in addition to triggering phagocytosis. Future tests of normal neutrophils with anti-CR_3 should allow precise definition of these and other possible CR_3-dependent functions.

5.4.4 Functions of H-R

(a) *Release of factor I*

Purified H triggered the release of factor I from B cells (Lambris *et al.*, 1980), neutrophils (Dobson *et al.*, 1981) and monocytes (Newman *et al.*, 1981). Anti-H-R (anti-anti-H idiotypic antibody) bound to B cell H-R in a similar manner to the normal H ligand, blocking [^{3}H]-H binding to B cells and triggering I release (Lambris and Ross, 1982). Insufficient amounts of anti-H-R were available to test neutrophils and monocytes for anti-H-R-stimulated I release.

(b) *B cell blastogenesis and monocyte respiratory burst*

Purified H has also been reported to stimulate B cell blastogenesis (Hammann *et al.*, 1981) and a monocyte respiratory burst (Schopf *et al.*, 1982). Future investigations of these functions with anti-H-R are planned to determine the role of H-R in B cell and monocyte function.

REFERENCES

Arnaout, M.A., Melamed, J., Tack, B.F. and Colten, H.R. (1981), *J. Immunol.*, **127,** 1348–1354.

Ault, K.A. and Springer, T.A. (1981), *J. Immunol.*, **126,** 359–364.

Beller, D.I., Springer, T.A. and Schreiber, R.D. (1982), *J. Exp. Med.*, **156,** 1000–1009.

Berger, M., Gaither, T.A., Hammer, C.H. and Frank, M.M. (1981), *J. Immunol.*, **127,** 1329–1334.

Bianco, C., Griffin, F.M., Jr. and Silverstein, S.C. (1975), *J. Exp. Med.*, **141,** 1278–1290.

Breard, J., Reinherz, E.L., Kung, P.C., Goldstein, G. and Schlossman, S.F. (1980), *J. Immunol.*, **124,** 1943–1948.

Cooper, N.R. (1969), *Science*, **165,** 396–398.

Dobson, N.J., Lambris, J.D. and Ross, G.D. (1981), *J. Immunol.*, **126,** 693–698.

Dykman, T.R., Cole, J.L., Iida, K. and Atkinson, J.P. (1983), *Proc. Natl. Acad. Sci. U.S.A.*, **80,** 1698–1702.

Eddy, A., Newman, S.L., Cosio, F., Le Bien, T. and Michael, A.F. (1984), *Immunol. Clin. Immunopathol.*, (in press).

Ehlenberger, A.G. and Nussenzweig, V. (1977), *J. Exp. Med.*, **145,** 357–371.

Fearon, D.T. (1979), *Proc. Natl. Acad. Sci. U.S.A.*, **76,** 5867–5871.

Fearon, D.T. (1980), *J. Exp. Med.*, **152,** 20–30.

Fearon, D.T. and Collins, L.A. (1983), *J. Immunol.*, **130,** 370–375.

Fearon, D.T., Kaneko, I. and Thomson, G.G. (1981), *J. Exp. Med.*, **153,** 1615–1628.

Fishelson, Z. and Müller-Eberhard, H.J. (1982), *J. Immunol.*, **129,** 2603–2607.

Gerdes, J., Mason, D.Y. and Stein, H. (1982), *Immunology*, **45,** 645–653.

Gupta, S., Ross, G.D., Good, R.A. and Siegel, F.P. (1976), *Blood*, **48,** 755–763.

Hamman, K.P., Raile, A., Schmitt, M., Scheiner, O., Mussle, H.H., Peters, H. and Dierich, M.P. (1981), *Immunobiology*, **159,** 126 (abstract).

Hogg, N., Ross, G.D., Slusarenko, M., Jones, D., Walport, M.J. and Lachmann, P.J. (1984), *Eur. J. Immunol.*, (in press).

Iida, K. and Nussenzweig, V. (1981), *J. Exp. Med.*, **153,** 1138–1150.

Iida, K. and Nussenzweig, V. (1983), *J. Immunol.*, **130,** 1876–1880.

Iida, K., Mornaghi, R. and Nussenzweig, V. (1982), *J. Exp. Med.*, **155,** 1427–1438.

Kazatchkine, M.D., Fearon, D.T., Appay, M.D., Mandet, C. and Bariety, J. (1982), *J. Clin. Invest.*, **69,** 900–912.

Lambris, J.D. and Ross, G.D. (1982), *J. Exp. Med.*, **155,** 1400–1411.

Lambris, J.D., Dobson, N.J. and Ross, G.D. (1980), *J. Exp. Med.*, **152,** 1625–1644.

Lambris, J.D., Dobson, N.J. and Ross, G.D. (1981) *Proc. Natl. Acad. Sci. U.S.A.*, **78,** 1828–1832.

Lambris, J.D., Cohen, P.L., Dobson, N.J., Wheeler, P.W., Papamichail, M. and Ross, G.D. (1982), *Clin. Res.*, **30,** 514A (abstract).

Medicus, R.G., Melamed, J. and Arnaout, M.A. (1983), *Eur. J. Immunol.*, **13,** 465–470.

Medof, M.E., Iida, K., Mold, C. and Nussenzweig, V. (1982), *J. Exp. Med.*, **156,** 1739–1754.

Melamed, J., Arnaout, M.A. and Colten, H.R. (1982), *J. Immunol.*, **128,** 2313–2318.

Nishioka, K. and Linscott, W.D. (1963), *J. Exp. Med.*, **118,** 767–793.
Newman, S.L. and Johnston, R.B., Jr. (1979), *J. Immunol.*, **123,** 1839–1846.
Newman, S.L., Musson, R.A. and Henson, P.M. (1980), *J. Immunol.*, **125,** 2236–2244.
Newman, S.L., Dobson, N.J., Lambris, J.D., Ross, G.D. and Henson, P.M. (1981), *Fed. Proc. Fed. Am. Soc. Exp. Biol.*, **40,** 1017 (abstract).
Ortaldo, J.R., Sharrow, S.O., Timonen, T. and Herberman, R.B. (1981), *J. Immunol.*, **127,** 2401–2409.
Ross, G.D. (1979), *Blood*, **53,** 799–811.
Ross, G.D. (1980), *J. Immunol. Methods*, **37,** 197–211.
Ross, G.D. (1982), *Fed. Proc. Fed. Am. Soc. Exp. Biol.*, **41,** 3089–3093.
Ross, G.D. and Lachmann, P.J. (1983), *Immunobiology*, **164,** 160 (abstract).
Ross, G.D. and Lambris, J.D. (1982), *J. Exp. Med.*, **155,** 96–110.
Ross, G.D. and Polley, M.J. (1975), *J. Exp. Med.*, **141,** 1163–1180.
Ross, G.D., Polley, M.J., Rabellino, E.M. and Grey, H.M. (1973), *J. Exp. Med.*, **138,** 798–811.
Ross, G.D., Winchester, R.J., Rabellino, E.M. and Hoffman, T. (1978), *J. Clin. Invest.*, **62,** 1086–1092.
Ross, G.D., Lambris, J.D., Cain, J.A. and Newman, S.L. (1982), *J. Immunol.*, **129,** 2051–2060.
Ross, G.D., Newman, S.L., Lambris, J.D., Devery-Pocius, J., Cain, J.A. and Lachmann, P.J. (1983), *J. Exp. Med.*, **158,** 334–352.
Ross, G.D., Cain, J.A. and Lachmann, P.J. (1984a), *Fed. Proc. Fed. Am. Soc. Exp. Biol.*, (in press) (abstract).
Ross, G.D., Thomson, R.A., Walport, M.J., Ward, R.H.R. and Lachmann, P.J. (1984b), *J. Clin. Invest.*, (in press).
Ross, G.D., Walport, M.J., Parker, C.J., Lentine, A.F., Fuller, C.R., Yount, W.J., Moynes, B.L., Winfield, J.B. and Lachmann, P.J. (1984c), *Clin. Res.*, (in press) (abstract).
Schopf, R.E., Hammann, K.P., Scheiner, O., Lemmel, E.-M. and Dierich, M.P. (1982), *Mol. Immunol.*, **19,** 1401 (abstract).
Schreiber, R.D., Pangburn, M.K., Bjornson, A.B., Brothers, M.A. and Muller-Eberhard, H.J. (1982), *Clin. Immunol. Immunopathol.*, **23,** 335–357.
Tedder, T.F., Fearon, D.T., Gartland, G.L. and Cooper, M.D. (1983), *J. Immunol.*, **130,** 1668–1673.
Todd, R.F., III and Schlossman, S.F. (1982), *Blood*, **59,** 775–786.
Todd, R.F., III, Nadler, L.M. and Schlossman, S.F. (1981), *J. Immunol.*, **126,** 1435–1442.
Whaley, K. (1980), *J. Exp. Med.*, **151,** 505–516.
Wilson, J.G., Tedder, T.F. and Fearon, D.T. (1982a), *Fed. Proc. Fed. Am. Soc. Exp. Biol.*, **41,** 965 (abstract).
Wilson, J.G., Wong, W.W., Schur, P.H. and Fearon, D.T. (1982b), *N. Engl. J. Med.*, **307,** 981–986.
Wong, W., Wilson, J. and Fearon, D. (1983), *Fed. Proc. Fed. Am. Soc. Exp. Biol.*, **42,** 1235 (abstract).

6 Monoclonal Antibodies to β-Adrenergic Receptors and Receptor Structure

CLAIRE M. FRASER

Glossary of terms and abbreviations

β_1-, β_2-Receptors	The two major subtypes of adrenergic (β) receptors
IHYP	Iodohydroxybenzylpindolol: ligand for assaying β-receptors
Protein A	Protein from *Staphylococcus aureus* (usually Cowan strain 1). Binds Fc region of IgG and used to assay antibody binding

Acknowledgement

This work was supported by Grant AI-19346 from the National Institutes of Health and by Grant 82-881 from the American Heart Association with funds contributed in part by the Heart Association of Western New York. I wish to thank Dr J. Craig Venter for his invaluable help and discussions.

Monoclonal Antibodies to Receptors: Probes for Receptor Structure and Function
(*Receptors and Recognition*, Series B, Volume 17)
Edited by M. F. Greaves
Published in 1984 by Chapman and Hall, 11 New Fetter Lane, London EC4P 4EE

6.1 INTRODUCTION

Adrenergic receptors exist on the surface of essentially every cell in the body and modulate a wide range of physiological events which include the force and rate of cardiac contractility, relaxation or contraction of bronchial and vascular smooth muscle and numerous metabolic responses such as glycogenolysis and lipolysis. Epinephrine and norepinephrine as well as a number of sympathomimetic agents can elicit adrenergic responses; however, clear differences exist in the sensitivity of adrenergic receptors from different organs to both adrenergic agonists and antagonists.

The receptors for catecholamines were first subclassified into α- and β-receptors in 1948 by Ahlquist on the basis of the order of potency of a series of adrenergic agents in eliciting biological responses in a number of tissues (Table 6.1). β-Receptors were further subclassified into β_1 and β_2 subtypes in 1967 by Lands and co-workers. This distinction was based upon the difference in potency of epinephrine and norepinephrine in eliciting physiological responses in a number of tissues. At β_1-receptors, found primarily in the heart and adipocytes, epinephrine and norepinephrine are essentially equipotent in eliciting responses, whereas at β_2-receptors, found in the liver, airway and vascular smooth muscle and the uterus, epinephrine is much more potent than norepinephrine in eliciting responses (Furchgott, 1972).

Table 6.1 Adrenergic receptor classification based on pharmacological criteria (Ahlquist, 1948)

α-Adrenergic receptors:	(−)-epinephrine > (±)-epinephrine > (±)-norepinephrine > α-methylepinephrine > α-methylnorepinephrine > isoproterenol
β-Adrenergic receptors:	isoproterenol > (−)-epinephrine > α-methylepinephrine > (±)-epinephrine > α-methylnorepinephrine > norepinephrine

More recently, a number of laboratories have used adenylate cyclase activation (Mayer, 1972; Burges and Blackburn, 1972; Murad, 1973; Lefkowitz, 1975) or analysis of radioligand-binding data (Rugg *et al.*, 1978; Hancock *et al.*, 1979; Minneman *et al.*, 1979) to confirm and extend the initial subclassification of β-receptors as proposed by Lands *et al.* in 1967. However, each of these types of studies suffers from a lack of information as to the molecular properties underlying the observed pharmacological or biochemical differences between β_1- and β_2-receptor subtypes. Hence, the ultimate characterization of molecular properties of β-adrenergic receptors will only be obtained from the purification and amino acid sequencing of β_1- and β_2-receptor species.

To this end, our laboratory has been concerned with the solubilization, biochemical characterization and purification of β-adrenergic receptors from avian and mammalian tissues. This task has been complicated by the extremely low concentration of β-receptors on each cell (a few hundred to a few thousand/cell) and by the relative instability of these proteins following detergent solubilization from membranes. Nonetheless, much evidence has accumulated to date which suggests that there are at least two distinct molecular forms of the β-adrenergic receptor which correspond, for the most part, to the β_1- and β_2-receptor subtypes defined by pharmacological and biochemical studies.

Table 6.2 Solubilization of β-receptor–[^{125}I]IHYP complexes from liver and heart with different detergents*

	Liver		Heart				
Detergent†	β-Receptor (fmol/ml)	Protein (mg/ml)	β-Receptor (fmol/ml)	Protein (mg/ml)	HLB No.‡	Polyoxy-ethylene‡ (*n*)	Lipophile‡
Polyoxyethylene ether W1	4.1	2.15	0.0	1.05			
Brij 35	0.0	—	0.0	—	16.9	23	Lauryl alcohol
Brij 56	2.1	2.22	0.0	1.25	12.0	10	Cetyl alcohol
Brij 58	0.9	2.13	0.0	1.23	15.7	20	Cetyl alcohol
Brij 76	0.0	2.12	1.9	1.23	12.4	10	Stearyl alcohol
Brij 96	5.8	2.35	9.1	1.25	12.4	10	Olelyl alcohol
Brij 98	1.9	2.32	4.8	1.25	15.3	20	Olelyl alcohol
Polyoxyethylene 9 lauryl ether	2.4	2.5	0.0	1.6		9	Lauryl alcohol
Lubrol WX	1.1	1.95	0.0	0.7	14.9	17	Cetylstearyl
Tween 20	1.7	2.17	0.0	0.65	16.7	20	Monolaurate
Tween 40	0.0	12.3	0.0	0.36	15.6	20	Monopalmitate
Tween 60	1.8	2.00	0.0	0.60	14.9	20	Monostearate
Tween 80	0.0	1.00	0.0	0.56	15.0	20	Mono-oleate
Tween 85	0.0	1.30	0.0	0.32	11.0	20	Trioleate
Triton X-100	3.7	2.48	0.0	2.00	13.5	9.5	Octylphenyl
SDS	0.0	2.67	0.0	1.80			
Deoxycholate	1.6	2.52	0.0	1.85			
Digitonin	0.0	1.88	0.0	1.85			Digitogenin
Triton–digitonin	6.0	2.43	10.2	1.6			

* Heart and liver membranes were incubated with 250 pM IHYP for 30 min at 30 °C in the presence and absence of 10 M (−)propranolol prior to solubilization with detergents for 10 min at 30°C as described by Strauss *et al.* (1979).

† All detergents were used as 1% solutions (final concentration) with the exception of the Triton–digitonin combination which was 0.25% Triton X-100 and 1% digitonin.

‡ From the Merck index; (*n*) refers to the number of polyoxyethylene units per detergent molecule.

Strauss *et al.* (1979) demonstrated that canine heart β_1- and liver β_2-adrenergic receptors were differentially solubilized from purified plasma membrane preparations by a series of non-ionic and ionic detergents (Table 6.2). Hepatic β_2-adrenergic receptors were partially solubilized by Triton X-100, Lubrol, digitonin and some detergents of the Tween series while these agents were ineffective in solubilizing cardiac β_1-receptors. Both Brij 96 and a combination of Triton X-100 and digitonin gave the optimum degree of solubilization of β-adrenergic receptors from both cardiac and hepatic plasma membranes. It was concluded that the differences in β-receptor solubilization might reflect differences in the protein structure of the cardiac β_1- and liver β_2-receptor molecules or in the ability of detergents to solubilize membranes of different lipid composition.

Table 6.3 Molecular parameters of cardiac, lung and liver β-adrenergic receptors*

Parameter	Heart (β_1)	Turkey erythrocyte (β_1)	Lung (β_2)	Liver (β_2)
Stokes radius (nm)	4.2	4.3	5.8	5.8
Sedimentation coefficient ($S_{20,\,w}$)	3.69	3.70	3.78	3.67
Partial specific volume (g/ml)	0.73	—	0.73	0.73
Hydrodynamic molecular weight	65 000	65 000	91 000	90 000
Frictional ratio f/f_0	1.6	1.6	2.0	2.0
Isoelectric point	5.0	5.5	4.2	—

* Hydrodynamic properties determined from Sepharose 6B gel permeation chromatography (Stokes radius), and sucrose density gradient centrifugation (sedimentation coefficients).

More direct evidence for molecular differences between β_1- and β_2-adrenergic receptors was obtained from determination of the hydrodynamic properties of β-receptors from a number of sources (Table 6.3; also Fraser and Venter, 1980a). The Stokes radii for the various receptors, obtained from gel permeation chromatography, indicated that the canine lung and liver β_2-receptors were substantially larger (5.8 nm) than the canine heart or turkey erythrocyte β_1-receptors (4.2 nm). The avian and mammalian β-receptors were similar in their sedimentation velocities with calculated $S_{20,\,w}$ values of approximately 3.7. Under the appropriate experimental conditions, soluble mammalian receptors do not appear to bind considerable quantities of detergent as the partial specific volumes of the receptor proteins calculated in these experiments were identical with those of the marker proteins used and equal to

0.73 g/ml. Using these data, the hydrodynamic molecular weights of β_1- and β_2-receptors were calculated. Cardiac and turkey erythrocyte β_1-receptors with molecular weights of 65 000 and frictional coefficients of 1.6, appeared to be smaller, more globular proteins than either canine liver or lung β_2-receptors, with molecular weights of 90 000 and frictional coefficients of 2.0.

The notion that β_1- and β_2-adrenergic receptors represented distinct molecular species was also supported by the observed difference in the isoelectric point of turkey erythrocyte β_1-receptors (pI = 5.5), canine heart β_1-receptors (pI = 5.0) and canine lung β_2-adrenergic receptors (pI = 4.2).

Table 6.4 Molecular subclassification of β-adrenergic receptors*

Species	Cell or tissue	IHYP binding inhibition by 1 mM DTT (%)	Stokes radius (nm)	β-Receptor subclass based on molecular parameters
Dog	Liver	0	5.8	β_2 (100)
Rat	Liver	0	—	β_2 (100)
Cat	Liver	0	—	β_2 (100)
Frog	Erythrocytes	0	5.8	β_2
Mouse	Lymphoma (S49 cells)	—	6.4	β_2
Dog	Lung	16±6.4	5.8	β_1 (20) β_2 (80)
Human	Lung (VA$_4$ cells)	17±6.0	5.8	β_1 (21) β_2 (79)
Rat	Lung	32±1.1	—	β_1 (40) β_2 (60)
Rabbit	Lung	41±2.0	—	β_1 (52) β_2 (48)
Cat	Lung	43±8.0	—	β_1 (54) β_2 (46)
Dog	Adipocytes	41±1.7	—	β_1 (51) β_2 (49)
Rat	Glioma (C6 cells)	44±2.3	—	β_1 (55) β_2 (45)
Turkey	Erythrocytes	35±4.6	4.2	β_1 (44) β_2 (56)
Rat	Heart	65±6.1	4.8	β_1 (81) β_2 (19)
Dog	Heart	80±1.5	4.2	β_1 (100)

* β-Receptor classification based upon IHYP-binding data in the presence and absence of 1 mM DTT. The 80% inhibition of binding to dog heart β-receptors by DTT is assumed to represent the presence of only β_1-receptors. The lack of a DTT effect on liver β-receptors is assumed to represent the presence of only β_2-receptors. From Strauss (1980).

Strauss and Venter (Strauss, 1980) have demonstrated a differential sulfhydryl reagent sensitivity of canine cardiac and hepatic β-receptors. For example, preincubation of purified plasma membranes with 1 mM dithiothreitol (DTT), a potent disulfide reducing agent, produced an 80% reduction in the specific binding of β-adrenergic ligands to cardiac β_1-receptors yet had no effect on hepatic β_2-receptors. Scatchard analysis of saturation isotherms of [^{125}I]iodohydroxybenzylpindolol (IHYP) binding to cardiac β-receptors in the presence of 1 mM DTT indicated that the decrease in IHYP binding observed was due to a loss of β-adrenergic receptors with no change in receptor affinity for IHYP. The data suggested that the β_1-receptor contains a highly reactive disulfide bond which is either absent from or less accessible in the β_2-adrenergic receptor (Table 6.4). The lower reactivity of the disulfide group in the β_2-adrenergic receptor suggested that the β_1- and β_2-receptor proteins may differ. In similar experiments, ligand binding to β_1-receptors was inhibited 83% by preincubation of cardiac membranes with 1 mM *p*-chloromercuribenzoate (pCMB), and this effect could be prevented by prior occupation of the receptor with β-adrenergic antagonists. Ligand binding to β_2-receptors was also affected by pCMB; however, this effect was apparently due to a change in receptor affinity for radioligands in contrast to the reduction in β_1-receptor number produced by pCMB. These data indicated that β_1-, and possibly, β_2-receptors contain free sulfhydryl groups in or near the ligand-binding site of the receptor molecule but suggest that the role of free sulfhydryl groups in ligand-binding to β_1- and β_2-receptors may differ.

6.2 THE USE OF MONOCLONAL ANTIBODIES AND AUTOANTIBODIES IN β-RECEPTOR CHARACTERIZATION

The application of immunological techniques to hormone receptor characterization and purification has allowed rapid advances in this area. Autoantibodies to cell membrane receptors (Patrick *et al.*, 1973; Abramsky *et al.*, 1975; Drachman *et al.*, 1978; Smith and Hall, 1974; Manley *et al.*, 1974; Flier *et al.*, 1975; Kahn *et al.*, 1976; Venter *et al.*, 1980; Fraser *et al.*, 1981) and, more recently, monoclonal antibodies to hormone receptors (see other chapters in this volume and Tzartos and Lindstrom, 1980; Tzartos *et al.*, 1981) have shown great utility in the molecular characterization of related receptor molecules. The extent of molecular homology between receptors can be examined with radioimmunoassay techniques and cross-reactivity studies. Immunochemical analysis of hormone receptors reveals differences between receptor subtypes which are not apparent with other biochemical techniques for determination of gross protein conformation such as gel permeation chromatography or sodium dodecyl sulfate polyacrylamide gel electrophoresis (SDS-PAGE). The development of monoclonal antibodies to receptor molecules has enabled the study of structural homology between receptor subtypes to progress to the level of single antigenic determinants. Furthermore, as illustrated elsewhere in this volume, autoantibodies and monoclonal antibodies to hormone and neurotransmitter receptors have been demonstrated to be valuable reagents in the immunoaffinity purification of receptor proteins.

6.3 β_1-ADRENERGIC RECEPTOR STRUCTURE

6.3.1 Monoclonal antibodies to the turkey erythrocyte β_1-adrenergic receptor

We have developed monoclonal antibodies to turkey erythrocyte β_1-adrenergic receptors by immunizing Balb/c mice with digitonin-solubilized receptors partially purified by preparative isoelectric focusing (Fig. 6.1; Fraser

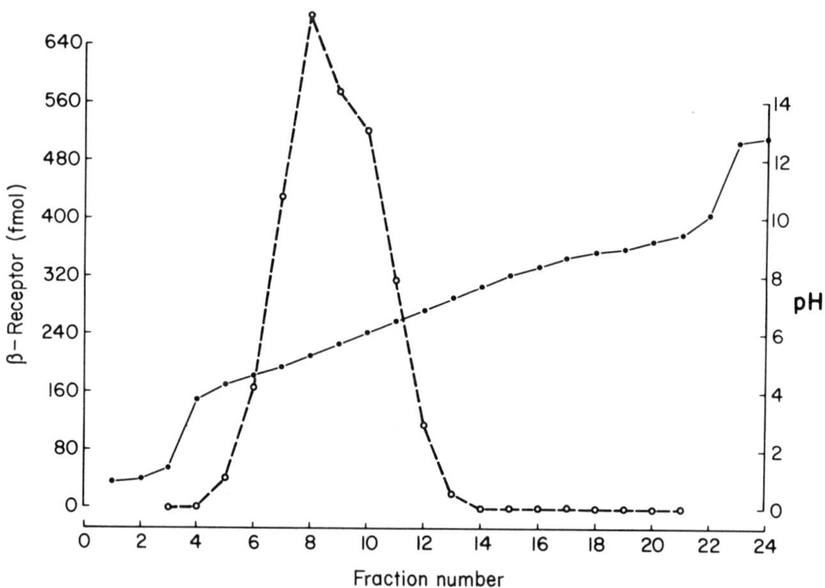

Fig. 6.1 Isoelectric focusing of turkey erythrocyte β-adrenergic receptors. Turkey erythrocyte β-adrenergic receptors were solubilized from erythrocyte 'ghosts' by treatment with 0.5% digitonin for 12 hours at 4 °C. Soluble β-adrenergic receptors were added to sucrose density gradients (0–50%) containing 1% ampholines (pH 3–10) for isoelectric focusing (110 ml column) for 16 hours at 4 °C (constant power, 15 watts; maximum voltage, 1600 volts). Fractions from isoelectric focusing columns were assayed for β-adrenergic receptors. The open circles represent the concentration of β-adrenergic receptors present in each fraction (4 ml) from the isoelectric focusing column and the closed circles represent the pH gradient. (From Fraser and Venter, 1981.)

and Venter, 1980b; Venter and Fraser, 1981). This technique results in a 2000–10 000-fold purification of the β-receptor as compared to the intact erythrocyte. Splenic lymphocytes obtained from mice immunized with this material (pI = 5.5) were fused with SP2/0 myeloma cells to generate hybridomas secreting β-receptor-specific monoclonal antibodies.

The screening assay which has been used to identify monoclonal antibodies to turkey erythrocyte β_1-receptors is an indirect immunoprecipitation assay (Fraser and Venter, 1980b; Venter and Fraser, 1981). As the digitonin-solubilized turkey erythrocyte β-receptor is relatively stable ($t_{1/2}$ = 3 days at 4 °C) in the absence of ligand, aliquots of unoccupied, partially purified receptor are incubated with a sample of culture medium from microtiter wells containing hybridomas followed by precipitation of receptor–monoclonal antibody complexes by addition of an excess of rabbit antiserum to mouse IgG. The supernates are assayed for loss of receptors as compared to samples containing control medium or, alternatively, the pelleted immune complexes

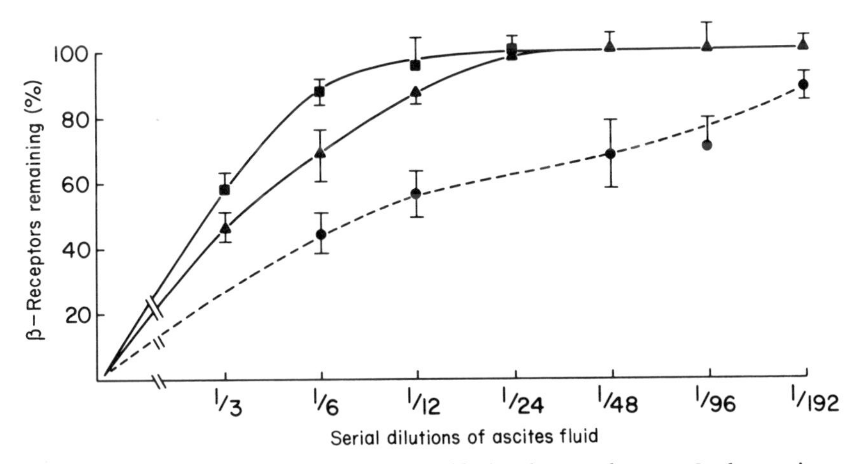

Fig. 6.2 Immunoprecipitation of partially purified turkey erythrocyte β-adrenergic receptors. β-Adrenergic receptors were solubilized from turkey erythrocyte membranes with 0.5% digitonin. Aliquots containing 5 fmol of soluble β-receptor were incubated with the indicated serial dilutions of ascites fluid containing monoclonal antibodies to β-receptors for 18 hours at 4 °C. Precipitation of β-receptor–antibody complexes was accomplished by incubation of the samples with 100 μg of rabbit anti-mouse IgG serum for 4 hours at 4 °C followed by centrifugation at 12 000 *g* for 5 minutes. The concentration of β-receptors remaining in solution was quantitated by labeling an aliquot of the supernates with IHYP in the presence or absence of 10 μM (−)-propranolol followed by precipitation of labeled receptors with 15% polyethylene glycol. The values represent the mean ± S.E.M. of triplicate determinations from two separate experiments. The closed squares indicate precipitation data obtained with monoclonal antibody 101, the closed circles with monoclonal antibody 103 and the closed triangles with monoclonal antibody 104.

can be assayed for the presence of immunoprecipitated receptor. This assay allows for identification of monoclonal antibodies to antigenic determinants on the β-receptor both within and outside of the ligand-binding site of the receptor molecule and does not rely solely on antibody effects on ligand binding to the receptor.

We have isolated three stable hybridomas producing monoclonal antibodies of the IgG class to turkey erythrocyte β-receptors and grown them as ascites tumors in BALB/c mice (Fraser and Venter, 1980b). Fig. 6.2 illustrates a concentration-dependent immunoprecipitation of partially purified turkey erythrocyte β_1-receptors by monoclonal antibodies 101, 103 and 104. The data indicate that each monoclonal antibody recognizes an antigenic determinant of the solubilized turkey erythrocyte and suggest that each antibody may differ in its affinity for the β-receptor.

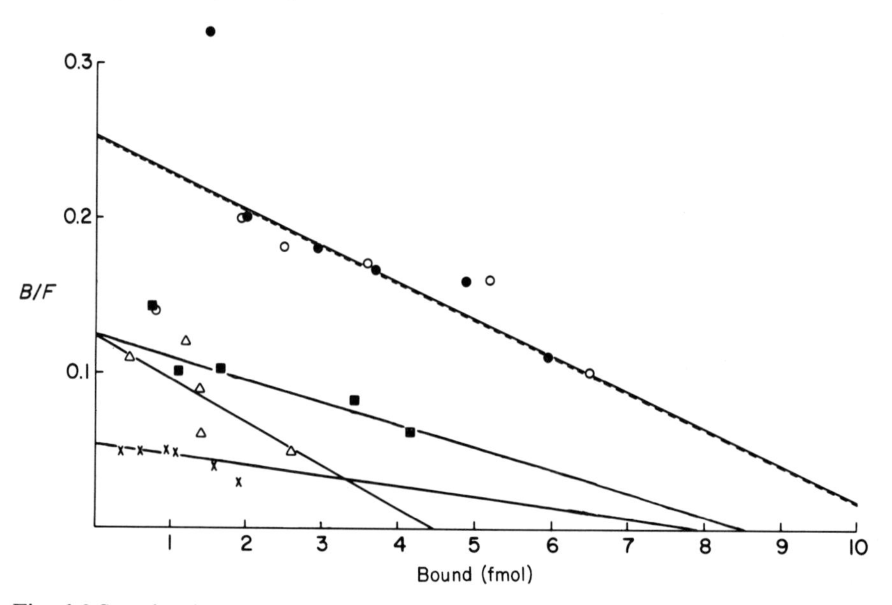

Fig. 6.3 Scatchard analysis of IHYP saturation isotherms to membrane-associated turkey erythrocyte β-receptors in the presence of monoclonal antibodies. Turkey erythrocyte membranes containing 10 fmol of β-receptor were preincubated with a 1:40 dilution of purified monoclonal antibody, ascites fluid containing monoclonal antibody or phosphate-buffered saline containing mouse IgG at the appropriate concentration for 60 minutes at 30 °C. Increasing concentrations of IHYP were added in the presence or absence of 10 μM (−)-propranolol and the samples were incubated for 30 minutes at 30 °C. Shown is Scatchard analysis of saturation isotherms obtained in the presence of purified monoclonal antibody 104 (closed squares), ascites fluid containing antibody 101 (open circles), 103 (open triangles), 104 (X) or control mouse IgG (closed circles). Each point represents the mean of triplicate determinations from two separate experiments. (From Venter and Fraser, 1983.)

In order to begin to elucidate whether the epitope(s) recognized by antibodies 101, 103 and 104 represented distinct regions of the turkey erythrocyte β_1-receptor, the ability of each of the monoclonal antibodies to affect β-adrenergic ligand binding was examined by analysis of IHYP-saturation isotherms obtained in the presence of monoclonal antibodies (Venter and Fraser, 1983). As shown in Fig. 6.3, when compared to controls, monoclonal antibody 101 did not affect β-adrenergic ligand binding to turkey erythrocyte β-receptors, suggesting that the determinant recognized by this monoclonal antibody was located outside of the ligand-binding site of the receptor. Monoclonal antibody 104 inhibited IHYP binding in a manner consistent with competitive antagonism (that is by altering the apparent affinity of the receptor for ligand without changing the total number of receptors), suggesting that the antigenic determinant recognized by antibody 104 was within the ligand-binding site of the β-receptor. Monoclonal antibody 103 inhibited IHYP binding non-competitively, that is by reducing the number of ligand-binding sites without changing the affinity of the receptor for ligand.

6.3.2 Immunological cross-reactivity between avian and mammalian β-receptors

It was of interest to determine whether or not the determinants recognized by monoclonal antibodies to turkey erythrocyte β-receptors were associated with β-adrenergic receptors from other sources or unique for the turkey erythrocyte receptor.

To this end, antibody 104 was assayed for its ability to inhibit β-adrenergic ligand binding to other β_1- and β_2-receptors. As illustrated in Fig. 6.4, monoclonal antibody 104 inhibited IHYP binding to membrane-associated and solubilized turkey erythrocyte β-receptors in a concentration-dependent fashion. More importantly, it was found that monoclonal antibody 104 equally inhibited IHYP binding to β_1-receptors from bovine heart and β_2-receptors from bovine lung and liver. These data indicated that some degree of structural homology existed between β_1- and β_2-receptors, and raised the possibility that the two types of receptor arose from a common ancestor by gene duplication (Fraser and Venter, 1980b). The determinant on the β-receptor recognized by monoclonal antibody 104 is not found on α_1-adrenergic or dopamine receptors, which makes receptor interconversion a highly unlikely event (Fraser and Venter, 1980b).

6.3.3 Immunoaffinity purification of turkey erythrocyte β_1-adrenergic receptors

We have produced an immunoaffinity column with antibody 104 covalently coupled to Sepharose. Because of the unique specificity of the antibody and its

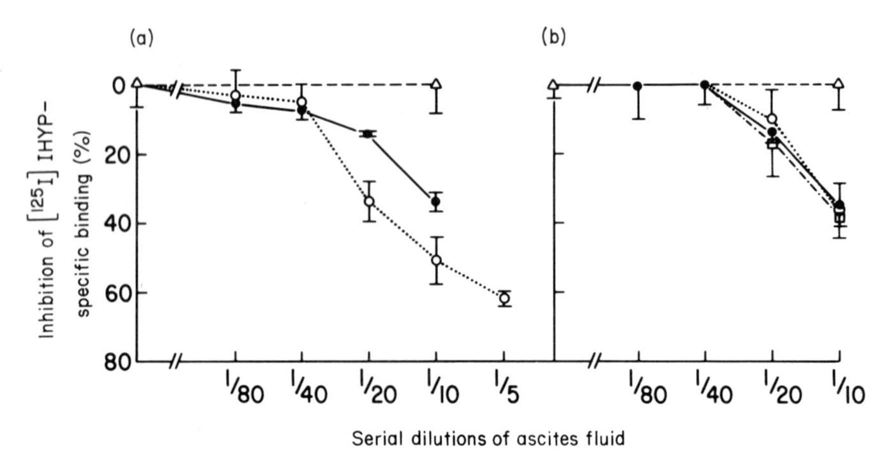

Fig. 6.4 Inhibition of IHYP-specific binding to soluble and membrane associated β-adrenergic receptors by monoclonal antibody 104. Turkey erythrocyte 'ghosts' ((a), solid circles), partially purified turkey erythrocyte β-receptors ((a), open circles), purified calf lung membranes ((b), open circles), purified calf heart membranes ((b), solid circles) or purified calf liver membranes ((b), open squares) at protein concentrations required to provide 1.6 fmol of β-receptor per assay were preincubated with serial dilutions of ascites fluid containing monoclonal antibody 104 for 60 minutes at 30°C in a final volume of 100 μl. IHYP-specific binding to membranes was determined by incubating samples with IHYP in the presence and absence of 10 μM (−)-propranolol for 30 minutes at 30°C followed by filtration on Whatman GF/C glass-fiber filters. Serial dilutions of mouse IgG served as the control for each experiment ((a) and (b), open triangles) The values are the means ± the S.E.M. of duplicate determinations from two separate experiments. (From Fraser and Venter, 1980.).

competitive binding to β-receptors of both β_1 and β_2 classes, it was possible to combine immunoaffinity chromatography with ligand-specific affinity chromatography by eluting β-receptor molecules from a monoclonal antibody affinity column with adrenergic ligands. The fact that the determinant recognized by monoclonal antibody 104 is within the adrenergic ligand-binding site of the receptor and common to both β_1- and β_2-receptor subclasses along with the relatively low affinity of monoclonal antibody 104 for β-receptors makes this antibody well suited for β-receptor affinity purification procedures (Fraser and Venter, 1980b).

When turkey erythrocyte β-receptors are labeled in the membrane with ^{131}I, solubilized, and purified by isoelectric focusing followed by monoclonal antibody affinity chromatography in which the bound receptor is eluted with sodium dodecyl sulfate and β-mercaptoethanol, proteins of 70000, 31000 and 22000 daltons are identified on SDS/polyacrylamide gels (Fig. 6.5). If the specificity of the immunoaffinity isolation is increased by elution of bound receptors with 10 μM (−)-propranolol, a single protein of 65000–70000 daltons is found on SDS-polyacrylamide gels (Fig. 6.6). The same 65000–70000-dalton protein is also obtained on SDS-polyacrylamide gels if the (−)-propranolol eluate from the immunoaffinity columns is iodinated by using lactoperoxidase or chloramine T. The identity of the 70 000-dalton protein as the intact turkey erythrocyte β-receptor was suggested by the excellent agreement of the SDS-PAGE molecular weight with the calculated hydrodynamic molecular weight of the turkey erythrocyte β_1-receptor and the canine heart β_1-receptor of

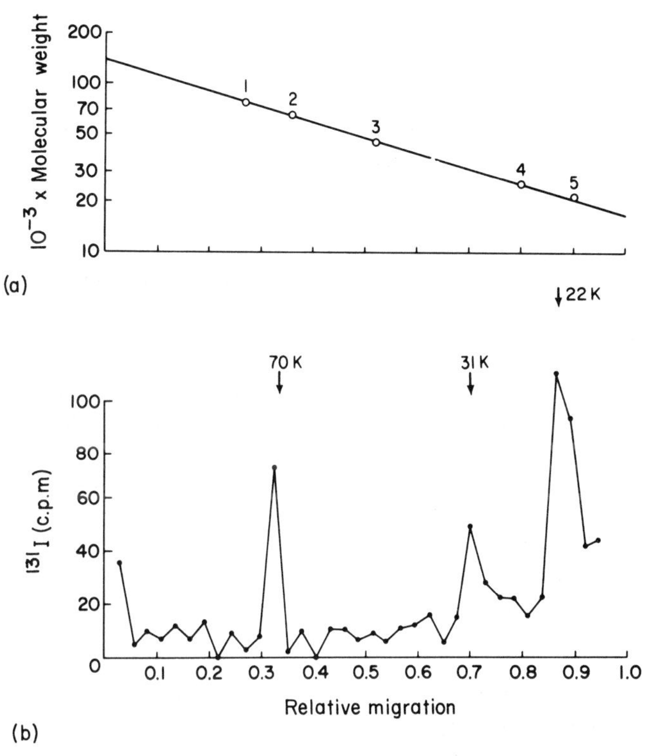

Fig. 6.5 Monoclonal antibody affinity purification of turkey erythrocyte β-adrenergic receptors. Turkey erythrocyte 'ghosts' were surface-labeled with [^{131}I]iodine. β-Receptors were solubilized from membranes with 0.5% digitonin and purified by preparative isoelectric focusing to a specific activity of 1.67 pmol/mg of protein. Partially purified receptors (2 ml) were incubated with monoclonal antibody 104–Sepharose 4B (2.5 ml) for 2 hours at 30 °C. The Sepharose beads were washed in a 10 ml column with 5 ml of phosphate buffer. β-Receptors were eluted from the antibody affinity columns with buffer containing 1% SDS, 2.5% mercaptoethanol and 5% glycerol. Samples were incubated at 100 °C for 5 minutes and analyzed on 10% SDS/polyacrylamide gels which were sliced and counted for radioactivity. (a) Molecular weight calibration curve for 10% SDS/polyacrylamide gels using: (1) Lactoperoxidase, 77 000 daltons; (2) Bovine serum albumin, 66 000 daltons; (3) Ovalbumin, 46 000 daltons; (4) Chymotrypsinogen A, 25 750 daltons and (5) Soy bean trypsin inhibitor, 21 500 daltons. (b) ^{121}I-labeled monoclonal antibody affinity column eluates of turkey erythrocyte β-receptors. Sizes marked as $M_r \times 10^{-3}$. ^{131}I marked as c.p.m. $\times 10^{-3}$. (From Fraser and Venter, 1980.)

65000. Furthermore, the 70000-dalton protein contains the β-adrenergic ligand-binding site, as it binds radioligands following removal of (−)-propranolol by ultrafiltration. Although the purified receptor is very unstable, the binding is identical with that found for detergent-solubilized β-receptors. The purified receptor protein has also been reconstituted into human erythrocyte acceptor membranes by the method previously published for detergent-solubilized β-receptors (Jeffrey *et al.*, 1980). Couraud *et al.* (1983) and Schreiber *et al.* (1980) have also reported a similar SDS-PAGE molecular

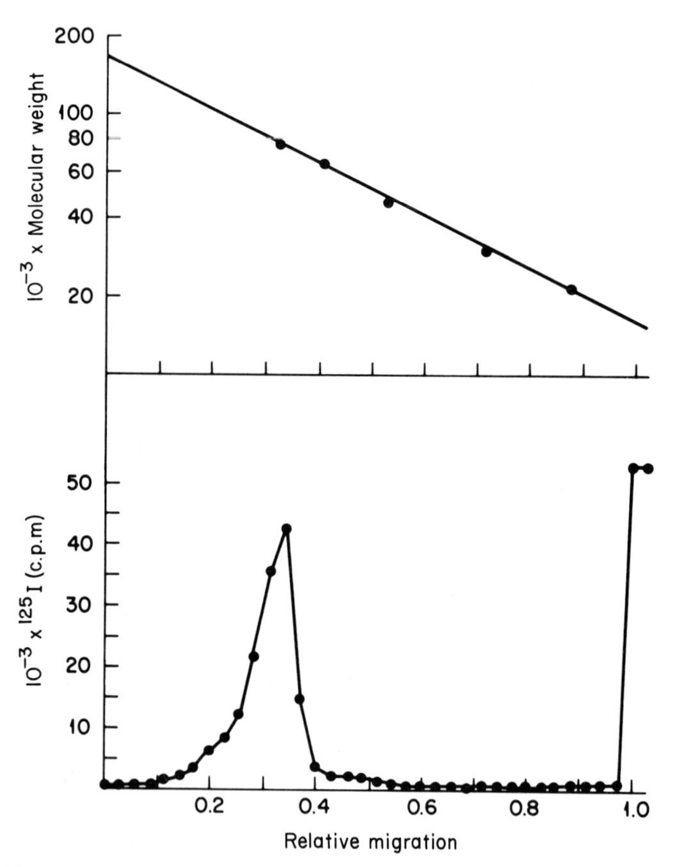

Fig. 6.6 Ligand-specific elution of turkey erythrocyte β-adrenergic receptors from immunoaffinity columns. Partially purified turkey erythrocyte β-adrenergic receptors (600 fmol) were applied to immunoaffinity columns of monoclonal antibody 104–Sepharose 4B (5 ml). Columns were washed with 10 ml of phosphate buffer, and adsorbed β-adrenergic receptors were eluted with 10 ml of 10 μM (−)-propranolol. The eluate was iodinated with Na ^{125}I and analyzed by 10% SDS-PAGE as described in Fig. 6.5. Molecular weight standards are as in Fig. 6.5. (From Venter and Fraser, 1981.)

weight for the turkey erythrocyte β_1-receptor isolated with anti-idiotypic antibodies having β-receptor specificity.

6.3.4 Target size analysis of β_1-adrenergic receptors

Two independent lines of evidence, immunoaffinity purification using monoclonal antibodies and hydrodynamic characterization of the β_1-receptor, suggested that the turkey erythrocyte β_1-receptor and the canine heart β_1-receptor are single polypeptides of molecular weight approximately 65 000–70 000. In order to determine whether or not the proteins isolated by gel permeation chromatography and SDS-PAGE represented the intact β-receptor *in situ,* we turned to target size analysis or radiation inactivation. Target-size analysis, first described in the 1950s (Pollard, 1953), is presently the only technique available for determination of the functional molecular size of a protein in the native membrane. According to the theory of target-size analysis, membranes are bombarded with high-energy radiation such that one high-energy electron has enough energy to destroy the physical structure and function of a target protein molecule. The amount of radiation required to inactivate a protein is inversely proportional to the size of the protein, that is, large proteins are statistically more likely to be hit and inactivated by a given dose of radiation than small proteins. Molecular-size calibrations can be performed by inclusion of enzymes of known molecular weight with the membrane preparations (Venter, 1983). Target-size analysis has recently been performed with a number of membrane receptor systems with considerable success (Lo *et al.*, 1982; Venter, 1983; Venter *et al.*, 1983; Harmon *et al.*, 1980; Doble and Iverson, 1982; Paul *et al.*, 1981; Fraser and Venter, 1982; Neilsen *et al.*, 1981; Lilly *et al.*, 1983).

When target-size analysis is applied to the canine heart β_1-receptor, a functional molecular size of 110 000–130 000 daltons is obtained for this protein as determined by loss of ligand-binding activity (Fig. 6.7). These data taken together with those obtained from SDS-PAGE have led us to propose the following model for the mammalian β_1-adrenergic receptor. The simplest explanation which incorporates the experimental findings to date is that, *in situ,* the β_1-receptor exists as a functional dimer of two identical 65 000–70 000 dalton subunits, each of which contains a catecholamine-binding site. Alternatively, the functional β_1-receptor may be composed of one receptor (65 000–70 000 dalton) subunit in association with the guanine nucleotide regulatory protein of adenylate cyclase or another yet to be identified subunit.

6.4 β_2-ADRENERGIC RECEPTOR STRUCTURE

The β_2-adrenergic receptor is of major physiological and clinical importance as it modulates vascular and airway smooth muscle diameter. Our initial studies

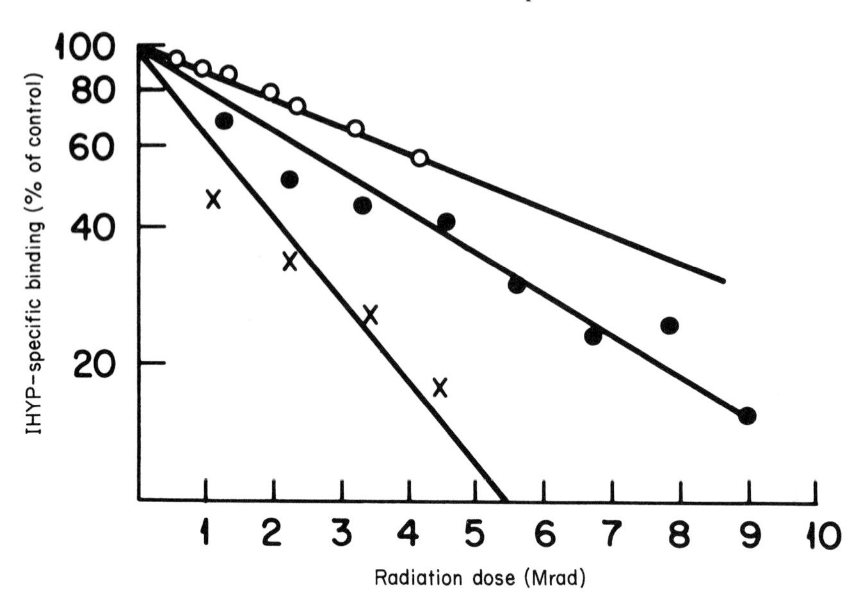

Fig. 6.7 Target-size analysis of canine heart β_1-adrenergic receptors. Canine heart membranes were frozen in thin layers on aluminium trays and subjected to high-energy electron bombardment in a Van de Graaf generator. Chamber temperature was maintained at −45 to −52 °C during the irradiation. Radiation-induced loss of the β_1-adrenergic receptor was assessed by measuring loss of IHYP-specific binding as a function of increasing doses of radiation (●). Molecular-size calibration was performed by inclusion of enzyme standards of known molecular weight as follows: liver alcohol dehydrogenase, 84 000 (O) and pyruvate kinase, 224 000 (×). The molecular size of the β_1-receptor was determined by comparing the slope ratio of inactivation of the receptor to the enzymes as described (Venter, 1983). Each point represents the mean of triplicate determinations from each experiment. Lines were drawn by least-squares linear regression.

on molecular properties of β-receptors have indicated clear differences in β_1- and β_2-receptors as determined by detergent solubilization, hydrodynamic parameters and thiol-reagent sensitivity. Immunological characterization of β-receptors has further advanced our understanding of the molecular differences between β_1- and β_2-receptors.

6.4.1 Monoclonal antibodies

In studies similar to those described for the turkey erythrocyte β_1-receptor, the β_2-adrenergic receptor from bovine lung was solubilized from membranes with Triton X-100, purified by immunoaffinity chromatography on monoclonal antibody 104–Sepharose, eluted from the immunoaffinity matrix with (−)-

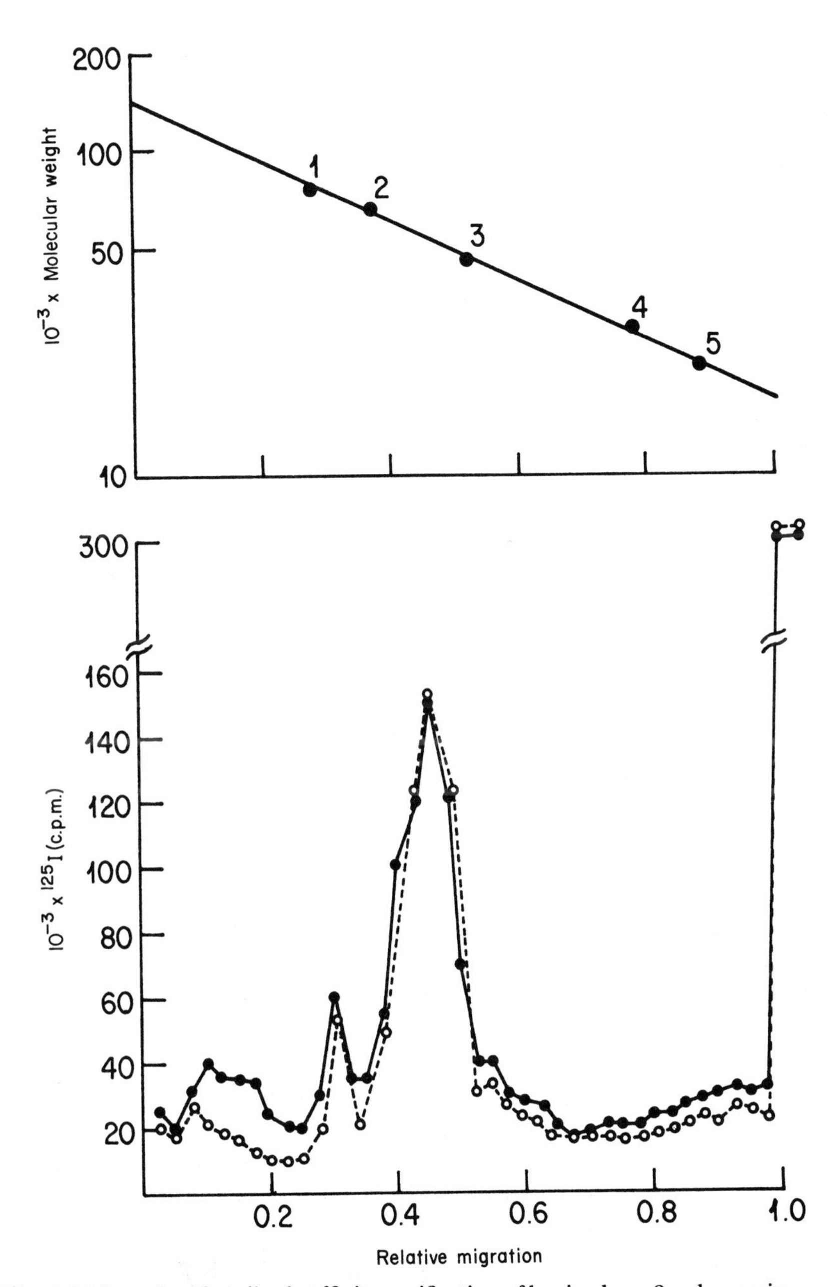

Fig. 6.8 Monoclonal antibody affinity purification of bovine lung β_2-adrenergic receptors. Triton X-100-solubilized β_2-adrenergic receptors, specifically labeled with [^{125}I]IHYP, were partially purified by gel permeation chromatography followed by immunoaffinity chromatography on monoclonal antibody 104–Sepharose 4B as described in Figs 6.5 and 6.6. The bottom panel represents an SDS/polyacrylamide gel (10%) of immunoaffinity-purified β_2-receptors under reducing (open circles) and non-reducing conditions (closed circles). The molecular weight standards in the top panel are identical with those described in the legend for Fig. 6.5. (From Fraser and Venter, 1982.)

propranolol and analyzed on SDS-polyacrylamide gels. As is illustrated in Fig. 6.8, the major protein obtained with this protocol is one of molecular weight 59000 (Fraser and Venter, 1982). When SDS-polyacrylamide gels are run under non-reducing conditions, a higher-molecular weight species of approximately 115000 daltons is also obtained. These findings suggest that the 59000-dalton protein may be derived from the 115000-dalton component. In order to address this question, gel slices containing the 115000-dalton protein were re-electrophoresed under reducing conditions. It was found that in the presence of β-mercaptoethanol, the 115000-dalton species disappeared and the radioactivity was found associated with a 59000-dalton protein (not shown). This indicated that the 59000-dalton protein is most likely a subunit of a larger 115000-dalton protein (Fraser and Venter, 1982).

The existence of a β_2-receptor species with a molecular weight greater than 59000 was also suggested by our earlier hydrodynamic molecular weight for the β_2-receptor from canine lung and liver.

6.4.2 Autoantibodies to β_2-adrenergic receptors

In 1980, we identified autoantibodies to β_2-adrenergic receptors in the sera of several patients with allergic respiratory disease including asthma, allergic rhinitis and cystic fibrosis (Venter *et al.*, 1980; Fraser *et al.*, 1981). The criteria we established for identification of autoantibodies to β_2-receptors were (1) the ability of serum to affect ^{125}I-labeled Protein A binding to membranes containing β_2-receptors, (2) the ability of serum to immunoprecipitate solubilized lung β_2-receptors in an indirect immunoprecipitation assay, and (3) the demonstration that serum factors responsible for such effects were immunoglobulins. Out of an initial patient population of ten asthmatic and atopic patients and four normal control subjects, we identified two asthmatics and one allergic rhinitic that possessed autoantibodies to β_2-adrenergic receptors according to the established criteria. As shown in Fig. 6.9, β-receptor autoantibodies cross reacted with β_2-receptors from a variety of species and tissues as determined by the degree of inhibition of IHYP binding. However, importantly, the autoantibodies had no detectable effects on IHYP binding to cardiac β_1-receptors. The lack of cross-reactivity between the β_2-receptor autoantibodies and β_1-receptors in the heart suggested that β_1- and β_2-receptors represent distinct protein species with structural differences in the ligand-binding sites of the molecules; furthermore, the apparent β_2-receptor specificity of the autoantibodies supported a possible pathogenic role for autoantibodies in reducing β-adrenergic responsiveness in allergic respiratory disease.

The autoantibodies were utilized in an immunoaffinity purification of mammalian lung β_2-adrenergic receptors. Triton X-100-solubilized bovine lung β_2-receptors labeled with the irreversible β-adrenergic affinity ligand,

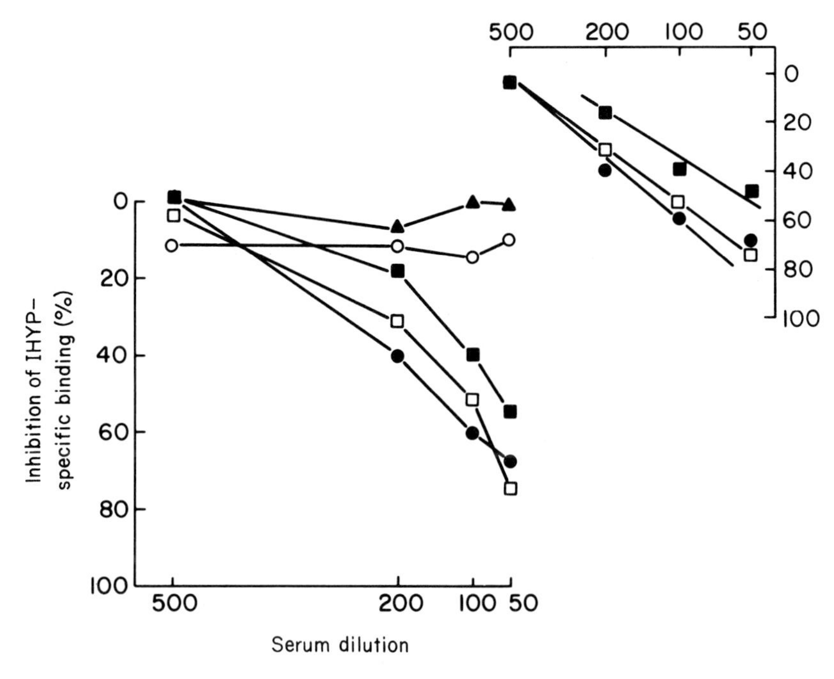

Fig. 6.9 Autoantibody inhibition of IHYP binding to β_2-adrenergic receptors. Purified membranes from canine heart (open circles), canine lung (open squares), calf lung (closed circles) and human placenta (closed squares) were preincubated with the indicated dilutions of serum from an allergic rhinitis patient with autoantibodies to β_2-adrenergic receptors or control serum for 60 minutes at 30 °C. Canine lung membranes were also preincubated under identical conditions with serum that had been depleted of gamma globulin (closed triangles). IHYP-specific binding to the membranes treated as described was determined by incubating samples with IHYP in the presence and absence of 10 μM (−)-propranolol for 30 minutes at 30 °C followed by filtration on Whatman GF/C glass-fiber filters. Percentage inhibition of IHYP-specific binding was determined from the ratio of IHYP bound in the presence of patient serum to that bound in the presence of control serum. Inset: Semi-logarithmic plot of data for IHYP binding to purified membranes. (From Venter *et al.*, 1980.)

[^{3}H]N-[2-hydroxy-3-(1-naphthoxy)propyl]-*N*-bromoacetylethylenediamine (NHNP-NBE), were partially purified by gel permeation chromatography and preparative isoelectric focusing followed by immunoaffinity chromatography on Protein A–Sepharose preincubated with autoimmune serum. When the eluate from the immunoaffinity column was analyzed on SDS-polyacrylamide gels it was found that the primary subunit of mammalian lung β_2-receptor containing the ligand-binding site had a molecular weight of 59 000 (Fig. 6.10), a value in excellent agreement with that obtained by immunoaffinity chromatography using monoclonal antibodies (Fraser and Venter, 1982).

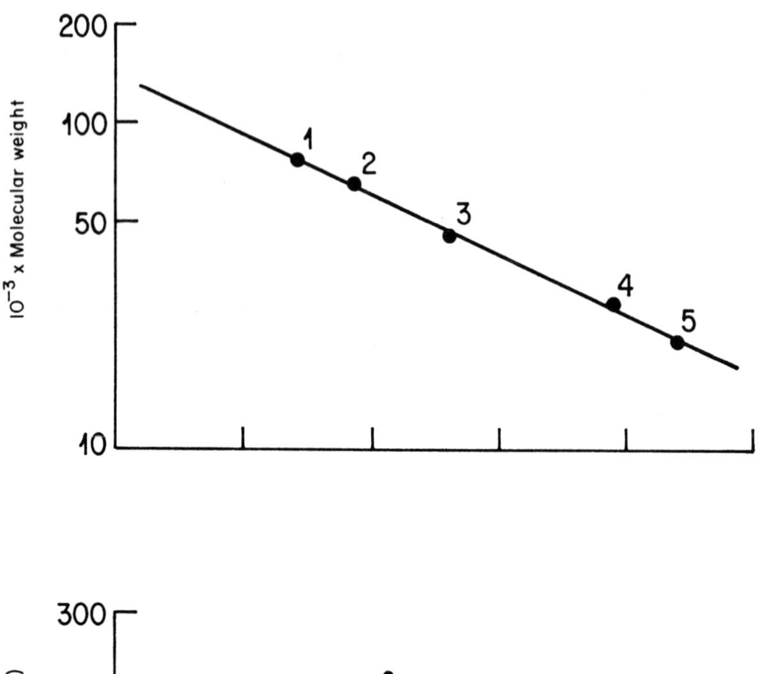

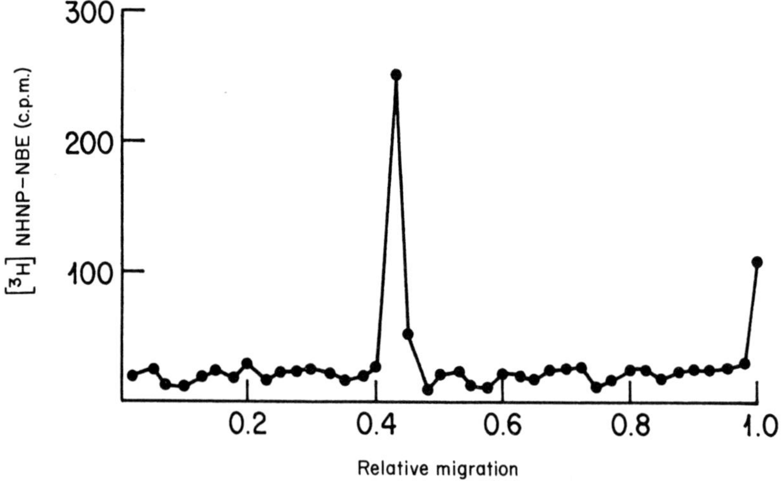

Fig. 6.10 Autoantibody affinity purification of bovine lung β_2-adrenergic receptors. Triton X-100-solubilized β_2-receptors, labeled with the irreversible affinity ligand [^{3}H]NHNP-NBE, were partially purified by gel-permeation chromatography and isoelectric focusing followed by immunoaffinity chromatography on autoantibody–Protein A–Sepharose. The bottom panel represents an SDS/polyacrylamide gel (10%) of immunoaffinity-purified β_2-receptors. The top panel is a linear plot of the molecular weight standards used for 10% SDS/polyacrylamide gels as described in Fig. 6.5. (From Fraser and Venter, 1982.)

6.4.3 Target-size analysis of β_2-adrenergic receptors

In an attempt to resolve the question of the molecular size of the β_2-receptor, we again used target-size analysis. Radiation inactivation of the β_2-receptor from canine lung revealed a functional molecular size for this receptor of 109 000 daltons, a value in very good agreement with that obtained for the larger protein species isolated by immunoaffinity chromatography and with the calculated hydrodynamic molecular weight for the β_2-receptor (Fraser and Venter, 1982).

The simplest model for the β_2-receptor consistent with the data obtained from immunoaffinity purification and target-size analysis is shown in Fig. 6.11. In the native membrane, the β_2-receptor appears to be composed of two 59 000-dalton, disulfide-linked proteins each of which contains a catecholamine-binding site.

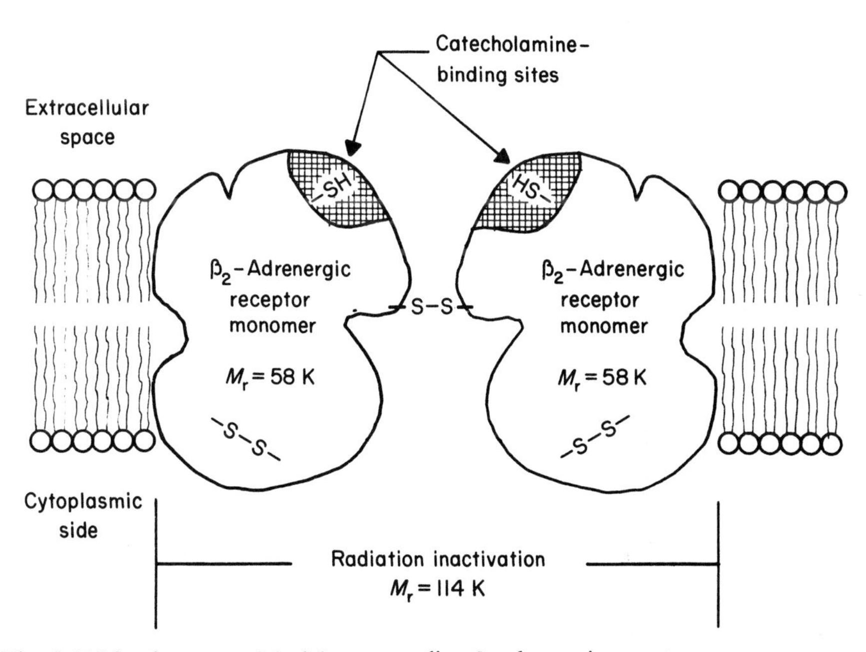

Fig. 6.11 Membrane model of the mammalian β_2-adrenergic receptor.

6.5 CONCLUSIONS

In summary, monoclonal antibodies to β-adrenergic receptors have been found to be of tremendous utility in the molecular characterization and immunoaffinity purification of β_1- and β_2-adrenergic receptors.

β_1-Adrenergic receptors from canine heart and turkey erythrocytes appear to be composed of subunits of 65 000–70 000 daltons as determined by hydrodynamic properties and immunoaffinity chromatography. Target-size analysis of canine heart β_1-adrenergic receptors indicates a functional molecular size for this receptor *in situ* of 130 000 daltons, suggesting that the β_1-receptor may exist as a dimer of two identical 65 000–70 000-dalton subunits in the native membrane.

β_2-Adrenergic receptors have been shown to contain at least one antigenic determinant in the ligand-binding site in common with β_1-receptors. Data derived from immunoaffinity purification of β_2-adrenergic receptors from mammalian lung indicate that the molecular weight of the β_2-receptor subunit is 59 000. Target-size analysis of β_2-receptors suggests that the β_2-adrenergic receptor may also exist as a functional dimer in the membrane, as evidenced by a functional molecular size of the β_2-receptor of 109 000 daltons.

The functional significance of the dimeric form of β-adrenergic receptors together with more detailed structural studies of each receptor subtype are currently underway in our laboratory. Monoclonal antibodies to β-receptors should prove to be valuable reagents in these studies.

REFERENCES

Abramsky, O., Aharonov, A., Webb, C. and Fuchs, S. (1975), *Clin. Exp. Immunol.*, **19,** 11–16.

Ahlquist, R.P. (1948), *Am. J. Physiol.*, **153,** 586–600.

Burges, R.A. and Blackburn, K.J. (1972), *Nature (London) New Biol.*, **235,** 249–250.

Couraud, P.O., Lo, B.Z., Schmutz, A., Durieu-Trautmann, O., Klutchko-Delavier, C., Hoebeke, J. and Strosberg, A.D. (1983), *J. Cell. Biochem.*, **21,** 187–193.

Doble, A. and Iversen, L.L. (1982), *Nature (London)*, **295,** 522–523.

Drachman, D.B., Angus, C.W., Adams, R.N., Michelson, J.D. and Hoffman, G.J. (1978), *N. Engl. J. Med.*, **298,** 1116–1118.

Flier, J.S., Kahn, C.R., Roth, J. and Bar, R.S. (1975), *Science*, **190,** 63–65.

Fraser, C.M. and Venter, J.C. (1980a), in *Membranes, Receptors and the Immune Response*, Alan R. Liss, New York, pp. 127–144.

Fraser, C.M. and Venter, J.C. (1980b), *Proc. Natl. Acad. Sci. U.S.A.*, **77,** 7034–7038.

Fraser, C.M. and Venter, J.C. (1982), *Biochem. Biophys. Res. Commun.*, **109,** 21–29.

Fraser, C.M., Venter, J.C. and Kaliner, M. (1981), *N. Engl. J. Med.*, **205,** 365–370.

Furchgott, R.F. (1972), *Handb. Exp. Pharmacol.*, **33,** 283–335.

Hancock, A.A., DeLean, A.L. and Lefkowitz, R.J. (1979), *Mol. Pharmacol.*, **16,** 1–9.

Harmon, J.T., Kahn, C.R., Kempner, E.S. and Schlegel, W. (1980), *J. Biol. Chem.*, **255,** 3412–3419.

Jeffrey, D.R., Charlton, R.R. and Venter, J.C. (1980), *J. Biol. Chem.*, **255,** 5015–5018.

Kahn, C.R., Flier, J.S., Bar, R.S., Archer, J.A., Gorden, A., Martin, M.M. and Roth, J. (1976), *N. Engl. J. Med.*, **294,** 739–745.

Lands, A.M., Arnold, A., McAuliff, J.P., Ludvena, F.P. and Brown, T.G. (1967), *Nature (London)*, **214,** 597–598.
Lefkowitz, R.J. (1975), *Biochem. Pharmacol.*, **24,** 583–590.
Lilly, L., Fraser, C.M., Jung, C.Y., Seeman, P. and Venter, J.C. (1983), *Mol. Pharmacol.*, (in press).
Lo, M.M.S., Barnard, E.A. and Dolly, J.O. (1982), *Biochemistry*, **21,** 2210–2217.
Manley, S.E., Bourke, J.R. and Hawker, R.W. (1974), *J. Endocrinol.*, **61,** 437–445.
Mayer, S.E. (1972), *J. Pharmacol. Exp. Ther.*, **181,** 116–125.
Minneman, K.P., Hegstrand, L.R. and Molinoff, P.B. (1979), *Mol. Pharmacol.*, **16,** 34–46.
Murad, F. (1973), *Biochim. Biophys. Acta*, **304,** 181–187.
Neilsen, T.B., Lad, P.M., Preston, S., Kempner, E., Schlegel, W. and Rodbell, M. (1981), *Proc. Natl. Acad. Sci. U.S.A.*, **78,** 722–726.
Patrick, J., Lindstrom, J., Culp, B. and McMillam, J. (1973), *Proc. Natl. Acad. Sci. U.S.A.*, **70,** 3334–3338.
Paul, S.M., Kempner, E.S. and Skolnick, P. (1981), *Eur. J. Pharmacol.*, **76,** 465–466.
Pollard, E.C. (1953), *Adv. Biol. Med. Phys.*, **3,** 153–189.
Rugg, E.L., Barnett, D.B. and Nahorski, S.R. (1978), *Mol. Pharmacol.*, **14,** 996–1005.
Schreiber, A.B., Couraud, P.O., Andre, C., Vray, B. and Strosberg, A.D. (1980), *Proc. Natl. Acad. Sci. U.S.A.*, **77,** 7385–7389.
Smith, B.R. and Hall, R. (1974), *Lancet*, **2,** 427–430.
Strauss, W.L. (1980), Doctoral Dissertation, State University of New York at Buffalo, Chapter 5.
Strauss, W.L., Ghai, G., Fraser C.M. and Venter, J.C. (1979), *Arch. Biochem. Biophys.*, **196,** 566–573.
Tzartos, S.J. and Lindstrom, J. (1980), *Proc. Natl. Acad. Sci. U.S.A.*, **77,** 755–759.
Tzartos, S.J., Rand, D.E., Einarson, B.L. and Lindstrom, J. (1981), *J. Biol. Chem.*, **256,** 8635–8645.
Venter, J.C. (1983), *J. Biol. Chem.*, **258,** 4842–4848.
Venter, J.C. and Fraser, C.M. (1981), in *Monoclonal Antibodies in Endocrine Research*, Raven Press, New York, pp. 117–134.
Venter, J.C. and Fraser, C.M. (1983), *Fed. Proc. Fed. Am. Soc. Exp. Biol.*, **42,** 273–278.
Venter, J.C., Fraser, C.M. and Harrison, L.C. (1980), *Science*, **207,** 1361–1363.
Venter, J.C., Fraser, C.M., Schaber, J.S., Jung, C.Y., Bolger, G. and Triggle, D.J. (1983), *J. Biol. Chem.*, (in press).

7 Cell Surface Structures Involved in Human T Lymphocyte Specific Functions: Analysis with Monoclonal Antibodies

COX TERHORST

Glossary of terms and abbreviations

β_2-m	β_2-Microglobulin, the small subunit of MHC class I antigens
Class I antigens	The products of HLA-A, -B and -C and H-2K, -D and -L loci
Class II antigens	The products of HLA-D and H-21 loci
CNBr	Cyanogen bromide
CTL	Cytotoxic T (thymus-derived) lymphocytes or killer T cells
Endo-F	Endo-β-*N*-acetylglycosaminidase-F
Endo-H	Endo-β-*N*-acetylglycosaminidase-H
Fab′	Monovalent fragment of immunoglobulin
HLA	Major histocompatibility complex of humans
H-2	Major histocompatibility complex of the mouse
IEF	Isoelectric focusing
INA	5-Iodonaphthyl-1-azide
MHC	Major histocompatibility complex
MLC	Mixed lymphocyte culture
NK cells	Mononuclear cells in blood, of uncertain lineage affiliation and which mediate 'natural killing' – the lysis of a variety of target cells – by what appears to be a non-immunological mechanism (as compared with lysis induced by antibody and complement or cytotoxic T cells)
SDS	Sodium dodecyl sulfate
SRBC	Sheep red blood cells
TCGF	T cell growth factor or interleukin-2
TL	A region of mouse chromosome 17 located to the right of the H-2 complex
T1, T3, T4, T6, T8, T11, T12	As yet 'unofficial' but widely used nomenclature for human T cell surface proteins identified by monoclonal antibodies
Lyt-1, Lyt-2, Lyt-3	Murine T cell surface proteins (Lyt-2 is probably the murine equivalent of human T8)
Clonotypic antibodies	Monoclonal antibodies specific for antigen-specific T cell clones. May identify the T cell receptor for antigen

Acknowledgements

I thank Drs Hergen Spits, Wil Tax and Jan de Vries for invaluable preprints, and Drs Carmelo Bernabeu, Peter van den Elsen, Bob Finberg, Hans Oettgen and Peter Snow for critical reading of the manuscript. The patience and expert secretarial help of Ms Donna Moschella is gratefully acknowledged. This work was supported by grants from the National Institutes of Health, the American Cancer Society and the Leukemia Society of America.

Monoclonal Antibodies to Receptors: Probes for Receptor Structure and Function
(*Receptors and Recognition*, Series B, Volume 17)
Edited by M. F. Greaves
Published in 1984 by Chapman and Hall, 11 New Fetter Lane, London EC4P 4EE

7.1 INTRODUCTION

Cell-mediated immunity encompasses a wide range of observable phenomena in a variety of regulatory and effector systems. The most common forms of cell-mediated responses studied, i.e., delayed type hypersensitivity, allograft rejection, tumor rejection, graft-versus-host reactions, all involve thymus-derived lymphocytes (T cells). Moreover, it has become increasingly clear from both animal and human studies that synergistic interactions of different T cells dictate the control mechanisms of the immune response. In the study of cell surface structures involved in T cell functions monoclonal antibodies have become invaluable tools. Particularly, since analysis of T cell interactions at the clonal level has become possible with the discovery of T-cell growth factor (TCGF; see Chapter 3 by Leonard *et al.*), monoclonal antibodies can be used for the delineation of T cell receptors for antigens.

Monoclonal antibodies have also been of great importance in preliminary studies of thymocyte differentiation within the thymus gland. As stem cells migrate into the thymus from the bone marrow, they differentiate and, as competent T lymphocytes, disperse from the thymus to the periphery. Several stages of thymic differentiation have been recognized with the use of monoclonal antibodies (Reinherz and Schlossman, 1980; Haynes, 1981; Goldstein *et al.*, 1982). Because of these new developments, selective expression of genes coding for cell surface structures, which govern associative recognition between co-operative T cell sets can now be studied.

This review summarizes briefly the available information about human thymic differentiation markers and T-cell-specific antigens, and focuses on cell surface structures implicated in specific T cell functions.

7.2 SUBPOPULATIONS OF HUMAN THYMUS—DERIVED LYMPHOCYTES

7.2.1 Monoclonal antibodies specific for T cells

Historically, the most efficient method of identifying human T cells involved rosetting with sheep red blood cells (SRBC). Monoclonal antibodies, which inhibit SRBC rosetting, are considered to be directed at the T cell surface receptor for SRBC. These antibodies (OKT11, Leu-5, 9.6, anti-D66) precipitate a glycoprotein (T11) with an apparent molecular weight of 50 K (Table 7.1) (Kamoun *et al.*, 1981; Howard *et al.*, 1981; Verbi *et al.*, 1982; Bernard *et al.*, 1982). Other so-called 'pan T cell markers' T3 (OKT3, Leu-4, UCHT1, SPV-T3, WT-31, WT-32) (Kung *et al.*, 1979; Ledbetter *et al.*, 1981; Beverley and Callard, 1981; Van Agthoven *et al.*, 1981; Spits *et al*, 1982a; Tax *et al.*, 1983a), 3A1 (Haynes *et al.*, 1980; Haynes, 1981) and T12 (Reinherz and Schlossman, 1980) are also present on all peripheral T cells (Table 7.1). While

Table 7.1 Monoclonal antibodies which detect T-cell surface-specific cell surface glycoproteins

Antigen	Molecular weight*	Monoclonal antibody	Percentage reactivity with: T cells	Thymocytes	NK cells
T3	20 K 25–28 K 37 K 44 K	OKT3 Anti-Leu-4 UCHT1 SPV-T3 WT-31 WT-32	100	30–60	0
T11	50 K	OKT11 9.6 Anti-D66	100	100	50
T12	120 K	Anti-T12	100	0	0
3A1	40 K	3A1	100	100	0
T4	60 K	OKT4 Anti-Leu-3	60	75	0
T8	34 K	OKT8 Anti-Leu-2a	30	75	50
Tac	55 K	Anti-Tac	Activated T cells only	0	0

* Determined under reducing conditions.

T11 and 3A1 are found on all thymocytes, T3 is present on only 30–70% of human thymocytes (dependent on the monoclonal antibody used) (Kung *et al.*, 1979; Ledbetter *et al.*, 1981). T3-bearing thymocytes appear to be functionally mature T cells, because they respond to antigenic stimuli (Reinherz and Schlossman, 1980). A very interesting monoclonal antibody (anti-Tac) is specific for activated T cell, but does not react with resting T cells (Uchiyama *et al.*, 1981a,b). Since anti-Tac detects the receptor for TCGF, this antibody is extremely useful in the study of T cell proliferation (Leonard *et al.*, 1982; Miyawaki *et al.*, 1982; Depper *et al.*, 1983 – see Chapter 3 by Leonard *et al.*).

The two cell surface markers T4 (Fig. 7.1) and T8 (Fig. 7.7) are not found on all peripheral T cells, but on two reciprocal subsets ($T4^+T8^-$ and $T4^-T8^+$). In contrast, the majority of human thymocytes are $T4^+T8^+$, whereas only a small percentage of thymocytes are either $T4^+T8^-$ or $T4^-T8^+$ (Reinherz and Schlossman, 1980) (Table 7.1). It is therefore tempting to speculate that during

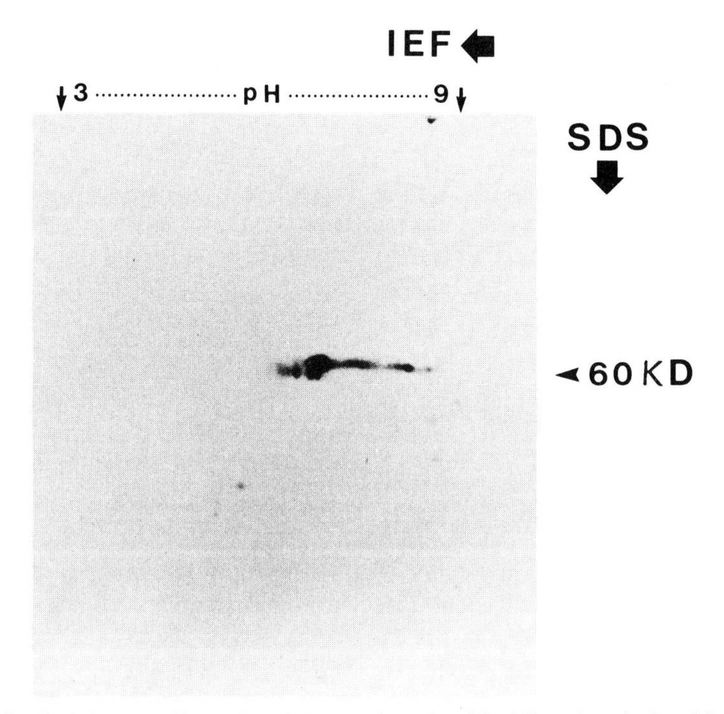

Fig. 7.1 Analysis by two-dimensional electrophoresis of the T4 antigen isolated from ^{125}I-labeled human T lymphocytes.

thymic differentiation a series of events occur resulting in the turning off of either T8 or T4 gene expression in the most mature T cells. Most recent reviews on human T lymphocyte antigens have focused on functional definitions of these peripheral T cell and thymocyte subsets.

As in the mouse (Cantor *et al.*, 1976; Cantor and Boyse, 1977), distinct functional programs have been assigned to human T cell subsets (T helper vs. T suppressor/cytotoxic) on the basis of the expression of one or two cell surface markers (Reinherz and Schlossman, 1980; Haynes, 1981; Goldstein *et al.*, 1982; Engelman *et al.*, 1981a). The idea that cell markers determine the cell's function is an oversimplification, for the same functions have been found in different subsets. Some recent observations by Swain (1981) have provided new insight into the possible overall role of the Lyt-2^{+} (T4^{-}T8^{+}) and Lyt-2^{-} (T4^{+}T8^{-}) T cell subsets. According to this author, the Lyt-2 antigen does not discriminate between cytotoxic and non-cytotoxic effector (helper) T cells, but rather reflects the *specificity* of cytotoxic T cells for H-2I-region or H-2K/D-region-associated (or 'restricted') antigen on target cells. The usual association of the Lyt-2 antigen and killer T cell function therefore reflects the tendency for CTLs to preferentially recognize H-2K/D determinants (MHC-class I).

Conversely, all functional interactions of Lyt-2$^-$ T cells are restricted by the presence of the H-2I-region antigens (MHC-class II).

In support of this notion are similar observations which have been made with limited sets of human CTL clones (Spits *et al.*, 1982a,b; Krensky *et al.*, 1982; Meuer *et al.*, 1982; Biddison *et al.*, 1982). Namely, most T4$^-$T8$^+$ CTL clones are directed at HLA-A/B/C antigens (MHC-class I), whereas most T4$^+$T8$^-$ CTL clones are directed at human MHC-class II structures. The analogy with the murine T cells is emphasized by structural similarities of the Lyt-2 antigen and T8 (Terhorst *et al.*, 1980; Ledbetter *et al.*, 1981). Most recently, Swain and colleagues generated human T cell cultures which responded to murine splenocytes (Swain *et al.*, 1983). CTLs in the T4$^-$T8$^+$ subset lysed H-2K/D-bearing mouse cells specifically. Human TCGF-producing T cells in the T4$^+$T8$^-$ subset responded specifically to mouse MHC-class II antigen. No evidence exists, however, which would support the conclusion that Lyt-2 (or T8) directly binds to a constant part of MHC-class I antigens or conversely that T4 interacts with a constant region of MHC-class II antigens.

However, some recent data do not fit the hypothesis of Swain and colleagues, thus challenging its universality. Namely, Miller and Stutman (1982) reported a Lyt-2$^+$ CTL clone directed at a determinant in the H-2I region (MHC-class II). More importantly, anti-Lyt-2 monoclonal antibody inhibited the cytotoxic function of that clone. Similarly, T4$^-$T8$^+$ human CTLs have been reported, which are specific for human MHC-class II antigens (Ball and Stastny, 1982; Spits *et al.*, 1983b). Whether these apparent exceptions are representative of a large percentage of the functional T cell repertoire remains to be determined.

Several monoclonal antibodies have been used to determine the location of T lymphocyte subpopulations in the thymus (see Section 7.2.2) or human lymph nodes by means of immunoperoxidase techniques (Poppema *et al.*, 1981; Cerf-Bensussan *et al.*, 1983). In lymph nodes the majority of cells in the paracortical area are T cells, most of which are T4$^+$T8$^-$ and only a few are T4$^-$T8$^+$. Unfortunately no conclusions can be derived about the functional state and interactions of these T cells.

One issue to be concerned with here is the specificity of the various monoclonal antibodies for T cells. For instance, natural killer cells (NK cells) express the T cell markers T8 and T11, but not T3 or T4 (Perussia *et al.*, 1983a,b). Conversely, T3$^+$ T cells can display NK-like activity (Hercend *et al.*, 1983; DeVries and Spits, 1984). The specificity of T3 as a T cell marker has been questioned, because some monoclonal anti-T3 antibodies have been shown to react with Purkinje cells in the cerebellum (Garson *et al.*, 1982). Attempts to identify the target antigen on these cells have failed so far and it is therefore unclear, whether monoclonal anti-T3 antibodies detect the same molecule or cross-react with an unrelated structure. Anti-T8 monoclonal antibodies have been reported to bind to cultured sheep oligodendrocytes by some investigators (Oger *et al.*, 1982), but not by others (Hirayama *et al.*, 1983). Recently, the T4 antigen defined by monoclonal antibodies has been detected on human monocytes and Langerhans cells (Moscicki *et al.*, 1983; Wood *et al.*, 1983). Although these observations need to be confirmed by a description of the target antigens on the non-T cells, it remains possible that monoclonal antibodies with absolute specificity for T cells may in fact not exist.

7.2.2 Monoclonal antibodies and thymocyte subpopulations

In addition to T3, T4 and T8, several antigens which are not specific for T cells have been very useful in dissecting thymocyte subpopulations (Table 7.2). For instance, anti-HLA-A/B/C antibodies (W6/32) stain medullary cells more strongly than cortical cells (Janossy *et al.*, 1980; Müller *et al.*, 1983). Immunohistological studies with monoclonal antibodies against HLA-A/B allospecificities revealed striking variations in the quantitative expression of certain HLA-A versus HLA-B locus determinants on cortical thymocytes (Müller *et al.*, 1983). A distinction between cortical and medullary thymocytes can clearly be made with monoclonal anti-T6 (Janossy *et al.*, 1980; Bhan *et al.*, 1983) and anti-M241. Most $T4^+T8^+$ thymocytes are $T6^+$ whereas $T4^+T8^-$ or $T4^-T8^+$ thymocytes are $T6^-$ (Reinherz and Schlossman, 1980).

Table 7.2 Monoclonal antibodies which detect thymocyte subpopulations

Antigen	Molecular weight (Daltons)*	Monoclonal antibody	Percentage reactivity with: T cells	Thymocytes	Reactivity with other cell types
T1	67 K	OKT1 anti-Leu-1 T-101	100	30–100	B cell lymphomas
T6	49 K/β_2-m	OKT6, NA1-34 IIC7	0	75	Langerhans cells in epidermis
M241	43 K/β_2-m	anti-M241	0	75	Dendritic cells in dermis
Transferrin receptor	95 K	OKT9 5E9 B3/25	0	10	All proliferating cells

* Determined under reducing conditions.

A monoclonal antibody with a tissue distribution similar to anti-T3 is anti-T1 (Kung *et al.*, 1979; Wang *et al.*, 1980). T1, which is borne by a 67-K dalton glycoprotein (Van Agthoven *et al.*, 1981) with a protein backbone of 58 K dalton (Bergman and Levy, 1982), is also present on B cell lymphomas. Some anti-T1 antibodies (OKT1) react with 30% of human thymocytes (Kung *et al.*, 1979), whereas others (e.g. Leu-1) stain all human thymocytes (Wang *et al.*, 1980). Yet, OKT1, as in the case of anti-T3, was used to purify the most mature thymocytes (Kung *et al.*, 1979; Umiel *et al.*, 1982; DeVries *et al.*, 1983).

Among other monoclonal antibodies which have been used in thymic differentiation studies, the anti-transferrin receptor reagents (OKT9, 5E9 and B3/25) (Trowbridge and Omary, 1981; Sutherland *et al.*, 1981) are worth mentioning (see also Chapter 10 by Trowbridge and Newman). The transferrin receptor, a (95 K dalton)$_2$ homodimer, is coded for by gene(s) on chromosome 3 (Goodfellow *et al.*, 1982; Van de Rijn *et al.*, 1983b). Only 10% of human thymocytes were found to be positive for OKT9 or 5E9. Those thymocytes are mainly large subcapsular cells which do not express T1, T3, T4, T6 and T8 (Reinherz and Schlossman, 1980). Because all metabolically active cells express the transferrin receptor, the T9-positive cells are most likely a set of actively dividing prothymocytes.

Thus, with the use of monoclonal antibodies prothymocytes and mature thymocytes can be separated from the majority of thymocytes. $T4^+T8^-$ and $T4^-T8^+$ thymocytes may have been derived from $T4^+T8^+$ cells during thymic differentiation. Whether this transition happened concomitantly with the disappearance of T6 and M241 from the thymocyte cell surface is uncertain. During that transition thymocytes become functionally active and gradually acquire the T3 antigen. Further understanding of thymic differentiation and of the number of cell lineages involved requires knowledge about migration of cells in and out of the thymus gland, about mobility of cells within the thymus and about the function of thymic epithelium.

(a) *Structure of T6 and M241*

Since anti-T6 and anti-M241 showed such a distinct staining pattern on human thymocytes and as no other cells with the exception of dendritic cells in the skin (see below) are stained by these antibodies, we have begun to define the target antigens T6 and M241. The T6 and M241 antigenic determinants are present on glycoproteins of molecular weights 49 K and 44 K, respectively (Terhorst *et al.*, 1981; Van Agthoven and Terhorst, 1982; Lerch *et al.*, 1983; Knowles and Bodmer, 1982; Van de Rijn *et al.*, 1983c). On thymocytes both M241 and T6 are associated with a small subunit (12 K dalton) β_2-microglobulin (β_2m). After cell surface radioiodination the heavy chains of M241 and T6 are labeled much more strongly than β_2-m. This is not the case when these protein complexes are labeled by reductive methylation with NaB[$^3H_4$] (Lerch *et al.*, 1983).

Strong evidence for the association of T6 with β_2-m exists. Namely:

1. When all β_2-m-associated proteins are isolated from ^{125}I-labeled human thymocytes by immunoadsorbent columns, this set contains T6 (Terhorst *et al.*, 1981).

2. ^{125}I-T6 antigen purified with a monoclonal anti-T6 antibody, can be re-precipitated with anti-β_2-m antibody (Lerch *et al.*, 1983).
3. The small subunit of T6 has the same isoelectric point as β_2-m (Van Agthoven and Terhorst, 1982).
4. The *N*-terminal amino acid sequence (25 steps) of the small subunit of T6 is identical with β_2-m (P. Lerch, J. Smart and C. Terhorst, unpublished.

T6 and M241 are also expressed on human T leukemic cells (Van Agthoven and Terhorst, 1982; Knowles and Bodmer, 1982); T6 from the cell line MOLT-4 is of particular interest. Although T6 was in part found associated with β_2-m, on MOLT-4 cells most of the T6 heavy chain is associated with another polypeptide (12 K dalton, pI = 7.0) (Van Agthoven and Terhorst, 1982). Previously, this polypeptide had been called βt (Ziegler and Milstein, 1979), for it did not cross react with β_2-m. We have recently determined, that βt, quite surprisingly is in fact bovine β_2-m. Since MOLT-4 cells were grown on fetal calf serum, which contains bovine β_2-m, we studied the exchange between human β_2-m and bovine β_2-m on the cell surface. Thus, it could be shown that the human β_2-m associated with T6 can be replaced completely by bovine β_2-m (Bernabeau *et al.*, 1983b). Exchange of β_2-m on the cell surface occurs with some, but not all, MHC-class I antigens (Bernabeu *et al.*, 1983b). Interestingly, β_2-m associated with M241 does not exchange at all. On the basis of their association with β_2-m, both T6 and M241 may be considered human MHC-class I antigens.

A more important difference between thymic T6 and MOLT-4 T6 (but not between thymic M241 and MOLT-4 M241) was found in their glycosylation patterns. On the surface of thymus cells and MOLT-4 cells a T6 species with four *N*-linked oligosaccharides was found (Van Agthoven and Terhorst, 1982; Lerch *et al.*, 1983; Van de Rijn, 1983c). In addition MOLT-4 cells express a T6 species with five *N*-linked oligosaccharide side chains (Van Agthoven and Terhorst, 1982; Lerch *et al.*, 1983). Whether this means that MOLT-4 T6 carries a mutation and is an 'alien histocompatibility antigen' has to be proven by comparative protein sequence analysis.

T6 and M241 differ from the classical MHC-class I antigens in glycosylation. After treatment with the enzyme Endo-F or with non-aqueous trifluoromethanesulfonic acid the protein backbone of T6 and M241 is 33 K daltons in contrast with the 40 K dalton backbone of HLA-A/B, murine H-2K/D or TL antigens (Lerch *et al.*, 1983; Van de Rijn *et al.*, 1983c). The difference in glycosylation between HLA-A/B (10%) and T6 (30–35%) could explain why biosynthetic labeling of T6 was found to be far more difficult than in the case of HLA-A/B antigens (Lerch *et al.*, 1983). Perhaps the post-translational modifications are a rate-limiting step in this process. On T-leukemic cell lines the expression of β_2-m-associated proteins, but not T6, can be enhanced by lymphocyte-conditioned medium containing interferon (Lerch *et al.*, 1983). Whether this is a reflection of the role of T6 in the thymocyte differentiation process remains to be determined. These considerable differences between T6 and HLA-A/B antigens were confirmed by absence of cross-reactivity between the two polypeptide chains (Lerch *et al.*, 1983).

The similarity in protein backbone size led us to compare the T6 and M241 glycoproteins by peptide mapping. For this purpose, the T6 and M241 glycoproteins were isolated from human thymocytes, denatured in SDS and radioiodinated using chloramine-T as an oxidizing agent. Tryptic peptide maps showed that T6 and M241 are different proteins, although some peptides were shared (Van de Rijn, *et al.*, 1983c).

Both M241 and T6 can be labeled by [^{125}I]iodonaphthylazide (Lerch *et al.*, 1983). T6 binds detergent as judged by charge-shift electrophoresis (P. Lerch, unpublished) and can be induced to form water-soluble protein micelles (Terhorst *et al.*, 1981). These combined data allow the conclusion that T6 and M241 are integral membrane proteins. If T6 and M241 have a similar domain structure to HLA-A/B/C antigens, one of the extracellular domains may have a shorter amino acid sequence (Fig. 7.2).

Although some similarities in tissue distribution between T6/M241 and murine TL antigens exist, two major differences should be emphasized. In addition to thymocytes and T cell leukemias, T6 and M241 are present on dendritic cells in the skin. T6 was found on Langerhans cells in the epidermis and M241 on indeterminate dendritic cells in the dermis (Van de Rijn *et al.*, 1983a). Murine TL antigens have only been found on thymocytes and T-leukemias (Flaherty, 1981). Also the difference in polypeptide chain backbone

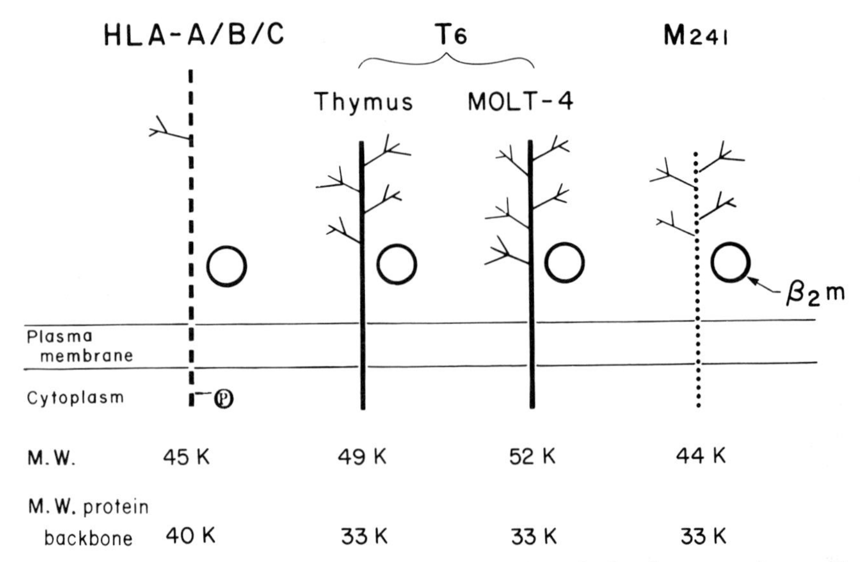

Fig. 7.2 Schematic diagram of human MHC-class I antigens in the plasma membrane. Ⓟ, Phosphorylation site; Ұ, oligosaccharide side chain.

indicates that T6 and M241 are not necessarily the human homologues of the murine TL antigens. Perhaps murine equivalents of T6 and M241 will be found among the many MHC-class I gene products coded for in the TL and Qa regions of the H-2 complex (Hood *et al.*, 1983).

The function of the T6 and M241 antigens is unknown. They may be receptors for thymic hormones or their structures may govern associative interactions between thymocytes. Since $T6^-M241^-$ thymocytes strongly express HLA-A/B antigens, whereas $T6^+M241^+$ thymocytes stain only weakly with fluoresceinated anti-HLA-A/B reagents (Janossy *et al.*, 1980; Müller *et al.*, 1983), these two MHC-class I antigens may fulfil a similar function in distinct stages of thymocyte differentiation.

7.3 HUMAN T CELL FUNCTIONS

Several of the cell surface glycoproteins identified by anti-T cell monoclonal antibodies, which were described in the previous sections, have been implicated in immunologic functions (Table 7.3). In addition, other antibodies have been selected on the basis of their ability to block *in vitro* T lymphocyte assays. Functional properties of human T cells can be tested *in vitro* by measuring:

1. Proliferative responses to cell-surface-bound antigens, soluble antigens, or lectins.
2. Cytolytic activity.
3. Production of antigen-specific factors or lymphokines.
4. Regulatory role (help or suppression) in antibody production by B cells or in proliferation, generation of CTL or production of factors by other T cells.

Table 7.3 Some functional properties of anti-T cell monoclonal antibodies

Monoclonal antibody	Mitogenicity	Inhibition of proliferation triggered by:			Inhibition of cytolysis by:	
		Soluble antigen	Cell surface antigen	Lectin	CTL	NK cells
Anti-T3	+	+	+	+	+	−
Anti-T11	−	+	+	+	+	+
Anti-LFA-1	−	+	+	+	+	+
Anti-Tac	−	+	+	+	−	−
Anti-T4	−	−	+*	−	+*	−
Anti-T8	−	−	+†	−	+†	−

* Inhibits only $T4^+$ cells.
† Inhibits only $T8^+$ cells.

The two assays used by most investigators to test for interference by monoclonal antibodies are specific T cell proliferation and cytolysis by CTLs. Monoclonal antibodies have also been tested for their ability to trigger DNA synthesis, i.e. mitogenicity. These assays involve very complex interactions, which need multipoint cell–cell contacts. Thus, blocking of an assay system by antibody does not necessarily interfere with the interactions between antigen and its T cell receptor.

7.3.1 T cell proliferation

(a) ***Mitogenic antibodies***

DNA synthesis in T lymphocytes is triggered in the context of antigen presenting cells (Rosenthal, 1978). In more artificial systems, a variety of chemical agents such as lectins, phorbol esters and the calcium ionophore A23187 have been shown to stimulate mitosis. The mechanism of transfer of signals across the plasma membrane and towards the nucleus is little understood. In assays for monoclonal antibodies that are mitogenic, DNA synthesis is simply measured by uptake of [^{3}H]thymidine into cells. In order to continue proliferating, T cells require the presence of TCGF, but TCGF does not support the growth of non-activated T lymphocytes (Gillis *et al.*, 1982). Mitogenicity induced by an antibody can therefore also be measured by assays for TCGF.

All monoclonal antibodies directed at the T3 surface antigen (OKT3, anti-Leu-4, UCHT1, WT-31, WT-32, SPV-T3 a,b,c) induce DNA synthesis in human peripheral blood cultures at a concentration of 10–20 ng/ml (Van Wauwe *et al.*, 1980; Burns *et al.*, 1982; Spits *et al.*, 1983a; Tax *et al.*, 1983a,b). The monovalent, Fab′ fragment of OKT3 is also mitogenic but 100 times less than the IgG. The mitogenic response is maximal after 3 days in culture and is terminated after 6 days (Van Wauwe *et al.*, 1980). Approximately one-third of the individuals do not respond to the IgG1 monoclonal antibodies anti-Leu-4 (Kaneoka *et al.*, 1983; Van Wauwe and Goossens, 1983), UCHT1 and WT-31 (Tax *et al.*, 1983b). This unresponsiveness was found to be due to the inability of autologous monocytes to exert their auxiliary function (Kaneoka *et al.*, 1983; Van Wauwe and Goossens, 1983). Perhaps an Fc receptor on monocytes that reacts with IgG1 is necessary for the mitogenic activation of T cells. This idea is contradicted, however, by the finding that immunoglobulin Fab fragments of OKT3 (Van Wauwe *et al.*, 1980), but not of anti-Leu-4 (Kaneoka *et al.*, 1983), can stimulate T cells.*

The mitogenic responses of T cells to monoclonal anti-T3 antibodies are similar to the responses to lectins [concanavalin A (Con A), phytohaemag-

*Editorial technical comment: only a very small contamination of the Fab fragments with aggregated fragments or whole IgG would be necessary to account for the mitogenic activity observed.

glutinin (PHA)]. Some investigators have indeed observed that Con A and PHA prevent the binding of anti-T3 antibodies (Palacios, 1982). The effect of anti-T3 antibodies is also reminiscent of the effects of antibodies to epidermal growth factor receptors on epidermal cells, which can induce protein kinase activity and epidermal-growth-factor-receptor clustering (Schreiber *et al.*, 1982; see Chapter 2 by Strosberg and Schreiber and Chapter 12 by Schlessinger *et al.*). Thus far, phosphorylation of the T3 antigens has not been detected (C. Terhorst, unpublished).

Recently, monoclonal antibodies have been developed which are specific for human CTL clones; these are referred to as anti-clonotypic monoclonal antibodies (Meuer *et al.*, 1983a). These antibodies, covalently bound to Sepharose beads, specifically induce proliferation of CTL clones as measured by TCGF production. The cell surface glycoproteins recognized by these antibodies appear to be associated with the T3 complex (Meuer *et al.*, 1983b).

(b) ***Monoclonal antibodies which inhibit T cell proliferation***

Upon activation, T cells become sensitive to TCGF (IL-2) which is produced by most T lymphocytes, albeit in variable amounts (Gillis *et al.*, 1982). Activated T cells express the TCGF receptor (see Chapter 3 by Leonard *et al.*). Once the responsiveness to TCGF is acquired, T cells need the presence of TCGF in order to proliferate and grow. Antibodies directed against the TCGF receptor therefore inhibit T cell proliferation (Leonard *et al.*, 1982; Miyawaki *et al.*, 1982; Depper *et al.*, 1983). The anti-Tac antibodies inhibit DNA synthesis only during a distinct stage of the activation, indicating the complexity of T cell activation (see Chapter 3 by Leonard *et al.*).

Other monoclonal antibodies, which *inhibit* proliferation of T cells induced by allogeneic antigens (i.e. MLC), lectins or by soluble antigens include anti-T3 (Reinherz *et al.*, 1980; Chang *et al.*, 1981; Burns *et al.*, 1982; Tax *et al.*, 1983a,b) and anti-T11 (Palacios and Martinez-Maza, 1982; Van Wauwe and Goossens, 1983; Martin *et al.*, 1983; Krensky *et al.*, 1983). The concentration of IgG necessary for blocking T cell proliferation is higher than for mitogenicity (Burns *et al.*, 1982). Anti-T3 antibodies appear to inhibit independently from the anti-T11 or anti-Tac antibodies. The anti-T3 monoclonal antibody UCHT1 also has a profound inhibitory effect on the proliferation of human T helper clones stimulated by influenza haemagglutinin (or haemagglutinin HA peptides) (Zanders *et al.*, 1983). Anti-LFA-1 antibodies, which react with T cells, B cells, monocytes, granulocytes and NK cells (Spits *et al.*, 1983a; see Chapter 5 by Ross), also block T cell proliferation induced by lectins or antigens (Sanchez-Madrid *et al.*, 1982). Several investigators have reported that some, but not all, monoclonal antibodies directed at the T cell subset markers T4 and T8 inhibit proliferation of the corresponding T4 or T8 positive subset (Engleman *et al.*, 1981b; Biddison *et al.*, 1983). Often very high concentrations of antibody were needed for these inhibitions, in contrast with the small amounts often required for inhibition by anti-T3, anti-T11 or anti-LFA-1.

It was recently shown that cyclosporin A, a fungal metabolite, suppresses proliferation of human T lymphocytes. This hydrophobic undecapeptide binds to specific sites on human T lymphocytes. Binding of ^{3}H-labeled cyclosporin A was partially inhibited by the monoclonal antibody OKT3 (Ryffel *et al.*, 1982). In addition, OKT3 binding was reduced by preincubation of T lymphocytes with cyclosporin A (Britton and Palacios, 1982). The latter could be explained by direct blocking of antibody binding or by dissociation of the T3 complex caused by cyclosporin A, which resulted in the loss of antibody binding (see below). Cyclosporin A does not appear to interfere with the expression of TCGF receptors upon lectin activation or with TCGF binding (Miyawaki *et al.*, 1983).

Both in the experiments described here, and in the inhibitions of cytolysis (see Section 7.3.2), the major question to be addressed is whether the monoclonal antibody is directed at a molecule which directly participates in the function which is studied. It is conceivable that the antibody perturbs a plasma membrane domain and thus indirectly influences a complex function such as onset of T cell proliferation. More detailed knowledge of the series of events which induce triggering of T cell proliferation and a better description of the interactions and lateral mobility of the glycoproteins detected by the antibodies described is required.

7.3.2 Monoclonal antibodies which inhibit cytolysis

With the exception of anti-Tac, all of the monoclonal antibodies which are known to block T cell proliferation also block cytotoxic effector function by T cells (Table 7.3). Cell-mediated lysis (CML) is measured by release of ^{51}Cr from a target cell by CTL. Anti-LFA-1 and anti-T11 inhibit the effector phase of CTL and NK cells (Spits *et al.*, 1983a; Martin *et al.*, 1983). In addition, anti-LFA-1 blocks the so-called lectin-dependent killing (Spits *et al.*, 1983a). Anti-T3, anti-LFA-1 and anti-T11 inhibit cytolysis by all CTLs, i.e. directed at MHC-class I and -class II antigens (Spits *et al.*, 1982a,b, 1983a; Meuer *et al.*, 1982). Anti-T8 antibodies inhibit cytolysis by T8$^+$ clones, whereas only some of the monoclonal anti-T4 antibodies inhibit cytolysis by T4$^+$ CTL (Spits *et al.*, 1982a,b, 1983a; Krensky *et al.*, 1982; Meuer *et al.*, 1982; Biddison *et al.*, 1982).

The inhibitory effect of the monoclonal anti-T3 and anti-T8 antibodies could not be attributed to a functional inactivation of the CTL clones, since Con A overcame the inhibition of cytotoxicity without affecting the binding of antibodies to the cells (Spits *et al.*, 1982a,b). These findings indicate that the monoclonal anti-T3 and anti-T8 antibodies do not block the cytolytic effector mechanism itself.

In an attempt to study the role of T3 and T8 in the CTL effector function in more detail, a series of experiments was conducted using proteolytic enzymes. Mild trypsinization of two T8$^+$ CTL clones resulted in a complete abrogation of the cytotoxic activity of these clones (Spits *et al.*, 1982b). In contrast, the cytotoxicity of T4$^+$ clones was not affected by trypsinization. The disappearance of cytotoxicity after trypsin treatment correlated with the removal of T8 determinants. T11 was removed on both T4$^+$ and T8$^+$ CTL clones, but the expression of LFA-1, T3 and T4 remained unchanged (Spits *et al.*, 1982b). The expression of T8 and T11 was restored after overnight incubation of the cells, as was the cytotoxic activity. The general ability to lyse 'target' cells was not lost by the trypsin treatment, since the lectin-dependent cytotoxicity was still found intact. These experiments demonstrate that the T8 antigen is involved in primary adhesion and that the functions of T3 and T8 can be uncoupled. The latter is clearly demonstrated by experiments using an assay for inhibition of CTL–target cell adhesion. Several groups (Landegren *et al.*, 1982; Tsoukas *et al.*, 1982) demonstrated that monoclonal anti-T8 antibodies inhibit the adhesion step and that anti-T3 antibody blocks at a later stage of cytolysis, possibly during programming for lysis (see Fig. 7.3). From studies in murine systems, anti-LFA-1 antibodies also seem to inhibit in the adhesion stage (Springer *et al.*, 1982). It remains to be determined at which stage anti-T11 inhibits. The complexity of CTL–target cell interactions is obvious. Recognition of MHC antigens occurs within the context of multipoint cell–cell interactions and all the T cell antigens described here may play a role in this process.

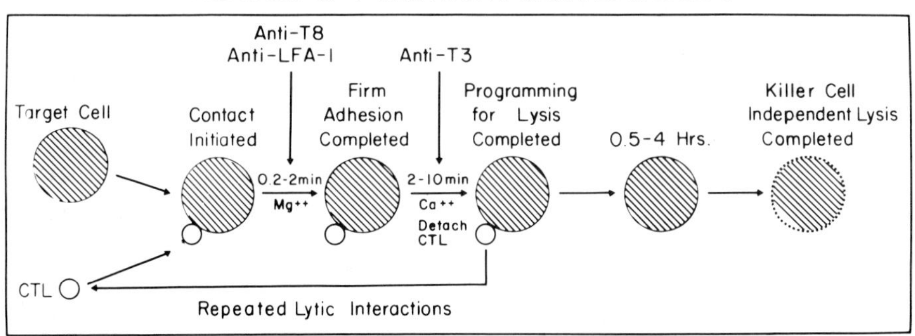

Fig. 7.3 Steps in T-lymphocyte-mediated cytolysis which are blocked by monoclonal antibodies to human T cell surface antigens. After Martz (1977).

The T3 complex plays a special role in the interactions between CTLs and target cells. The recently described anti-clonotypic monoclonal antibodies, which inhibit specific effector function of CTL clones, may be associated with the T3 antigens as judged by co-modulation* experiments (Meuer *et al.*, 1983a,b). This raises the interesting possibility that the T3 antigen is associated with receptors for MHC-class I and II antigens which are detected by the anti-clonotypic antibodies. The preliminary structural data on the T3 protein ensemble support the possibility of such an associative system.

7.4 STRUCTURE OF THE T3 COMPLEX

As discussed above, the T3 antigen appears to be involved in several T-cell-specific functions. The most plausible explanation for the pleiotropic effect of the anti-T3 antibody is that T3 consists of several subunits. Our protein chemistry studies support this notion. The main antigenic determinants (epitopes) detected by the anti-T3 reagents (which bind to 3×10^4–5×10^4 sites per cell; Van Wauwe *et al.*, 1980; Burns *et al.*, 1982) are borne by a 20 K-dalton glycoprotein (Borst *et al.*, 1983a). In addition, glycoproteins of 25–28, 37 and 44 K-daltons are found in anti-T3 immunoprecipitates derived from surface-labeled cells (Fig. 7.4). The four antigens remain associated upon electrophoresis in the presence of several detergent solutions (Borst *et al.*, 1983a). Peptide maps prepared from ^{125}I-labeled 20 K-dalton T3 and the 25–28 K-dalton protein appeared to differ greatly, and since the molecular masses and isoelectric points of the four polypeptides were distinct after removal of all *N*-linked oligosaccharides, they must differ in primary structure. It is however possible that the four chains have sequence homologies, as is the case with the subunits of the acetylcholine receptor (Raftery *et al.*, 1980; see Chapter 8 by Fuchs *et al.*). Since these subunits may not be in a permanent arrangement, but associate only to fulfill a given function, this T3 complex can be considered as a set of *'protein ensembles'*. Studies with different anti-T3 monoclonal antibodies show that these protein ensembles also exist on the cell surface (see below).

Radioiodinated 20 K-dalton T3 and the 37 K-dalton molecules, but not the 25–28 K-dalton and 44 K-dalton molecules, were susceptible to Endo-H digestion (Borst *et al.*, 1982). This indicates that the mature 20 and 37 K-dalton glycoproteins contain high-mannose-type carbohydrates. In contrast, all four glycoproteins are susceptible to digestion with Endo-F, which removes both complex and high-mannose-type *N*-linked sugars. Endo-F treatment reduced the molecular mass of the 20 K-dalton T3 to 14 K-daltons. The 25–28 K-dalton protein was digested to 16 K-daltons; the 37 and 44 K-dalton proteins were digested to 32 and 34 K-daltons, respectively. The Endo-F digestion products have no charge heterogeneity, indicating that most of the complexity in the

*Editor's note: Co-modulation is seen as co-incident *re*distribution of two different determinants on the cell surface into the same discrete aggregate (i.e. patch or polarized cap).

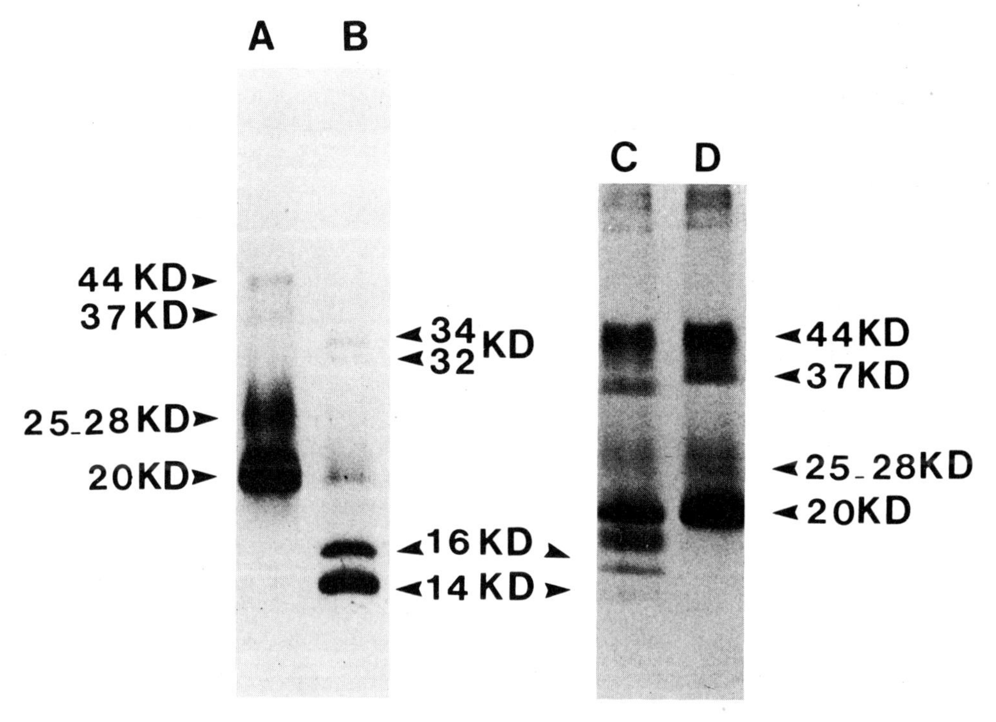

Fig. 7.4 Analysis by SDS/polyacrylamide gel electrophoresis of Endo-F- and Endo-H-treated ^{125}I-labeled T3 antigens isolated from human T cells. Samples were incubated without endoglycosidases (lanes A and D), with Endo-F (lane B) or Endo-H (lane C) and analyzed under reducing conditions.

isoelectric focusing pattern is caused by *N*-linked, as opposed to *O*-linked, carbohydrates (Borst *et al.*, 1983a) (Fig. 7.4).

Biosynthetic labeling with [^{35}S]-methionine and -cysteine shows that only part of the 20 K-dalton T3 was susceptible to Endo-F and Endo-H treatment. In contrast to the ^{125}I-labeled 20 K-dalton T3, the ^{35}S-labeled T3 could never be completely converted into the 14 K-dalton species. This indicates that a 20 K-dalton T3 molecule exists which does not carry any *N*-linked carbohydrates. Experiments using the inhibitor of *N*-linked glycosylation, tunicamycin, are consistent with the existence of such a form, since even at a very high tunicamycin concentration (5 μg/ml) a 20 K-dalton T3 molecule was synthesized (Borst *et al.*, 1983b).

Firm evidence that two 20 K-dalton T3 species exist was provided by labeling experiments using the hydrophobic label [^{125}I]iodonaphthylazide (INA). The majority of the INA label is found in one 20 K-dalton species, which is not susceptible to Endo-F, Endo-H or neuraminidase. This 'hydrophobic

20 K-dalton T3' shows little heterogeneity upon isoelectric focusing. In contrast, the 20 K-dalton T3 which is a glycosylated form of the 14 K-dalton T3 is very heterogeneous in IEF and is susceptible to neuraminidase. These observations led us to speculate that the two 20 K-dalton T3 species may exist together in the membrane (Borst *et al.*, 1983b,c). A similar case was reported for the IgE receptor on mast cells (Holowka *et al.*, 1981), where a 50 K-dalton glycoprotein was found to be associated with a 30 K-dalton INA-labeled subunit. Thus, the INA-labeled 20 K-dalton T3 may be the anchor for the glycosylated 20 K-dalton T3 form and for the 25–28, 37 and 44 K-dalton glycoproteins.

On the basis of the relative incorporation of ^{125}I label, the 25–28, 37 and 55 K-dalton glycoproteins on normal human T cells and the T leukemic cell line HPB-ALL are present in different concentrations. The 20 K-dalton T3 forms are more abundant than the other T3 glycoproteins. Careful salt washes indicate that monoclonal antibodies OKT3 and anti-Leu-4 react with the 20 K-dalton T3 species (Borst *et al.*, 1983a). This idea is supported by modulation experiments, which show that after incubation with anti-Leu-4 an antigen–antibody complex with the 20 K-dalton component bound to antibody was detected (Rinnooy-Kan *et al.*, 1983). These modulation experiments and preliminary cross-linking experiments suggest the existence of a complex between

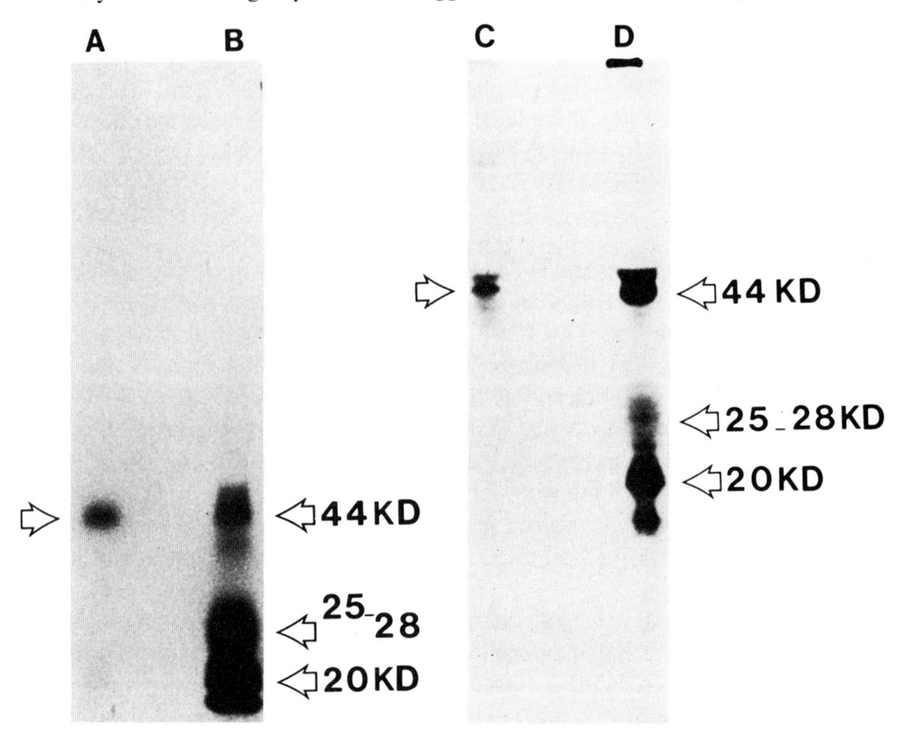

20 and 25–28 K-dalton T3 (C. Brezin and C. Terhorst, unpublished). Thus, independent protein ensembles of 20 K-dalton T3 with the other components may exist (see Fig. 7.6).

Evidence for the existence of a 20/44 K-dalton T3 ensemble was derived the use of a monoclonal antibody WT-31 (Tax *et al.*, 1983a,b). As

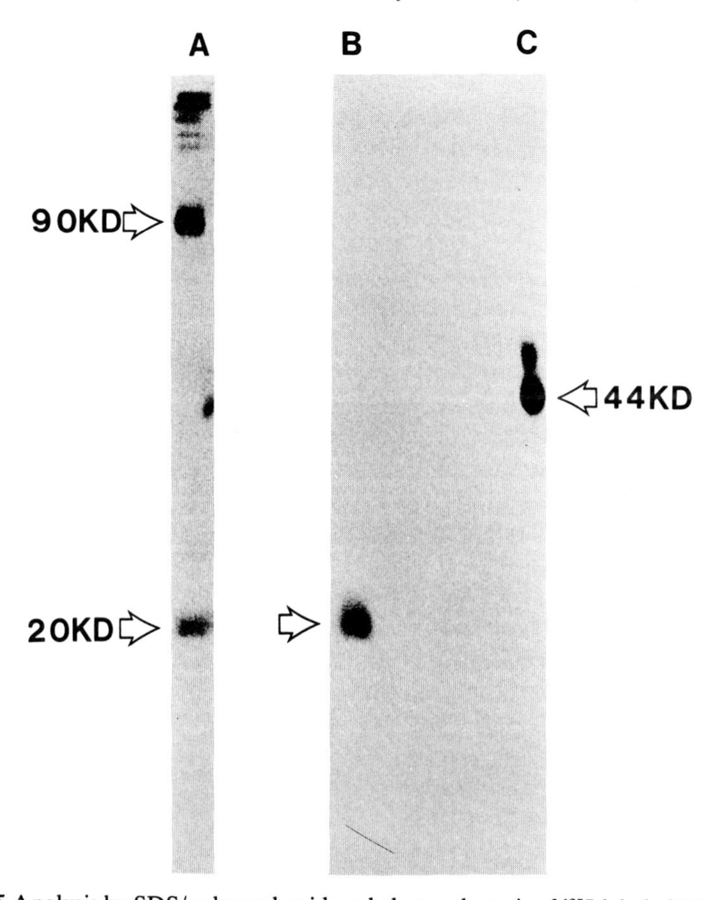

Fig. 7.5 Analysis by SDS/polyacrylamide gel electrophoresis of ^{125}I-labeled T3 antigens from human thymocytes and human T leukemic cells. Fig. 7.5.1 (*opposite*) A, Immunoprecipitate made with monoclonal antibody WT-31 from human thymocytes. B, Immunoprecipitate made with monoclonal antibody anti-Leu-4 from human thymocytes. C, Immunoprecipitate made with monoclonal antibody WT-31 from leukemic cells. D, Immunoprecipitate made with monoclonal antibody anti-Leu-4 from leukemic cells. All samples were analyzed under reducing conditions. Fig. 7.5.2 (*above*) Immunoprecipitate made with monoclonal antibody anti-Leu-4 from leukemic cells from a patient with Sézary syndrome. A, non-reducing conditions; B, the 20 K-dalton band in lane A was excised and analyzed under reducing conditions. C, the 90 K-dalton band in Lane A was excised and analyzed under reducing conditions.

discussed above (Section 7.3.1), WT-31 is mitogenic for human T cells and blocks CML. Purified IgG from WT-31 ascites, labeled with ^{125}I, binds to all human T lymphocytes. This binding is completely blocked by OKT3 (Tax *et al.*, 1983a). Conversely binding of monoclonal anti-T3 reagents OKT3 and WT-32 is only blocked partially by WT-31. In addition, modulation of the T3 complex with OKT3 removed reactivity with WT-31 from CTL clones. These results indicate that WT-31 may react with only one of the possible protein T3 ensembles. Immunoprecipitation experiments confirm that notion (Fig. 7.5): WT-31 precipitates the 37 K-dalton and/or 44 K-dalton forms of T3 from normal T lymphocytes, thymocytes or leukemic cells. When radioiodinated cells from a T-leukemic cell are used for immunoprecipitations, anti-Leu-4 detects the whole T3 complex, with a strong 44 K-dalton band (Fig. 7.5). Again, WT-31 precipitates only the 44 K-dalton band although in some instances a very faint 20 K-dalton T3 was found. Under non-reducing conditions the 44 K-dalton band ran as a dimcr of 90 K daltons (Fig. 7.5). Taken together these data provide evidence for the existence of a non-covalently associated 44/20 K-dalton T3 ensemble (Fig. 7.6). More protein chemical information is required to prove that these various T3 protein ensembles exist on the cell surface and that the hydrophobic 20 K-dalton T3 is the anchor protein of these ensembles.

Several groups have recently described the process of monoclonal antibody-induced cell surface modulation of T3. Anti-Leu-4 and OKT3 induce disappearance of T3 from the cell surface within 7 hours (Rinnooy-Kan *et al.*, 1983). WT-31 co-modulates with OKT3 on T lymphocytes and CTL clones (H. Spits, personal communication). Some investigators find that antibody–antigen complexes are internalized (Rinnooy-Kan *et al.*, 1983), whereas others have reported that both antibody and antigen are shed (Reinherz *et al.*, 1982). It remains to be determined whether antigenic modulation is a particular

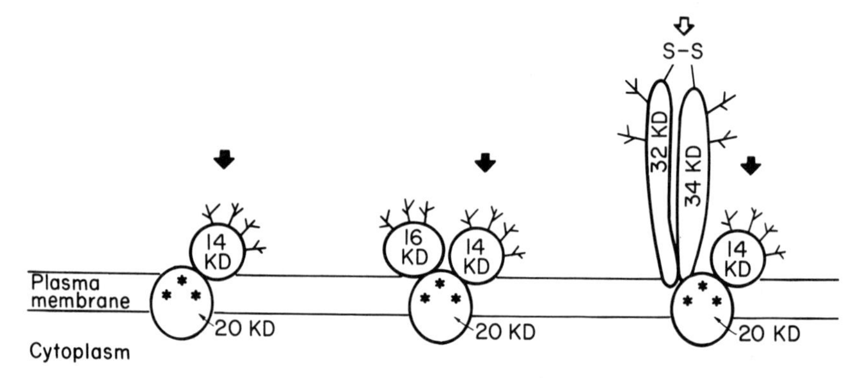

Fig. 7.6 Schematic diagram of possible T3 protein ensembles in the plasma membrane. *, Site of ^{125}INA label. Solid arrow, target antigen of OKT3, anti-Leu-4, SPV-T3 and WT-32. Open arrow, target antigen of WT-31. ⚲, oligosaccharide side chain.

feature of T3 or an artefact caused by some antibodies. Evidence that modulation could be a functionally relevant property of T3 was provided by studies with specific peptide antigens and T helper clones (Zanders *et al.*, 1983). At high concentrations (>20 μg/ml) of a peptide antigen the specific T helper clone has become unresponsive (i.e. tolerant) to induction of proliferation by the peptide. At these tolerizing concentrations reactivity with monoclonal anti-T3 (UCHT1) had disappeared.

Several investigators have suggested that the T3 complex may contain antigen-recognition structures on human T cells. Recent experiments by Meuer and colleagues (Meuer *et al.*, 1983a,b) strongly support this idea. Anti-clonotypic antibodies specifically inhibit cytolysis by CTL clones and proliferative responses by those clones. The anti-clonotypic monoclonal antibodies detect a 90 K-dalton heterodimer, which consists of disulfide bonded 49 and 43 K-dalton chains. Co-modulation experiments indicate that the clonotypic structures are associated with the T3 complex. The 43 and 49 K-dalton most likely belong to a family of clonotypic structures, which also include the 37 and 44 K-dalton T3. WT-31 probably recognizes a common determinant on all these structures. Interestingly, binding of the anti-clonotypic monoclonal antibody was not blocked by OKT3, whereas binding of WT-31 was. Although the clonotypic structures on CTLs are the best candidates for the T cell receptors for MHC-class I and class II antigens, experimental proof for that notion is still needed. Similar structures have been described in murine systems (Allison *et al.*, 1982; Haskins *et al.*, 1983). Comparisons of the 20 K-dalton T3 species from different individuals (Borst *et al.*, 1982) and different CTL clones (Spits *et al.*, 1982b) by IEF and amino acid sequencing suggest that little heterogeneity among these forms exists. The idea that part of the T3 complex is composed of constant elements (20, 25–28 K daltons) and that the 37/44 K-dalton chains are variable portions is therefore very appealing. Upon recognition of antigen by the variable structures signal transfer to other T cell components could take place via the constant part of the T3 ensembles. These signals could give rise to either cell proliferation or cytolysis, perhaps dependent on a different mode of ensemble formation.

7.5 STRUCTURE OF T8

Monoclonal antibodies specific for T8 inhibit cytolysis by CTL clones in the adhesion phase. Both structural (Terhorst *et al.*, 1980; Ledbetter *et al.*, 1982) and functional studies suggest that T8 may be the human homologue of the murine Lyt-2,3 antigens.

The T8 molecule is a 34 K-dalton glycoprotein under reducing conditions if isolated from human T cells, CTL-clones or T leukemic cell lines (Snow *et al.*, 1983). Under non-reducing conditions T8 forms a series of multimers (Fig.

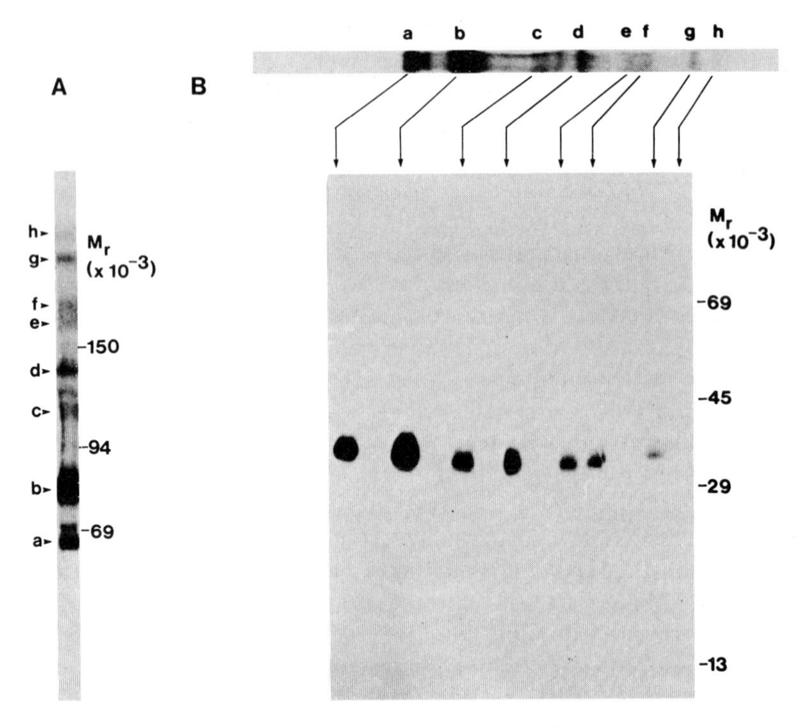

Fig. 7.7 Analysis by SDS/polyacrylamide gel electrophoresis of the T8 antigen isolated from ^{125}I-labeled peripheral blood T lymphocytes. Sequential non-reduced and reduced SDS/polyacrylamide gel analysis of T8 immunoprecipitates from ^{125}I-labeled peripheral T lymphocytes. Analysis under non-reducing conditions on a 5–15% SDS/polyacrylamide gradient gel (A). The bands indicated as a–h in (A) were visualized by autoradiography, excised, and equilibrated with sample buffer containing 5% 2-mercaptoethanol. The reduced proteins were subsequently analyzed on a 10–15% SDS/polyacrylamide gradient gel.

7.7). Most notably two dimer forms, of 67 and 76 K-daltons exist. These species are most likely generated by differences in internal disulfide bonding (Snow *et al.*, 1983). When analyzed under non-reducing conditions, identical homo-multimers are precipitated by three different monoclonal antibodies (OKT8, OKT5 and anti-Leu-2A). Although the antigens precipitated appear identical, the determinants recognized by the antibodies seem to differ. The OKT8 epitope is not destroyed by reduction and alkylation, whereas those bound by the OKT5 and anti-Leu-2A antibodies are sensitive to this treatment (Snow *et al.*, 1983).

Cleavage with CNBr of the 34 K-dalton T8 gives two fragments (9 and 34 K daltons). The 9 K-dalton *N*-terminal segment is not involved in interchain disulfide bridges, whereas the 24 K-dalton *C*-terminal fragment is (Snow and

Terhorst, 1983). The 24 K-dalton CNBr fragment contains the oligosaccharide side chain, which probably contains *O*-linked carbohydrates. The T8 antigen can be labeled by [^{125}I]iodonaphthylazide, which indicates the presence of a hydrophobic segment. Most integral membrane proteins are anchored in the lipid bilayer via a stretch of hydrophobic amino acids. It was therefore suggested that T8 is an integral membrane protein. The [^{125}I]INA label was found in the 24 K-dalton segment. Taken together these data support the model for the T8 dimer and multimers, as shown in Fig. 7.9.

On thymocytes the same series of homomultimers was found as described above. In addition, a series of heteromultimers was detected (see Fig. 7.8). The 34 K-dalton T8 is disulfide-linked to a 46 K-dalton protein (Snow and Terhorst, 1983). Preliminary analysis showed that the 46 K-dalton T8 structure was not related to the 34 K-dalton T8 molecule. This conclusion is based upon analysis of peptide maps, two-dimensional gel electrophoresis, and CNBr cleavage patterns. The carbohydrate content of the two subunits was also found to be dissimilar, on the basis of the differential cleavage susceptibilities of the two molecules to Endo-F. This showed that the 46 K-dalton molecule contained several *N*-linked oligosaccharides, whereas the 34 K-dalton molecule did not. Finally, the 46 K-dalton form does not contain a determinant recognized by an anti-T8 monoclonal antibody.

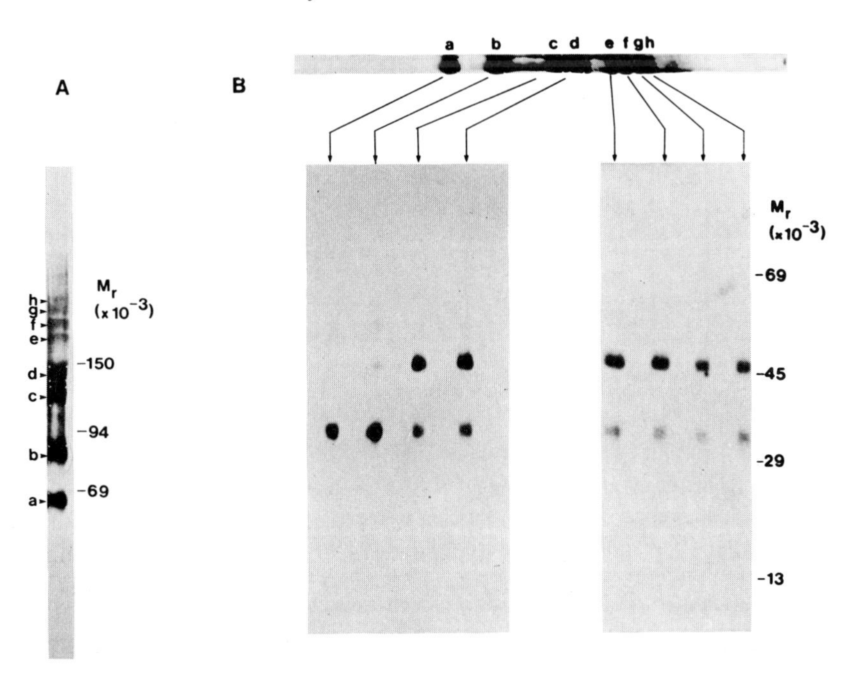

Fig. 7.8 Analysis by SDS/polyacrylamide gel electrophoresis of the T8 antigen isolated from ^{125}I-labeled human thymocytes. See legend to Fig. 7.7.

Experiments using CNBr cleavage and SDS-polyacrylamide gel electrophoresis suggest the model presented in Fig. 7.9. The other multimeric forms of T8 are derived from the two complexes shown here. The functional significance of the disulfide-induced multimerization of the T8 molecule on CTL is at present unclear. In this regard, it has been found that sulfhydryl reagents have an inhibitory effect on cytolysis by CTL (Redelman and Hudig, 1980). Although there is no evidence that such compounds bind directly to T8 at the cell surface, the fact that T8 has been shown to be involved in cytolysis, as well as to be heavily dependent upon sulfhydryl cross-linking for its structure, make this an attractive possibility. Additional investigation aimed at determining the

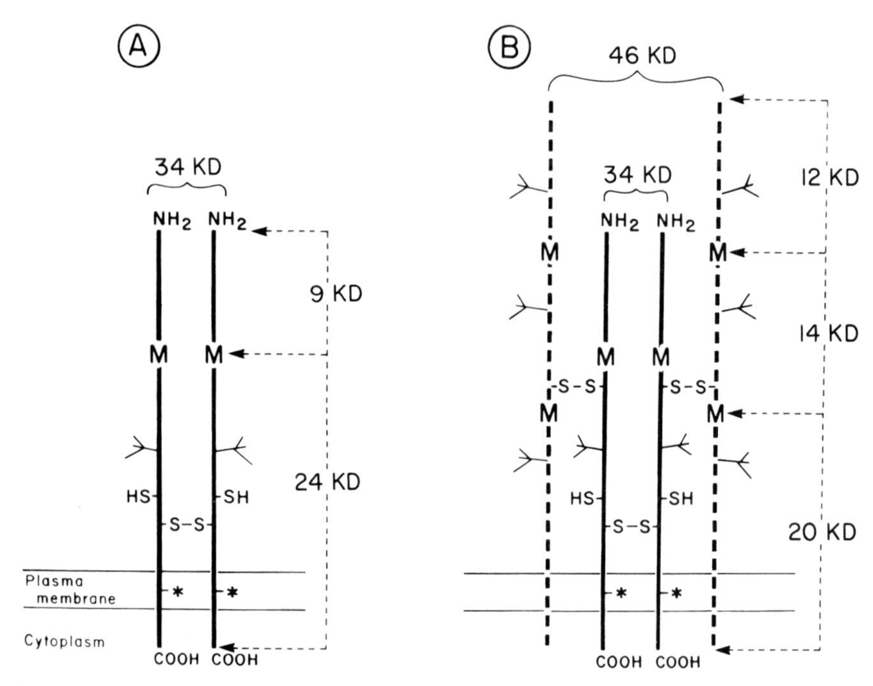

Fig. 7.9 Possible models for the disposition of the T8 molecule on T lymphocytes and thymocytes. (A) This structure represents the dimeric form of the T8 molecule on peripheral T lymphocytes and thymocytes. The binding sites for INA and the CNBr fragments are indicated in their relationship to the plasma membrane. *, Site of [^{125}I]INA label. (B) The simplest heteromultimeric structure composed of the 43 and 34 K-dalton subunits as postulated on thymocytes. The CNBr fragments of both subunits are indicated in their putative disposition with respect to the cell membrane. In both figures, M signifies methionine, SH represents free sulfhydryl groups of cysteines, and S-S represents disulfide-linked cystine residues. Higher-order multimers are presumably formed through the free sulfhydryls indicated in the figure. COOH and NH_2 signify the *C*- and *N*-termini, respectively. Y, oligosaccharide side chain.

effects of mild reduction and alkylation on the action of CTL may aid in the elucidation of this question.

Questions concerning the differences between the T8 molecules on thymus and peripheral blood lymphocytes (PBL) are intriguing. For example, it would be interesting to investigate at what stage of maturation in the thymus, the 46/34 K-dalton heteropolymers are lost and replaced by the 34 K-dalton homopolymers. This transformation during the final stages of T lymphocyte differentiation is of particular interest with respect to its potential importance in the acquisition of functional competence by thymocytes.

7.6 STRUCTURE OF LFA-1

Anti-LFA-1 monoclonal antibodies react with thymocytes, T cells, B cells, monocytes, granulocytes and NK cells. LFA-1 antigens were first described by Springer and colleagues in the mouse (Springer *et al.*, 1982), but have recently been found on the surface of human cells (Spits *et al.*, 1983a; Sanchez-Madrid *et al.*, 1982; Hildreth *et al.*, 1983). The anti-LFA-1 reagents SPV-L1 and SPV-L5 react more strongly with T cells and proliferative T cells than with monocytes and granulocytes (Spits *et al.*, 1983a). The monoclonal antibodies, FK14.1 (J. Borst and C. Terhorst, unpublished) and TA-1 (LeBien *et al.*, 1983), however, react more strongly with monocytes than with T cells. All anti-LFA-1 monoclonal reagents inhibit the cytolysis by CTL clones. SPV-L1 and SPV-L5 also inhibit lectin-dependent cytolysis of CTL clones directed at a large variety of non-specific target cells (Spits *et al.*, 1983a). In addition, these reagents inhibit cytolysis by NK cells. Anti-LFA-1 reagents inhibit lectin-induced and cell-surface- and soluble-antigen-induced T cell proliferation (Spits *et al.*, 1983a; Krensky *et al.*, 1983; Sanchez-Madrid *et al.*, 1983).

Anti-LFA-1 monoclonal antibodies precipitate heterodimers of 95 and 160 K-daltons. Upon cell-surface radioiodination, the heavy chain is much more strongly labeled than the light chain (Spits *et al.*, 1983a, Sanchez-Madrid *et al.*, 1983). Both chains are glycoproteins, which can be labeled by $NaB[^3H_4]$ after mild oxidation with sodium periodate. Both chains most likely carry both *N*-linked and *O*-linked oligosaccharides (J. Borst and C. Terhorst, unpublished).

The LFA-1 dimer is reminiscent of the 95/160 K-dalton Mo-1 antigen, which is found on monocytes, granulocytes and NK cells (Todd *et al.*, 1982). Anti-Mo-1 antibodies specifically inhibit rosetting between phagocytes and erythrocytes coated with C3bi, a fragment of the complement component C3b (Arnaout *et al.*, 1983; see Chapter 5 by Ross). Like LFA-1, the Mo-1 antigen is relatively resistant to trypsin (Todd *et al.*, 1983). Peptide-mapping studies revealed that the 95 K-dalton subunit of Mo-1 and LFA-1 are identical, but that the 160 K-dalton chains are distinct (J. Borst and C. Terhorst, unpublished).

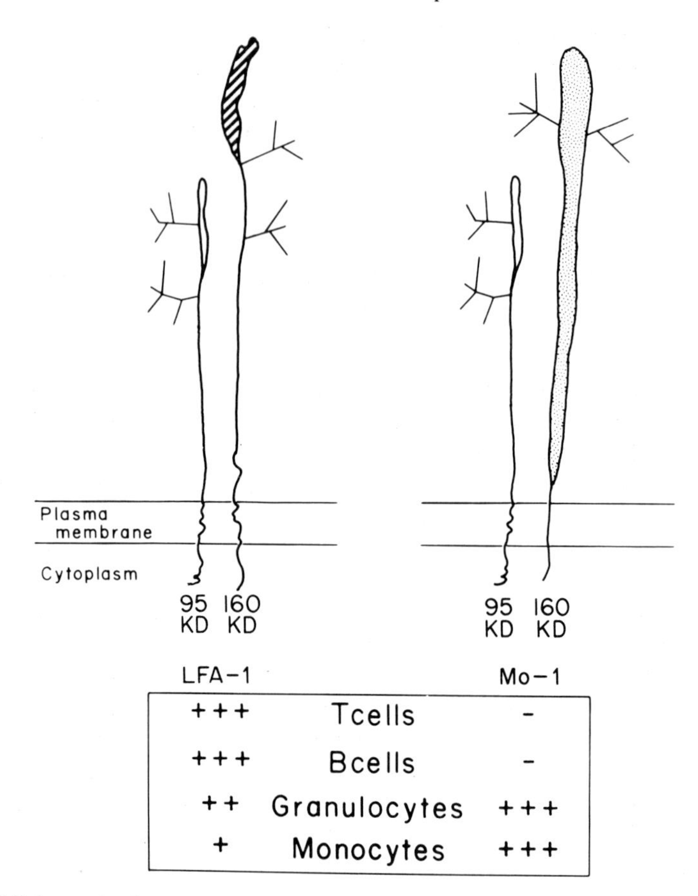

Fig. 7.10 Schematic diagram of human LFA-1 and Mo-1 antigens in the plasma membrane. , oligosaccharide side chain.

This conclusion (Fig. 7.10) is supported by the finding that anti-Mo-1 monoclonal antibodies bind to the 160 K-dalton chain only (Arnaout *et al.*, 1983). Interestingly, one patient with recurrent bacterial infections has been described who lacks both Mo-1 and LFA-1 (A. Arnaout, personal communication). A similar relationship to that between Mac-1 and LFA-1 has been described in the mouse (Springer *et al.*, 1982; see Chapter 5 by Ross for discussion of this topic).

Taken together these data strongly suggest that Mo-1 and LFA-1 belong to a family of cell adhesion molecules. Whether these glycoproteins are somehow related to the so-called cell adhesion molecules (CAMs) on neuronal and liver cells (Edelman, 1983) remains to be determined. It is plausible that LFA-1- and Mo-1-like molecules provide for stabilization in adhesion in T–T cell, T–B cell and T cell–macrophage interactions. Anti-LFA-1 monoclonal antibodies appear to block the effector CTL but not at the target cell level, because some target cells such as K562 and melanoma cells do not carry the target antigens for anti-LFA-1 monoclonal antibodies (DeVries and Spits, 1984). However, the counter structures of LFA-1 on these target cells, may have LFA-1-like properties without being recognized by the particular monoclonal anti-LFA-1 antibodies available.

7.7 FUNCTION-RELATED T CELL ANTIGENS AND THE T CELL RECEPTOR

Most T lymphocytes recognize foreign antigens only when these antigens are associated on cell surfaces with the membrane glycoproteins encoded by genes in the MHC. For instance, CTLs recognize viral antigens in association with MHC-class I antigens on target cells and helper T cells respond to antigen only if associated with MHC-class II products on antigen-presenting cells. Monoclonal antibodies have already been of paramount importance in defining these 'antigen driven' cell–cell adhesions. In particular, since T cell clones have been generated, the identification and biochemical description of antigen-specific receptors on T cells is now within our reach.

In this chapter an attempt was made to summarize the most salient features of cell surface structures which are involved in human T lymphocyte functions. The clonotypic structures found on human CTL clones are at the moment the best candidates for T cell receptors which recognize MHC-class I and -class II antigens (Meuer *et al.*, 1983a,b). These variable glycoproteins may form *protein ensembles* with the invariable proteins of the T3 complex. Recognition of MHC-class I or -class II antigens by allogeneic CTLs also requires participation of LFA-1, T8 or T4 and T11. Each of these T cell antigens must have a complementary structure on the target cell. In other words, MHC-class I or -class II antigens on the target cell are recognized by the T cell receptor, if sufficient cell–cell adhesions can be established through multipoint interactions. Speculations have been made that the counter structures for T4 and T8 are constant regions of class II and class I antigens, respectively. But more experimental evidence is needed for that idea. The counter structures for LFA-1 and T11 on the target cell are also unknown. It is plausible that depending on relative affinities of various receptor and acceptor molecules not every molecular interaction may be required in each T cell–cell contact. The notion that MHC antigens are recognized by CTLs in concert with other cell–cell interactions is supported by the observation that human MHC-class I antigens expressed on mouse L cells are not recognized by human CTLs (Bernabeu *et al.*, 1983a). This absence of recognition is not corrected by the introduction of human β_2-m together with the human class I antigen. Most likely, mouse L cells do not carry the correct recognition structures for T8, LFA-1 and T11.

The observation that the multipoint interactions between CTL and target cells can be inhibited by a monoclonal antibody specific for only one of the components suggests a multistep mechanism. On the basis of the inhibition studies of CTL–target cell interactions with the monoclonal antibody described, the following order of events can be proposed.

1. Adhesion via interactions of T8 (or T4) and LFA-1 on the CTL with their counterparts on the target cell.
2. Association between T cell receptor (T3 variable elements) and MHC-class I or -class II antigen.
3. Triggering of the constant part of T3, which initiates lethal hit delivery.

At which point the monoclonal anti-clonotype and anti-T11 block CTL–target cell interaction has not yet been reported. A similar order of events could be described with respect to interactions between antigen-presenting cells and helper T lymphocytes. In the latter case a constant element of the T3 complex may interact with the proliferating machinery of the T cell.

In summary, monoclonal antibodies have been invaluable tools in describing the structures which are involved in human T cell functions. It is logical to assume that in the near future additional antibodies will be found which allow a full description of all the participants in T-cell-mediated interactions, the temporal order of such interactions, the mode of action of the receptor for antigen and the mechanism of T cell proliferation.

REFERENCES

Allison, J., McIntyre, B.W. and Bloch, D. (1982), *J. Immunol.*, **129,** 2293–2299.

Arnaout, A.A., Todd, R.F., Dana, N., Melamed, J., Schlossman, S.F. and Colten, H. (1983), *J. Clin. Invest.*, **72,** 171–179.

Ball, E.J. and Stastny, P. (1982), *Immunogenetics,* **2,** 157–163.

Bergman, Y. and Levy, R. (1982), *J. Immunol.*, **128,** 1334–1340.

Bernabeu, C., Van de Rijn, M., Finlay, D., Maziarz, R., Biro, A., Spits, H., DeVries, J. and Terhorst, C. (1983a), *J. Immunol.*, **131,** 2032–2037.

Bernabeu, C., Van de Rijn, M., Lerch, P.G. and Terhorst, C. (1983b), *Nature (London),* (in press).

Bernard, A., Gelin, C., Raynal, B., Pham, D., Gosse, C. and Boumsell, L. (1982), *J. Exp. Med.*, **155,** 1317–1333.

Beverley, P.C.L. and Callard, R.E. (1981), *Eur. J. Immunol.*, **11,** 329–334.

Bhan, A.K., Mihm, M.C. and Dvorak, H.F. (1982), *J. Immunol.*, **129,** 1578–1583.

Biddison, W.E., Rao, P.E., Talle, M.A., Goldstein, G. and Shaw, S. (1982), *J. Exp. Med.*, **156,** 1065–1083.

Biddison, W.E., Rao, P.E., Talle, M.A., Goldstein, G. and Shaw, S. (1983), *J. Immunol.*, **131,** 152–157.

Borst, J., Prendiville, M.A. and Terhorst, C. (1982), *J. Immunol.*, **128,** 1560–1565.

Borst, J., Alexander, S., Elder, J. and Terhorst, C. (1983a), *J. Biol. Chem.*, **258,** 5135–5141.

Borst, J., Prendiville, M.A. and Terhorst, C. (1983b), *Eur. J. Immunol.*, **13,** 576–580.

Borst, J., Spits, H., DeVries, J.E., Pessano, S., Rovera, G. and Terhorst, C. (1983c), *Hybridoma,* **2,** 265–274.

Britton, S. and Palacios, R. (1982), *Immunol. Rev.*, **65,** 5–22.

Burns, G.F., Boyd, A.W. and Beverley, P.C. (1982), *J. Immunol.*, **129,** 1451–1452.

Cantor, H. and Boyse, E.A. (1977), *Cold Spring Harbor Symp. Quant. Biol.*, **41,** 23–32.

Cantor, H., Shen, F.W. and Boyse, E.A. (1976), *J. Exp. Med.*, **143,** 1391–1401.

Cerf-Bensussan, M., Schneeberger, E.E. and Bhan, A.K. (1983). *J. Immunol.*, **130,** 2615–2622.

Chang, T.W., Kung, P.G., Gingras, S.P. and Goldstein, G. (1981), *Proc. Natl. Acad. Sci. U.S.A.*, **78,** 1805–1810.

Depper, J.M., Leonard, W.J., Robb, R.J., Waldmann, T.A. and Greene, W.C. (1983), *J. Immunol.*, **131,** 690–696.

DeVries, J.E. and Spits, H. (1984), *Immunol.*, **132,** 510–519.

DeVries, J.E., Vyth-Dreese, F.A., Figdor, C.G., Spits, H., Leemans, J.M. and Bont, W.S. (1983), *J. Immunol.*, **131,** 201–206.

Edelman, G.M. (1983), *Science,* **219,** 450–457.

Engleman, E.G., Benike, C.J., Grumet, F.C. and Evans, R.L. (1981a), *J. Immunol.*, **127,** 2124–2129.

Engleman, E.G., Benike, C.J., Glickman, E. and Evans, R.L. (1981b), *J. Exp. Med.*, **153,** 193–198.

Flaherty, L. (1981), in *The Role of the Major Histocompatibility Complex in Immunobiology* (M.E. Dorf, ed.), Garland Press, New York, pp. 33–57.

Garson, J.A., Beverley, P.C.C., Coakham, H.B. and Harper, E.I. (1982), *Nature (London),* **298,** 375–377.

Gillis, S., Mochizuki, D.Y., Conlon, P.J., Hefeneider, S.H., Ramthum, C.A., Gilles, A.E., Frank, M.B., Henney, C.S. and Watson, J.D. (1982), *Immunol. Rev.*, **63,** 166–209.

Goldstein, G., Lifter, J. and Mittler, R. (1982), in *Monoclonal Antibodies and T cell Products* (D.H. Katz, ed.), CRC Press, pp. 71–89.

Goodfellow, P.N., Banting, G., Sutherland, R., Greaves, M., Solomon, E. and Povey, S. (1982), *Som. Cell. Genet.*, **8,** 197–206.

Haskins, K., Kubo, R., White, J., Pigeon, M., Kappler, J. and Marrack, P. (1983), *J. Exp. Med.*, **157,** 1149–1169.

Haynes, B.F. (1981), *Immunol. Rev.*, **57,** 127–161.

Haynes, B.F., Mann, D.L., Hemler, M.E., Schroer, J.A., Schelhamer, J.H., Eisenbarth, G.S., Strominger, J.L., Thomas, C.A., Mostowski, H.S. and Fauci, A.S. (1980), *Proc. Natl. Acad. Sci. U.S.A.*, **77,** 2914–2919.

Hercend, T., Reinherz, E.L., Meuer, S., Schlossman, S.F. and Ritz, J.E. (1983) *Nature (London),* **301,** 158–160.

Hildreth, J.E.K., Gotch, F.M., Hildreth, P.D.K. and McMichael, A.J. (1983), *Eur. J. Immunol.*, **13,** 202–208.

Hirayama, M., Lisak, R.P., Kim, S.U., Pleasure, D.E. and Silberberg, D.H. (1983), *Nature (London),* **301,** 153–158.

Holowka, D., Gitler, C., Bercovici, T. and Metzger, H. (1981), *Nature (London)*, **289**, 206–209.

Hood, L., Steinmetz, M. and Malissen, B. (1983), *Annu. Rev. Immunol.*, **1**, 529–568.

Howard, F.D., Ledbetter, J.A., Wong, J., Bieber, C.P., Stinsor, E.B. and Herzenberg, L.A. (1981), *J. Immunol.*, **126**, 2117–2122.

Janossy, G., Thomas, J.A., Bollum, F.J., Granger, S., Pissolo, G., Bradstock, K.F., Wong, L., McMichael, A., Ganeshrager, K. and Hoffbrand, A.V. (1980), *J. Immunol.*, **125**, 202–210.

Kamoun, M., Martin, P.J., Hanse, J.A., Brown, M.A., Siadak, A.W. and Nowinski, R.C., (1981), *J. Exp. Med.*, **153**, 207–212.

Kaneoka, H., Perez-Rojas, G., Sasasuki, T., Benike, C.J. and Engleman, E.G. (1983), *J. Immunol.*, **131**, 158–164.

Knowles, R.W. and Bodmer, W.F. (1982), *Eur. J. Immunol.*, **12**, 676–681.

Krensky, A.M., Reiss, C.S., Mier, J.W., Strominger, J.L. and Burakoff, S.J. (1982), *Proc. Natl. Acad. Sci. U.S.A.*, **79**, 2365–2370.

Krensky, A.M., Sanchez-Madrid, F., Robbins, E., Nagy, J.E., Springer, T.A. and Burakoff, S.J. (1983), *J. Immunol.*, **131**, 611–616.

Kung, P., Goldstein, G., Reinherz, E. and Schlossman, S.F. (1979), *Science*, **206**, 347–349.

Landegren, U., Ramstedt, U., Axberg, I., Ullberg, M., Jardal, M. and Wigzell, H. (1983), *J. Exp. Med.*, **155**, 1579–1584.

LeBien, T.W., Bradley, J.G. and Koller, B. (1983), *J. Immunol.*, **130**, 1833–1839.

Ledbetter, J.A., Evans, R.L., Lipinski, M., Cunningham-Rundles, C., Good, R.A. and Herzenberg, L.A. (1981), *J. Exp. Med.*, **153**, 310–323.

Leonard, W.J., Depper, J.M., Uchiyama, T., Smith, K.A., Waldmann, T.A. and Greene, W. (1982), *Nature (London)*, **300**, 267–269.

Lerch, P.G., Van de Rijn, M., Schrier, P. and Terhorst, C. (1983), *Human Immunol.*, **6**, 13–30.

Martin, P., Longton, G., Ledbetter, J.A., Newman, W., Braun, M.P., Beatty, P.G. and Hansen, J.A. (1983), *J. Immunol.*, **131**, 180–185.

Martz, E. (1977), *Contemp. Top. Immunobiol.*, **7**, 301–356.

Meuer, S.C., Hussey, R.E., Hodgdon, J.C., Hercend, T., Schlossman, S.F. and Reinherz, E.L. (1982), *Science*, **218**, 471–473.

Meuer, S.C., Acuto, O., Hussey, R.E., Hodgdon, J.C., Fitzgerald, K.A., Schlossman, S.F. and Reinherz, E.L. (1983a), *Nature (London)*, **303**, 808–810.

Meuer, S.C., Fitzgerald, K.A., Hussey, R.E., Hodgdon, J.C., Schlossman, S.F. and Reinherz, E.L. (1983b), *J. Exp. Med.*, **157**, 705–719.

Miller, R.A. and Stutman, O. (1982), *Nature (London)*, **296**, 76–78.

Miyawki, T., Yachie, A., Uwandana, N., Ohzeki, S., Nagaoki, T. and Tanaguchi, N. (1982), *J. Immunol.*, **129**, 2474–2478.

Miyawaki, T., Yachie, A., Ohzeki, S., Nagaoki, T. and Tanaguchi, N. (1983), *J. Immunol.*, **130**, 2737–2742.

Moscicki, R.A., Amento, E.P., Krane, S.M., Kurnick, J.T. and Colvin, R.B. (1983), *J. Immunol.*, **131**, 743–748.

Müller, C., Stein, H., Ziegler, A. and Wernet, P. (1983), *Eur. J. Immunol.*, **13**, 414–418.

Oger, J., Szuchet, S., Antel, J. and Arnason, B. (1982), *Nature (London)*, **295**, 66–68.

Palacios, R. (1982), *Immunol. Rev.*, **63,** 73–110.
Palacios, R. and Martinez-Maza, O. (1982), *J. Immunol.*, **129,** 2479–2485.
Perussia, B., Acuto, O., Terhorst, C., Faust, J., Lazarus, R., Fanning, V. and Trinchieri, G. (1983a), *J. Immunol.*, **130,** 2142–2148.
Perussia, B., Fanning, V. and Trinchieri, G. (1983b), *J. Immunol.*, **131,** 223–231.
Poppema, S., Bhan, A.K., Reinherz, E.L., McCluskey, R.T. and Schlossman, S.F. (1981), *J. Exp. Med.*, **153,** 30–41.
Redelman, D. and Hudig, D. (1980), *J. Immunol.*, **124,** 870–878.
Reinherz, E.L. and Schlossman, S.F. (1980), *Cell,* **19,** 821–827.
Reinherz, E.L., Hussey, R.E. and Schlossman, S.F. (1980), *Eur. J. Immunol.*, **10,** 758–762.
Reinherz, E.L., Meuer, S.C., Fitzgerald, K.A., Hussey, R.E., Leving, H. and Schlossman, S.F. (1982), *Cell,* **30,** 735–743.
Raftery, M.A., Hunkapillar, M.W., Schrader, C.D. and Hood, L. (1980), *Science,* **208,** 1454–1457.
Rinnooy-Kan, E.A., Wang, C.Y., Wang, L.C. and Evans, R.L. (1983), *J. Immunol.*, **131,** 536–539.
Rosenthal, A.S. (1978), *Immunol. Rev.*, **40,** 136–157.
Ryffel, B., Gotz, U. and Heuberger, B. (1982), *J. Immunol.*, **129,** 1978–1982.
Sanchez-Madrid, F., Krensky, A.M., Ware, C.F., Robbins, E., Strominger, J.L., Burakoff, S.J. and Springer, T.A. (1982), *Proc. Natl. Acad. Sci. U.S.A.*, **79,** 7489–7493.
Schreiber, A.B., Libermann, T.L., Lax, I., Yarden, Y. and Schlessinger, J. (1982), *J. Biol. Chem.*, **256,** 846–853.
Snow, P. and Terhorst, C. (1983), *J. Biol. Chem.*, **258,** 14 675–14 681.
Snow, P., Spits, H., DeVries, J. and Terhorst, C. (1983), *Hybridoma,* **2,** 187–199.
Spits, H., IJssel, H., Terhorst, C. and DeVries, J.E. (1982a), *J. Immunol.*, **128,** 95–99.
Spits, H., Borst, J., Terhorst, C. and DeVries, J.E. (1982b), *J. Immunol.*, **129,** 1563–1569.
Spits, H., Keizer, G., Borst, J., Terhorst, C., Hekman, A. and DeVries, J.E. (1983a), *Hybridoma*, (in press).
Spits, H., IJssel, H., Thompson, A. and DeVries, J.E. (1983b), *J. Immunol.*, **131,** 678–683.
Springer, T.A., Davignon, D., Ho, M.K., Kurzinger, K., Martz, E. and Sanchez-Madrid, F. (1982), *Immunol. Rev.*, **68,** 111–139.
Sutherland, R., Delia, D., Schneider, C., Newman, R., Kemshead, J. and Greaves, M. (1981), *Proc. Natl. Acad. Sci. U.S.A.*, **78,** 4515–4519.
Swain, S.L. (1981), *Proc. Natl. Acad. Sci. U.S.A.*, **78,** 7101–7105.
Swain, S.L., Dutton, R.W., Schwab, R. and Yamamoto, J. (1983), *J. Exp. Med.*, **157,** 720–729.
Tax, W.J.M., Leeuwenberg, H.F.M., Willems, H.M., Capel, P.J.A. and Koene, R.A.P. (1983a), in *Human Leukocyte Markers Detected by Monoclonal Antibodies* (C. Boumsell and A. Bernard, eds), Springer, Berlin, (in press).
Tax, W.J.M., Willems, H.W., Reekers, P.P.M., Capel, P.J.A. and Koene, R.A.P. (1983b), *Nature (London)*, **304,** 445–447.
Terhorst, C., Van Agthoven, A., Reinherz, E.L. and Schlossman, S.F. (1980), *Science,* **209,** 520–521.

Terhorst, C., Van Agthoven, A., LeClair, K., Snow, P., Reinherz, E.L. and Schlossman, S.F. (1981), *Cell*, **23**, 771–780.

Todd, R.F., Agthoven, A., Schlossman, S.F. and Terhorst, C. (1982), *Hybridoma*, **1**, 329–337.

Trowbridge, I. and Omary, M.B. (1981), *Proc. Natl. Acad. Sci. U.S.A.*, **78**, 3039–3043.

Tsoukas, C.D., Carson, D.A., Fong, S. and Vaughan, J.H. (1982), *J. Immunol.*, **129**, 1421–1425.

Uchiyama, T., Broder, S. and Waldmann, T. (1981a), *J. Immunol.*, **126**, 1393–1397.

Uchiyama, T., Nelson, D.C., Fleisher, T.A. and Waldmann, T.A. (1981b), *J. Immunol.*, **126**, 1398–1403.

Umiel, T., Daley, J.F., Bhan, A.K., Levey, R.H., Schlossman, S.F. and Reinherz, E.L. (1982), *J. Immunol.*, **129**, 1054–1061.

Van Agthoven, A. and Terhorst, C. (1982), *J. Immunol.*, **128**, 426–430.

Van Agthoven, A., Terhorst, C., Reinherz, E.L. and Schlossman, S.F. (1981), *Eur. J. Immunol.*, **11**, 18–21.

Van de Rijn, M., Geurts van Kessel, A.H.M., Kroezen, V., Van Agthoven, A.J., Terhorst, C., Bootsma, D. and Hilgers, J. (1983a), *Cytogenet. Cell Genet.*, **36**, 525–536.

Van de Rijn, M., Lerch, P.G., Knowles, R.W. and Terhorst, C. (1983b), *J. Immunol.*, **131**, 851–855.

Van de Rijn, M., Lerch, P.G., Knowles, R.W., Bhan, A.K. and Terhorst, C. (1983c), *Hum. Immunol.*, (in press).

Van Wauwe, J.P. and Goossens, J.G. (1983), *Cell. Immunol.*, **77**, 23–29.

Van Wauwe, J.P., DeMey, J.E. and Goossens, J.G. (1980), *J. Immunol.*, **124**, 2708–2713.

Verbi, W., Greaves, M.F., Schneider, C., Koubek, K., Janossy, G., Stein, H., Kung, P. and Goldstein, G. (1982), *Eur. J. Immunol.*, **12**, 81–86.

Wang, C.Y., Good, R.A., Ammirati, P., Dymbort, G. and Evans, R.L. (1980), *J. Exp. Med.*, **151**, 1539–1551.

Wood, G., Warner, N.L. and Warnke, R.A. (1983), *J. Immunol.*, **131**, 212–216.

Zanders, E.D., Lamb, J.R., Feldmann, M., Green, N. and Beverley, P.C.L. (1983), *Nature (London)*, **303**, 625–627.

Ziegler, A. and Milstein, C. (1979), *Nature (London)*, **279**, 243–245.

8 Antibodies to the Acetylcholine Receptor

SARA FUCHS, MIRY C. SOUROUJON and DARIA MOCHLY-ROSEN

Glossary of terms and abbreviations

anti-dAChR	Antibodies to non-structural (or denaturation-resistant) determinants of the acetylcholine receptor
AChR	Acetylcholine receptor
EAMG	Experimental autoimmune myasthenia gravis
MG	Myasthenia gravis
RCM-AChR	Reduced and carboxy-methylated receptor
SDS	Sodium dodecyl sulfate
α-Bgt	α-Bungarotoxin, a potent acetylcholine receptor antagonist (and neurotoxin) from snakes
T-AChR	Trypsin-digested acetylcholine receptor
PA-AChR	Polyalanyl-derivatized acetylcholine receptor
nit-AChR	Nitrated acetylcholine receptor
anti-nAChR	Antibodies to conformational (denaturation-sensitive) or 'native' determinants of acetylcholine receptor
Torpedo californica, *Electrophorus electricus*	Species of electric fish providing a rich source of acetylcholine receptors (– from the electroplax organ)
JR	Junctional acetylcholine receptor (present at high density in adult, mammalian skeletal neuromuscular junctions)
EJR	Extrajunctional acetylcholine receptor (present over the entire surface of embryonic or denervated adult muscle fibres)
Mepp	Miniature end plate potential
$F(ab')_2$	Pepsin-digested, divalent fragments of antibody

Acknowledgements

The work from our laboratory cited in this review was supported in part from The Muscular Dystrophy Association of America, The Los Angeles Chapter of the Myasthenia Gravis Foundation and from the United States–Israel Binational Science Foundation (BSF).

Monoclonal Antibodies to Receptors: Probes for Receptor Structure and Function
(*Receptors and Recognition*, Series B, Volume 17)
Edited by M. F. Greaves
Published in 1984 by Chapman and Hall, 11 New Fetter Lane, London EC4P 4EE

8.1 INTRODUCTION

The nicotinic acetylcholine receptor (AChR) is one of the key proteins governing the function of the neuromuscular junction. It is the first neurotransmitter receptor to be identified as a molecular entity and to be isolated and purified in an active form. One advantage of this receptor over other receptor systems is the availability of a rich and convenient source for AChR in electric organs of electric fish (eels or rays). This, in addition to the fact that α-neurotoxins from some Elapid snakes bind specifically and with very high affinity to the AChR, has facilitated and enabled the isolation and purification of this receptor and its thorough analysis.

As soon as the first purified preparations of AChR became available in the early seventies, scientists set forth to produce antibodies against this receptor, to be applied as immunochemical tools for structural and functional analysis of the receptor. In a way it was found as a result of a mere serendipity that immunization of animals with purified AChR leads to the development of clinical symptoms which are very similar to those observed in the human neuromuscular disease, myasthenia gravis. The model disease induced in animals was termed experimental autoimmune myasthenia gravis (EAMG).

In the last decade the research of the immunological properties of AChR in relation to myasthenia gravis (MG) has been a very attractive topic for immunologists and neurologists. The detailed study of the immune response to AChR and of EAMG have contributed enormously to the elucidation of the nature and molecular origin of myasthenia gravis and possibly of other receptor-associated autoimmune diseases. It is now established that myasthenia gravis is an autoimmune disease where AChR is the major autoantigen. This has been, and still is, a major contribution of basic research to the understanding and improved management of a human disease. This particular immunological aspect of the AChR has been the topic of many publications and review articles (for reviews see Fuchs, 1979a; Drachman, 1978; Lindstrom, 1979; Lindstrom and Dau, 1980; Vincent, 1980) and will not be dealt with in depth in this review. Aside from that, antibodies to the AChR (polyclonal and monoclonal) proved to be a powerful tool for following and studying various structural parameters and biological functions of the AChR. Antibodies to AChR elicited in various animal species, and found spontaneously in patients with myasthenia gravis, have been widely applied in different studies of AChR such as: immunochemical analysis of the receptor in various species, organs and excitable membranes; localization, orientation, biosynthesis and development of AChR; metabolism, channel properties and other physiological functions, etc. It is to these aspects of antibodies to AChR that this review is devoted.

8.2 THE ACETYLCHOLINE RECEPTOR

A detailed biochemical analysis of the nicotinic AChR is now available. Most structural studies were performed on the AChR purified from electric organs (of *Torpedo* and *Electrophorus electricus*), but the AChR from various mammalian muscles was also isolated and characterized. The molecular and functional properties of the AChR in its membrane-bound and detergent-solubilized forms have been summarized in several recent review articles (Karlin, 1980; Changeux, 1981; Conti-Tronconi and Raftery, 1982).

The *Torpedo* AChR molecule is a transmembrane glycoprotein of apparent molecular weight 250 000. It exists in a monomeric form of 9 S and in a disulfide bridged dimeric form of 13 S (Sobel and Changeux, 1977; Chang and Bock, 1977; Hamilton *et al.*, 1977; Aharonov *et al.*, 1977; Vandlen *et al.*, 1979). It is composed of four different subunits termed α, β, γ and δ. The apparent molecular weights of these subunits as determined by gel electrophoresis are about 40 K, 48 K, 55 K and 64 K for α, β, γ and δ respectively (Weill *et al.*, 1974; Chang and Bock, 1977; Aharonov *et al.*, 1977; Froehner and Rafto, 1979; Vandlen *et al.*, 1979). Recently, the exact amino acid sequence of each subunit was established and the real molecular weight of the subunits is somewhat different (see below).

Two major ligand-binding sites have been distinguished in AChR-rich membranes: the acetylcholine (ACh) site(s), which binds agonists and antagonists, and the site for local anesthetics. The local anesthetic drugs inhibit the ion-channel opening without inhibiting the binding to the ACh site. The coupling of the binding of ACh to the opening of the channel occurs through conformational changes. The binding of ACh to one site affects the affinity of the binding to the other. Three membrane-bound receptor states have been defined: a resting state, an active state and desensitized state (Changeux, 1981).

The stoichiometry of the various chains in the 9 S form is $\alpha_2\beta\gamma\delta$ (Reynolds and Karlin, 1978; Raftery *et al.*, 1980). The α subunit was identified to be associated with the ACh-binding site, as was demonstrated by affinity labeling with cholinergic ligands (Weill *et al.*, 1974; Sobel and Changeux, 1977; Damle *et al.*, 1978; Moore and Raftery, 1979; Lyddiatt *et al.*, 1979). The functional significance of the other subunits was uncertain for some years. However, it was unequivocally demonstrated recently that all the subunits are intrinsic parts of the receptor molecule.

The peptide maps of the four subunits were found to be different (Froehner and Rafto, 1979; Nathanson and Hall, 1979; Lindstrom *et al.*, 1979), and earlier studies did not reveal immunological cross-reactivity between the chains (Lindstrom *et al.*, 1978; Claudio and Raftery, 1977). However, recent studies with polyclonal antibodies (Fuchs and Bartfeld, 1983; Mehraban *et al.*, 1982), and with monoclonal antibodies (Tzartos and Lindstrom, 1980; Souroujon *et al.*, 1982) revealed antigenic homologies among the receptor subunits.

Recently, the homology between the four receptor subunits was clearly demonstrated, when first the *N*-terminal sequence analysis of the four subunits was performed (Raftery *et al.*, 1980) and later the whole amino acid sequence was identified using the recombinant DNA technology (Ballivet *et al.*, 1982; Sumikawa *et al.*, 1982; Giraudat *et al.*, 1982; Noda *et al.*, 1982, 1983; Claudio *et al.*, 1983). The homology observed strongly suggests that the genes encoding the four subunits descended from a single common ancestor which underwent gene duplications. From the amino acid sequence the real molecular weight of the subunits was found to be 50 K, 54 K, 57 K and 57.5 K for the α, β, γ and δ subunits respectively, with a total of 437, 469, 489 and 501 amino acids in the respective chains (Noda *et al.*, 1983).

The amino acid sequence of the receptor subunits makes it possible to deduce how the receptor is folded and positioned in the membrane. It has been suggested that all the subunits are oriented in a pseudosymmetric fashion with respect to the membrane (Noda *et al.*, 1983). Each subunit has several strongly hydrophobic regions with secondary structures which may represent transmembrane segments, or may be involved in intersubunit interaction (Giraudat *et al.*, 1982; Claudio *et al.*, 1983; Noda *et al.*, 1983). The possible location of the ACh-binding site, of the ionic channel, of the glycosylated amino acids, the location of the disulfide bond responsible for the dimerization of the receptor, the main antigenic determinants and the whole transmembrane topology of the AChR molecule have been proposed on the basis of these studies (Noda *et al.*, 1983).

The amount of AChR in mammalian skeletal muscle is about three orders of magnitude lower than in the *Torpedo* electric organ (Potter, 1973; Nathanson and Hall, 1980). In normal, innervated muscle the receptor molecules are localized at the nerve–muscle junction, leaving the rest of the muscle membrane practically free of AChR. Denervation of the muscle leads to a diffuse distribution of the receptors over the extrajunctional surface, accompanied by a 20-fold increase in total receptor sites (Miledi, 1960). Embryonic muscle fibers also have a diffuse distribution of the AChR (Hartzell and Fambrough, 1972), and during development the number of extrajunctional AChRs decreases, and the levels and distribution of the adult receptor are found (Hartzell and Fambrough, 1972; Bevan and Steinbach, 1977).

Similar to the electric fish AChR, the muscle receptor is a transmembrane glycoprotein (Brockes and Hall, 1975), composed of several subunits which correspond to the *Torpedo* AChR subunits (Froehner *et al.*, 1977a; Shorr *et al.*, 1978; Nathanson and Hall, 1979; Fambrough, 1979; Kemp *et al.*, 1980; Einarson *et al.*, 1982). One or two of the subunits bear the ACh-binding site as was demonstrated by affinity labeling with cholinergic ligands (Froehner *et al.*, 1977b; Lyddiatt *et al.*, 1979).

8.3 POLYCLONAL ANTI-AChR ANTIBODIES

8.3.1 Immunization and characterization of anti-AChR antibodies

Immunization of rabbits (Patrick and Lindstrom, 1973; Sugiyama *et al.*, 1973; Heilbronn and Mattson, 1974; Aharonov *et al.*, 1975a), rats (Lennon *et al.*, 1975), guinea pigs (Lennon *et al.*, 1975; Tarrab-Hazdai *et al.*, 1975a), monkeys (Tarrab-Hazdai *et al.*, 1975b), mice (Fuchs *et al.*, 1976), goats (Gordon *et al.*, 1977) and frogs (Nastuk *et al.*, 1979) with AChR isolated from electric fish results in the formation of anti-AChR antibodies. Earlier works utilized AChR from *Electrophorus electricus* for immunization and, later, AChR from *Torpedo* has been extensively used for immunization and for immunochemical and immunopathological studies.

Immunochemical analysis of AChR was carried out mostly on rabbit and rat antisera produced against AChR purified from electric organs. Different methods were applied for characterizing anti-AChR antibodies. These included direct quantitative precipitin reactions (Aharonov *et al.*, 1975a; Bartfeld and Fuchs, 1977; Green *et al.*, 1975; Sugiyama *et al.*, 1973; Penn *et al.*, 1976), immunodiffusion and immunoelectrophoresis (Aharonov *et al.*, 1975a, 1977), microcomplement fixation (Aharonov *et al.*, 1975a), passive microhemagglutination (Fuchs *et al.*, 1976) and radioimmunoassays. For radioimmunoassay the antigen (AChR) was either radiolabeled directly, using e.g. ^{125}I-AChR (Aharonov *et al.*, 1977), [^{3}H]acetyl AChR (Patrick *et al.*, 1973), [^{3}H]4-(*N*-maleimido)benzyltrimethylammonium (MBTA)-AChR (Karlin *et al.*, 1976), or a complex of the antigen with radiolabeled toxin (Patrick *et al.*, 1973) was used. In all radioimmunoassay procedures the antigen–antibody complexes formed were precipitated with goat anti-immunoglobulin sera or with *S. aureus* protein A.

There is no cross-reactivity between AChR and the closely related membrane protein, acetylcholinesterase (Sugiyama *et al.*, 1973; Aharonov *et al.*, 1975a). On the other hand, AChR is highly conserved during evolution and there is an immunologic cross-reactivity between AChR of different species (Fuchs *et al.*, 1976; Tarrab-Hazdai *et al.*, 1977; Aharonov *et al.*, 1977; Prives *et al.*, 1979) (see Section 8.3.5).

Anti-AChR antibodies bind to ^{125}I-toxin-labeled AChR, indicating that antibodies directed against determinants other than the acetylcholine-binding site are present (Aharonov *et al.*, 1977). Moreover, the relationship between the antigenic and toxin-binding sites of AChR was studied by experiments measuring the inhibition by cholinergic ligands and α-toxins of the binding of ^{125}I-AChR to anti-(*Torpedo* AChR) antibodies. No inhibition of the binding was observed by any of the following materials: carbamoylcholine, *d*-tubocurarine, decamethonium, hexamethonium, atropine, *Naja naja siamensis* α-toxin or α-bungarotoxin. Thus, these experiments with polyclonal anti-

AChR antibodies suggested that no antibodies were produced against the toxin-binding or acetylcholine-binding site of the receptor (Aharonov *et al.*, 1977). Although anti-AChR antibodies seemed not to be directed against the toxin-binding site they could nevertheless block both toxin-binding (Aharonov *et al.*, 1977) and the physiologic activity of the receptor (Patrick *et al.*, 1973; Sugiyama *et al.*, 1973; Green *et al.*, 1975). This blockage may be caused by antibodies to antigenic determinants located close to the physiologically active site, e.g. on an adjacent part of the same polypeptide chain, or brought into proximity either by folding or by juxtaposition of sites on different subunits. Alternatively the blockage may be of an allosteric nature by combination of antibodies at a site distant from the acetylcholine-binding site. In addition, antibodies may block the physiologic activity of the receptor by interfering with the regulation and/or function of the ionophore moiety of the receptor molecule. It should be pointed out here that later studies with monoclonal antibodies (see Section 8.4.2) have clearly demonstrated that the cholinergic binding site of the receptor can be immunogenic and thus some of the blocking effects observed with polyclonal antibodies might be due to such an antibody specificity.

The immunochemical properties of the detergent-purified AChR molecule appear to be indistinguishable from those of the receptor in the crude extract (Aharonov *et al.*, 1977), in a water-soluble preparation of AChR not exposed to detergents (Aharonov *et al.*, 1975a), or in AChR-rich membrane fraction. Immunization of rabbits with AChR-rich membrane fraction (Cohen *et al.*, 1972), in which probably only part of the AChR molecule is immunogenically exposed, gives rise to antibodies of similar specificity to those obtained by immunization with detergent-purified AChR and also leads to the development of experimental autoimmune myasthenia gravis (Tarrab-Hazdai and Fuchs, unpublished data; Berti *et al.*, 1976).

8.3.2 Antibodies against AChR derivatives

The AChR is a high-molecular-weight multi-subunit molecule. In order to analyze and classify its various antigenic determinants leading to various antibody specificities, antibodies were elicited against various derivatives of the receptor such as denatured receptor preparations, proteolyzed AChR, chemically modified AChR and the individual receptor subunits. Antibodies against denatured AChR were obtained in our laboratory following immunization with an irreversibly denatured AChR. Such an AChR derivative was prepared by complete reduction and carboxymethylation of *Torpedo* AChR in 6 M guanidine hydrochloride (Bartfeld and Fuchs, 1977). The reduced and carboxymethylated receptor (RCM-AChR) is devoid of the pharmacologic activity of the receptor and does not bind α-bungarotoxin (α-Bgt). RCM-AChR also did not induce any clinical signs of experimental myasthenia in rabbits. Nevertheless, anti-RCM-AChR sera cross-react with AChR (Bartfeld and Fuchs, 1978; Fuchs, 1979b). Comparison of the specificity of anti-AChR with that of anti-RCM-AChR has indicated some qualitative differences between the two antisera. By both immunodiffusion and radioimmunoassay measurements an identity was observed between the reaction of anti-RCM-AChR with RCM-AChR and AChR. However, anti-AChR serum showed only a partial cross-reactivity with RCM-AChR. It was also observed that whereas anti-AChR antibodies block the *in vitro* binding of ^{125}I-α-Bgt to AChR effectively, anti-RCM-AChR antibodies blocked this binding only to a very limited extent (Bartfeld and Fuchs, 1977; Fuchs, 1979a,b).

Analysis of both the intact and denatured receptor suggested that some antigenic determinants in the AChR molecule were abolished by the denaturation procedure. However, no additional determinants that were not expressed in the intact molecule became immunopotent after reduction and carboxymethylation. The altered antigenic specificity of antibodies to RCM-AChR along with their altered effect in blocking toxin binding to AChR led us to propose that the denaturation of AChR destroyed some antigenic determinant(s) (conformation-dependent in nature) that is (are) important for the induction of EAMG and may be located close to the toxin-binding site. The cross-reactivity between AChR and RCM-AChR and the non-pathogenicity of the latter was shown to be crucial in governing the immunosuppressive and therapeutic effects of RCM-AChR on EAMG (Bartfeld and Fuchs, 1978; Fuchs, 1979a,b).

A similar pattern of cross-reactivity between AChR and denatured AChR was also obtained by employing SDS-denatured AChR preparation (Karlin *et al.*, 1976; Claudio and Raftery, 1977). In agreement with our results, Claudio and Raftery (1980) have suggested that the antigenic determinants on AChR which are responsible for the induction of experimental myasthenia are lost upon denaturation and that there is a correlation between disease and the inhibition of α-Bgt binding.

Whereas denatured AChR represents the non-conformational antigenic determinants of AChR, trypsin-digested AChR (T-AChR) represents the conformation-dependent antigenic determinants which are resistant to proteolytic digestion (Bartfeld and Fuchs, 1979a). Tryptic digestion of AChR did not change the pharmacological specificity and the pathological myasthenic activity of the receptor molecule. Antibodies elicited against T-AChR are quite similar in their binding properties to anti-AChR antibodies and are different from anti-RCM-AChR antibodies. The binding of AChR to anti-RCM-AChR, which is inhibited by AChR and RCM-AChR to an equal extent, is inhibited to a very limited extent by T-AChR. T-AChR has been employed to detect or screen for anti-AChR antibodies directed against conformation-dependent antigenic determinants (Fuchs *et al.*, 1981a,b; Mochly-Rosen *et al.*, 1979).

Further immunochemical analysis of AChR was achieved by immunization with chemically modified AChR preparations. Chemical modification of free amino groups of the receptor was achieved by reacting the receptor with *N*-carboxy-DL-alanine anhydride to form polyalanyl AChR (PA-AChR) (Tarrab-Hazdai *et al.*, 1980). PA-AChR bound toxin to the same extent as the unmodified receptor but it did not induce myasthenia. Polyalanylation resulted in a change in the immunogenicity but not in the antigenic specificity of AChR. Anti-PA-AChR antibodies were directed mainly to the polyalanine determinants. On the other hand, the reactivity of PA-AChR with anti-AChR antibodies was as efficient as that of AChR (Tarrab-Hazdai *et al.*, 1980).

Modification of tyrosine residues of AChR was achieved by nitration with tetranitromethane (Fuchs *et al.*, 1981a, 1982; Mochly-Rosen and Fuchs, unpublished data). Nitrated AChR (nit-AChR) is similar in its pharmacological and immunological activities to denatured AChR (RCM-AChR); it does not bind α-Bgt and other cholinergic ligands and does not induce myasthenia. Antibodies elicited by immunization with nit-AChR bind also to AChR and RCM-AChR. However, they do not inhibit the binding of α-Bgt to solubilized AChR. The binding of ^{125}I-AChR to anti-nit-AChR antibodies is inhibited by nit-AChR, AChR and RCM-AChR to an equal extent, and is inhibited only partially by T-AChR. On the other hand, the binding of ^{125}I-AChR to anti-AChR antibodies is inhibited only partially by nit-AChR as is also the case with RCM-AChR (Fig. 8.1).

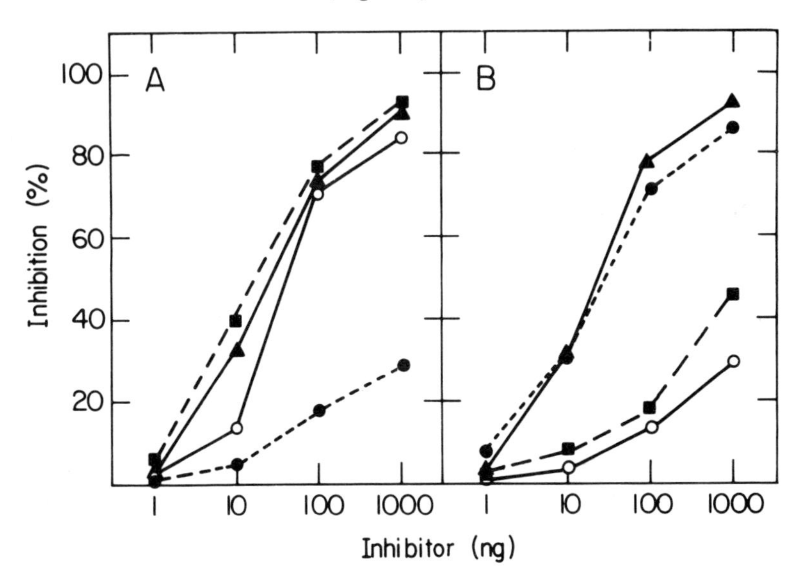

Fig. 8.1 Antigen specificity of nit-AChR. Inhibition of the binding of ^{125}I-AChR to anti-nit-AChR serum (A) and anti-AChR serum (B) by AChR (▲—▲), nit-AChR (O—O), RCM-AChR (■— —■) and T-AChR (●- - -●).

8.3.3 Antibody purification and antibody fractionation

Specific anti-AChR antibodies can be purified from anti-AChR sera by immunoadsorption to and elution from AChR–Sepharose columns (Martinez *et al.*, 1977; Fuchs *et al.*, 1981a,b). In some cases, it was preferred to employ for antibody purification an immunoadsorbent in which α-Bgt was first covalently attached to Sepharose and the resulting toxin–Sepharose was then complexed with AChR (Schwartz *et al.*, 1979).

As mentioned above, immunization with AChR results in a heterogeneous immune response representing varying levels of different antibody specificities. By specific immunoadsorptions, anti-AChR antibodies can be fractionated according to their antigenic specificities. For instance, anti-AChR antibodies were fractionated on an RCM-AChR–Sepharose column (Bartfeld and Fuchs, 1979b). In such a fractionation antibodies directed against conformational antigenic determinants in AChR and designated anti-native AChR (nAChR) do not absorb to RCM-AChR. On the other hand, the antibody fraction which adsorbs to RCM-AChR represents antibodies against non-structural determinants of AChR and are designated anti-dAChR antibodies (Bartfeld and Fuchs, 1979b; Fuchs *et al.*, 1981a,b). Binding experiments demonstrated that denatured receptor (RCM-AChR) or any of the individual subunits (see below) bind only to anti-dAChR antibodies. On the other hand, T-AChR binds to anti-nAChR, and binds weakly to anti-dAChR. By further fractionation of anti-dAChR on T-AChR it is possible to achieve further fractionation according to the proteolytic sensitivity of the antigenic determinants (Fuchs *et al.*, 1981a,b).

Anti-AChR antibodies were also fractionated using a covalently cross-linked receptor–toxin immunoadsorbent lacking free toxin-binding site (Barkas *et al.*, 1982). The majority of the antibodies which were adsorbed to such a column are not directed against the toxin-binding site; nevertheless, they could inhibit the binding of toxin to AChR in solution. The minor population of antibodies which did not bind to the affinity resin are those directed against the toxin-binding site, and, of course, inhibited toxin binding.

8.3.4 Antibodies to subunits

The *Torpedo* AChR is composed of four different subunits (α, β, γ and δ) with a stoichiometry of $\alpha_2\beta\gamma\delta$ in the 9 S monomer. The four subunits were suggested earlier to be distinct from each other from experiments with peptide maps (Froehner and Rafto, 1979; Nathanson and Hall, 1979; Lindstrom *et al.*, 1979). However, recent experiments of amino acid sequencing (Raftery *et al.*, 1980) and of DNA sequencing (Noda *et al.*, 1983) revealed a high degree of homology between the various subunits.

Both polyclonal and monoclonal antibodies were employed for studying the relationship and homology between the various receptor subunits. By using polyclonal antibodies against the isolated denatured subunits of *Torpedo* AChR Claudio and Raftery (1977) and Lindstrom *et al.* (1978) demonstrated that the subunits are immunologically distinct from one another. On the other hand, Mehraban *et al.* (1982) and Fuchs and Bartfeld (1983) observed antigenic similarities between the subunits of the receptor, as was demonstrated later also with monoclonal antibodies (see Section 8.4.3).

Rabbit (Claudio and Raftery, 1977) and rat (Lindstrom *et al.*, 1978) antibodies against the individual subunits of *Torpedo californica* AChR each only reacted with the immunizing subunit. Using these anti-subunit antibodies immunological cross-reactivity with AChRs from other *Torpedos* as well as from rat and human muscle AChR was observed, suggesting that the receptor subunits are conserved through evolution. Mehraban *et al.* (1982) raised antibodies in rabbits against isolated and *renatured* polypeptides of *Torpedo marmorata* AChR. These antibodies were shown to recognize determinants on the heterologous chains to appreciable extents, and also to cross-react with AChR from avian and mammalian muscles. It is possible that the renaturation step of the polypeptides before immunization was necessary for revealing the cross-reactivity between the various polypeptide chains, as many immunogenic determinants of the native receptor are lost upon denaturation for the isolation of the chains. Rabbit antibodies elicited against the 27 K subunit derived from AChR following tryptic digestion (Bartfeld and Fuchs, 1979a) bound the α subunit of the receptor as well as they bound the 27 K subunit. The β subunit bound weakly to these antibodies whereas the γ and δ subunits did not bind (Fuchs *et al.*, 1981a).

Our approach was to study the immunological relationship between the various subunits using antibodies elicited against *intact* AChR or with antibody subpopulations fractionated from these antibodies (Fuchs *et al.*, 1981a; Fuchs and Bartfeld, 1983). The individual subunits of AChR, like the whole denatured receptor, bind to both anti-AChR and to anti-RCM-AChR sera. However, in anti-AChR antibodies, the subunits bind only to that fraction of antibodies which is directed against non-structural antigenic determinants (anti-dAChR) and do not bind to anti-nAChR) (Fig. 8.2).

The antigenic specificity of the isolated subunits was analyzed by inhibition experiments. Each of the four radiolabeled subunits were reacted with anti-dAChR and the displacement of each particular binding by the homologous unlabeled subunits or by the other three unlabeled subunits was measured (Fig. 8.3). By this assay it is possible to select from a certain antibody population only antibodies which bind to one particular subunit, even if they represent a minor antibody fraction, and to analyze their specificity. Although in all cases the highest displacement was achieved by the homologous unlabeled subunit, indicating the presence of some distinct antigenic determinants specific for the individual subunits, a significant cross-reactivity between the various subunits

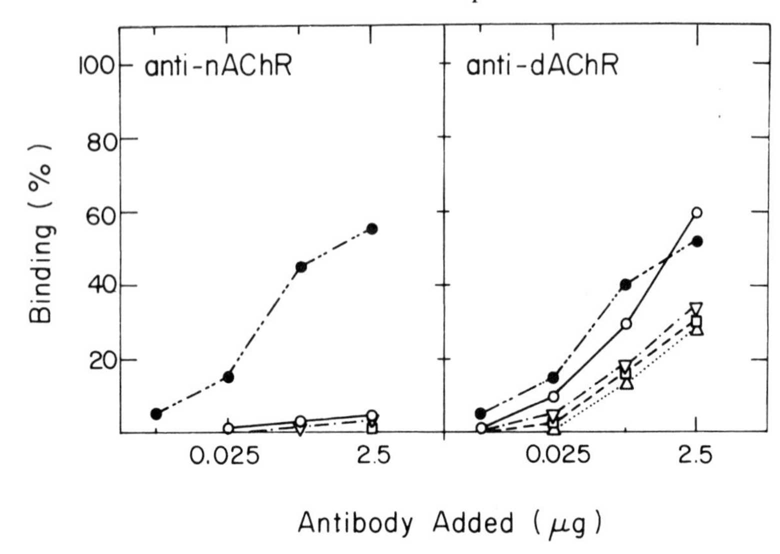

Fig. 8.2 Binding of ^{125}I-labeled AChR (●), α(O), β(▽), γ(□) and δ(△) subunit to anti-nAChR and anti-dAChR antibodies. (Fuchs and Bartfeld, 1983.)

was observed (Fig. 8.3). The cross-inhibition among the various subunits may be interpreted either by some shared immunogenic determinants between the various subunits expressed at the level of antibody production, or by antigenic cross-reactivity between common or similar determinants which can bind to a certain extent to antibodies raised by a particular subunit. The α-subunit is most likely the most immunogenic of all the receptor subunits when immunization with intact receptor is performed. From analysis of the reactivity of the isolated subunits with additional antibody fractions (Fuchs and Bartfeld, 1983), it was concluded that most of the antigenic determinants which are distinct for the various receptor chains are present in the exposed regions of the molecule which are sensitive to trypsin, and that the undigested portion of the molecule contains the determinants which are more cross-reactive either on the level of immunogenicity or antigenic specificity.

8.3.5 Antibodies to muscle AChR

There are several reports on the production of anti-muscle AChR antibodies by immunization with AChR preparations of skeletal muscle. Granato *et al.* (1976) immunized mice with AChR purified from rat denervated muscle. These antibodies showed 60% cross-reactivity with mouse receptor. Lindstrom *et al.* (1976a) immunized rats with syngeneic rat muscle AChR preparation and found that the anti-muscle AChR titers were not different from those obtained upon immunization with *Torpedo* AChR.

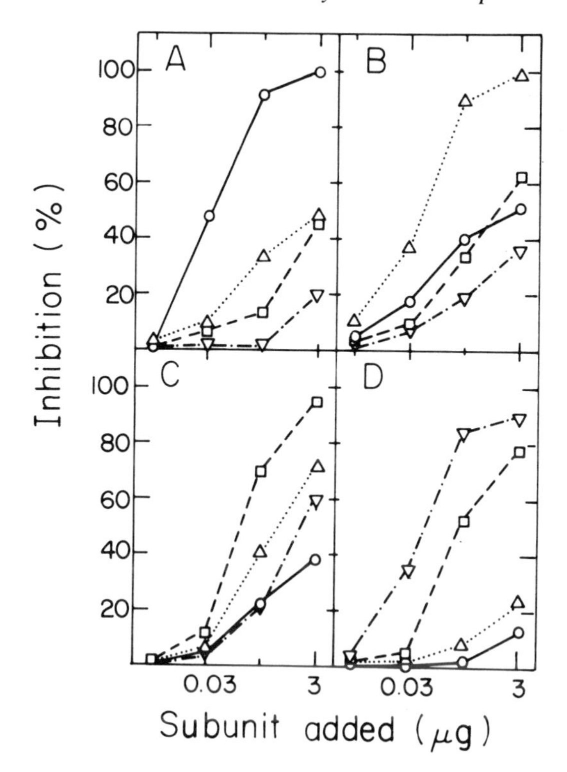

Fig. 8.3 Antigenic specificity of the receptor subunits. Inhibition of the binding of A, ^{125}I-α, B, ^{125}I-β, C, ^{125}I-γ and D, ^{125}I-δ to anti-d AChR antibodies by unlabeled subunits α(O), β(△), γ(□) and δ(▽). (Fuchs and Bartfeld, 1983.)

Another source for anti-muscle AChR antibodies is provided by sera of patients with myasthenia gravis, where spontaneous antibodies against human AChR are present. The anti-AChR antibodies in myasthenic patients are directed against various antigenic determinants of the AChR molecule and cross-react with AChR from different species (Almon *et al.*, 1974; Aharonov *et al.*, 1975b; Brenner *et al.*, 1978; Lindstrom *et al.*, 1976b). There seems to be no direct correlation between the antibody titers and the severity of the disease (Brenner *et al.*, 1978; Drachman, 1978; Lindstrom *et al.*, 1976b).

AChR is highly conserved during evolution and there is an immunologic cross-reactivity between the nicotinic AChRs of different species (Fuchs *et al.*, 1976; Lindstrom *et al.*, 1976b; Tarrab-Hazdai *et al.*, 1977; Aharonov *et al.*, 1977; Prives *et al.*, 1979). Owing to this cross-reactivity, antibodies elicited against electric organ AChR as well as sera of myasthenic patients have been widely applied as probes for structural and functional analysis of muscle AChR (see Section 8.5). The degree of cross-reactivity between AChR from various sources depends on both the phylogenetic distance between receptors tested as well as on the sensitivity of the assay used. Antibodies against eel receptor cross-react with *Torpedo* and with chicken receptor (Sugiyama *et al.*, 1973). Cross-reactivity between *Torpedo* and eel AChR was reported also by others (Penn *et al.*, 1976; Karlin *et al.*, 1976; Fulpius *et al.*, 1976). Immunologic cross-reactivity was also observed between fish and mammalian skeletal muscle AChR (Green *et al.*, 1975; Lindstrom *et al.*, 1976b; Fuchs *et al.*, 1976; Aharonov *et al.*, 1977). The cross-reactivity between fish and mammalian AChR is a crucial requirement for the induction of experimental myasthenia by immunization with fish AChR.

Antibodies raised against the isolated subunits of *Torpedo californica* AChR were also used to demonstrate cross-reactivity with AChR from various *Torpedo* fish (Claudio and Raftery, 1977) and with eel AChR, as well as with rat, calf and human skeletal muscle AChR (Lindstrom *et al.*, 1978).

Antibodies elicited against fish AChR were shown to cross-react with the nicotinic receptor also in tissues of non-muscle origin such as for instance, the nervous system (Patrick and Stallcup, 1977; Tarrab-Hazdai and Edery, 1980; Tarrab-Hazdai, 1981; Wonnacott *et al.*, 1982) and the thymus (Aharonov *et al.*, 1975c; Fuchs *et al.*, 1980a). Such cross-reactive antibodies were employed for studying the receptor from these non-muscle origins (see Section 8.5.8).

8.4 MONOCLONAL ANTI-AChR ANTIBODIES

8.4.1 Production of anti-AChR monoclonal antibodies

Since 1979 the technology of lymphocyte hybridization and monoclonal antibodies (Kohler and Milstein, 1975) has been extensively applied for the AChR system (Mochly-Rosen *et al.*, 1979; Gomez *et al.*, 1979; James *et al.*, 1980; Tzartos and Lindstrom, 1980). The advantages in employing monoclonal antibodies are obvious since they are monospecific and available in unlimited amounts. In the AChR system they can be applied as probes for various sites in the molecule for which no other specific markers are available. There are many reports in the literature describing the application of monoclonal anti-AChR antibodies as probes for studying AChR from various organs and species (see Section 8.5 below). Most monoclonal antibodies reported were derived from hybridomas for which electric organ AChR was employed as the original immunogen. Some anti-AChR antibodies were obtained from hybridization of AChR-sensitized mouse spleen cells with a mouse myeloma cell (Mochly-Rosen *et al.*, 1979; James *et al.*, 1980), whereas others produced xenogeneic rat–mouse hybridomas by fusing immune rat cells with a mouse myeloma line (Gomez *et al.*, 1979; Tzartos and Lindstrom, 1980; Lennon and Lambert, 1980). The advantage of monoclonal antibodies derived from mouse–mouse hybridizations is that they can be propagated and maintained easily in the form of ascitic fluids.

Monoclonal anti-AChR antibodies were obtained in our laboratory following a fusion of spleen cells of C57BL/6 mice immunized with *Torpedo californica* AChR and exhibiting myasthenic symptoms, with the P3-NSI/1-Ag4-1 (NS1) non-secreting plasmacytoma line (Mochly-Rosen *et al.*, 1979). We have characterized our monoclonal antibodies by measuring their binding in radioimmunoassay to various preparations and derivatives of *Torpedo* and mammalian AChR and have verified their restricted specificities. Some monoclonal antibodies (about 50%) were shown to be directed against conformation-dependent antigenic determinants and did not bind to a denatured AChR preparation or to the isolated subunits. On the other hand, they bound to trypsinated AChR (T-AChR), where the conformational regions were retained following proteolysis. Another group of antibodies were shown to be directed against antigenic determinants which are non-conformation-dependent and bound to denatured receptor (RCM-AChR). Only antibodies with this specificity could be employed for studies of subunit specificity (see Section 8.4.3). It is of interest to point out that all the anti-AChR antibodies which were derived from cells obtained from mice immunized with detergent-solubilized AChR bound also to membrane-bound AChR. This indicates that the antibodies are directed against antigenic determinants which are exposed on the membrane (Mochly-Rosen *et al.*, 1979).

James *et al.* (1980) have produced mouse–mouse hybridoma anti-AChR monoclonal antibodies by fusing spleen cells from BALB/c mice immunized with *Torpedo marmorata* AChR with the non-secreting B-lymphocytoma cell line SP2-OAg14. Rat–mouse (-hetero) hybridoma anti-AChR monoclonal antibodies were produced by hybridization of spleen cells from rats immunized with *Torpedo* AChR, with a mouse myeloma cell line (Gomez *et al.*, 1979; Tzartos and Lindstrom, 1980; Lennon and Lambert, 1980).

8.4.2 Anti-site monoclonal antibodies

As mentioned above (Section 8.3) studies with polyclonal antibodies have demonstrated that the binding of α-Bgt or other cholinergic ligands did not interfere with the binding of antibodies to the AChR, and it was suggested that the binding site of AChR is not immunogenic. However, by using monoclonal antibodies it was demonstrated that antibodies against the receptor-binding site can be elicited. The ability to select for and augment a particular antibody specificity, even one which is extremely minor, is one of the advantages offered by the lymphocyte hybridization method.

We have studied in detail in our laboratory one monoclonal antibody designated mcAb 5.5 which was shown to be directed against the cholinergic-binding site of AChR (Mochly-Rosen and Fuchs, 1981; Mochly-Rosen *et al.*,

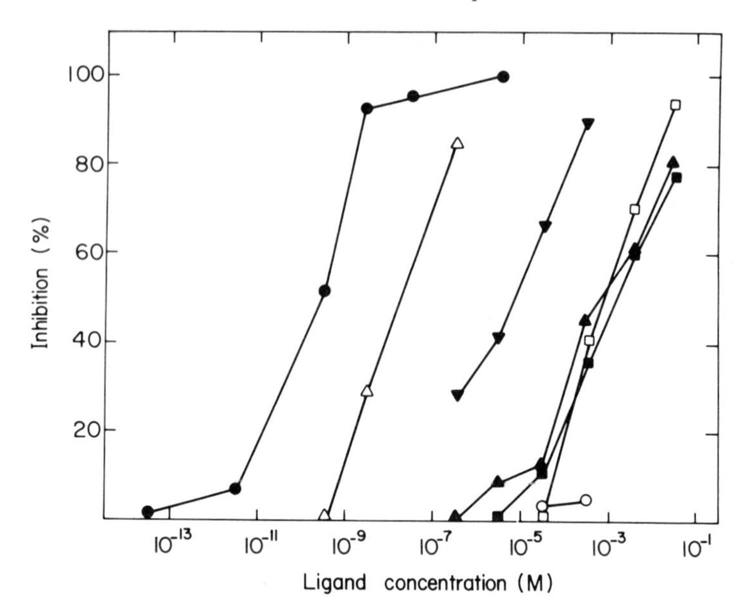

Fig. 8.4 Inhibition of the binding of ^{125}I-AChR to mcAb 5.5 by α-Bgt (●), Naja naja siamesis α-neurotoxin (△), *d*-tubocuranine (▼), decamethonium (▲), carbamoylcholine (■), acetylcholine (□) and atropine (O). (Mochly-Rosen and Fuchs, 1981.)

1981). Of our monoclonal antibodies, mcAb 5.5 is the only one which did not bind to AChR when complexed with α-Bgt. This antibody is directed against a conformation-dependent determinant of AChR and does not bind to denatured AChR. We have demonstrated that not only α-Bgt but also other cholinergic agonists and antagonists inhibit the binding of AChR to mcAb 5.5 (Fig. 8.4). The concentrations of various ligands necessary to give 50% inhibition of the binding of AChR to mcAb 5.5 compare quite well with the concentrations of the same ligands needed to produce 50% inhibition of the binding of α-Bgt to AChR. These data strongly indicate that mcAb 5.5 is directed against the cholinergic-binding site on the AChR, or at least against a site very closely associated with it.

McAb 5.5 was shown to react with AChR from various sources tested such as muscle of chick and mammals (Mochly-Rosen and Fuchs, 1981; Mochly-Rosen *et al.*, 1981; Souroujon *et al.*, 1983a), brain homogenates of rat, mouse, bovine, *Drosophila* and goldfish (Mochly-Rosen and Fuchs, 1981, 1983, and unpublished data), as well as neuroblastoma N1E-115 cells, pheochromocytoma PC12 cells and chromaffin cells of the bovine adrenal medulla (Mochly-Rosen and Fuchs, unpublished data). This wide cross-reactivity indicates that the cholinergic-binding site of AChR is highly conserved and makes this antibody a powerful probe for studying functional and structural properties of AChR in various sources (see Section 8.5).

Monoclonal antibodies which do not bind to AChR in the presence of α-Bgt were studied also by others (Gomez *et al.*, 1979; James *et al.*, 1980; Watters and Maelicke, 1983). It seems however that these antibodies are not identical in their specificity. The binding of the monoclonal antibody described by Gomez *et al.* was not inhibited by ligands other than α-Bgt. One of the monoclonal antibodies described by James *et al.* was inhibited also by *d*-tubocurarine and carbamoylcholine. Watters and Maelicke have studied in detail a series of anti-(*Torpedo marmorata* AChR) monoclonal antibodies which were directed against the ligand-binding regions. Only some of these antibodies cross-react with AChR from other species. From competition binding studies with cholinergic ligands they divide the antibodies directed against the ligand-binding regions into three groups. They suggest a model for the organization of the ligand-binding site of the receptor, according to which there are two ligand-binding regions differing in their antigenic properties (Watters and Maelicke, 1983).

8.4.3 Monoclonal antibodies to the receptor subunits

Monoclonal antibodies which react with the different AChR subunits were elicited by immunization with native AChR from *Torpedo* and eel or with the isolated subunits (Tzartos and Lindstrom, 1980; Souroujon *et al.*, 1982). Several methods were used to study the subunit specificity of these various antibodies. The first involves the use of ^{125}I-labeled SDS-denatured subunits of the receptor (Tzartos and Lindstrom, 1980; Souroujon *et al.*, 1982) and their corresponding proteolytic peptides as antigens (Gullick *et al.*, 1981). However, SDS denaturation, which is the only known method for separating the AChR subunits, alters or destroys most conformationally dependent antigenic determinants on AChR (Bartfeld and Fuchs, 1978; Fuchs, 1979a). As a result the subunit specificity of only those antibodies which react with denatured receptor can be tested by this method. The immunoblot technique ('Western' blotting) was also applied for analyzing the subunit specificity of monoclonal antibodies. Using this procedure some antibodies which were shown to be directed against conformation-dependent sites reacted also with the subunits on the gel. Another method that can be employed for mapping the subunit specificity of monoclonal antibodies directed against conformation-dependent determinants involves the competition among antibodies for antigenic determinants on native AChR.

It was demonstrated that the α subunit of the AChR is the most immunogenic and that the majority of the monoclonal antibodies raised against native AChR were directed against this subunit (Tzartos and Lindstrom, 1980; Souroujon *et al.*, 1982). In competition experiments Tzartos and Lindstrom

(1980) have shown that an anti-α AChR subunit monoclonal antibody can prevent the binding of about 50% of the antibodies in an antiserum to native AChR. Monoclonal antibodies directed to the main immunogenic region of the α subunit cross-react rather well with AChR of many other species, indicating that this determinant is highly conserved.

Immunological cross-reactivity between various AChR subunits was demonstrated by monoclonal antibodies. We have several monoclonal antibodies that bind both α and β subunits. However, in most cases the antibody titer towards the β subunit is tenfold lower than towards the α subunit. Antigenic cross-reactivity between the α and δ subunit was demonstrated also by one antibody (1.39) whereas antibody 1.17 bound exclusively to the δ subunit (Souroujon *et al.*, 1982; Mochly-Rosen, 1983). The γ subunit elicited a very weak response as demonstrated both by polyclonal serum and by monoclonal antibodies. One monoclonal antibody (5.30) exhibits a low binding to the γ subunit, to an antigenic site which appears to be cross-reactive also with the α and β subunits. It is interesting to point out that the titers of some of our monoclonal antibodies observed with the isolated subunits were higher than with the intact native form of the AChR. That may indicate binding to a site which is not exposed unless the subunits are separated.

Tzartos and Lindstrom (1980) have also reported on an antigenic cross-reactivity between the α and β subunits and between the γ and δ subunits. Fine-scale mapping of the antigenic determinants of the receptor subunits was achieved by analyzing the binding of monoclonal antibodies to proteolytically digested subunits (Gullick *et al.*, 1981). The immunological cross-reactivity between the various AChR subunits is compatible with the structural homology between the subunits as was discussed earlier in this review (Sections 8.2 and 8.3.4).

8.4.4 Monoclonal antibodies to muscle AChR

As with polyclonal anti-AChR antibodies, monoclonal anti-AChR antibodies elicited against electric organ AChR also exhibit a certain degree of cross-reactivity with AChR of various species and organs. Several groups have established that given antigenic determinants are common to AChR from different sources and monoclonal antibodies with a known subunit specificity bind to the corresponding subunit of different AChRs (Lennon and Lambert, 1980; Tzartos and Lindstrom, 1980; Mochly-Rosen and Fuchs, 1981; Souroujon *et al.*, 1982, 1983,a,b; Gullick and Lindstrom, 1982, 1983). Many of the monoclonal antibodies raised against native *Torpedo* and eel AChR are species specific. The least species-specific region is located on the α subunit.

Anti-AChR monoclonal antibodies were demonstrated to cross-react with muscle AChR from rat, fetal calf, mouse, human, chicken, guinea pig, dog and cat (Lennon and Lambert, 1980; Tzartos and Lindstrom, 1980; Mochly-Rosen and Fuchs, 1981; Souroujon *et al.*, 1983a). The highest degree of cross reactivity was found between *Torpedo* and *Electrophorus* AChR. It was also demonstrated that AChR purified from the muscle of *Electrophorus electricus* contained α, β, γ and δ subunits which correspond immunochemically to those of AChR from the electric organ of *Electrophorus* and *Torpedo*. Combined peptide mapping and immunochemical assays suggest that these two receptors differ only in small post-translational modifications (Gullick and Lindstrom, 1983).

Some anti-AChR monoclonal antibodies which cross-react with muscle AChR were also demonstrated to cross-react with the putative nicotinic AChR in rat brain (Mochly-Rosen and Fuchs, 1983) and chick brain (Swanson *et al.*, 1983) and were employed for studying this receptor in the nervous system (see Section 8.5.8).

There is only one described report on the elicitation of monoclonal antibodies by immunization with muscle AChR (Vernet der Garabedian and Morel, 1983). These authors have used for immunization partially purified AChR from human muscle. They described two anti-human AChR antibodies which cross-react with mouse but not with *Torpedo* or porcine AChR.

8.5 APPLICATIONS OF ANTI-AChR ANTIBODIES

Polyclonal and monoclonal anti-AChR antibodies directed against different antigenic determinants in the AChR molecule provide a powerful tool in the study of various structural and functional aspects of this receptor. In this section we have listed several directions in the research of AChR where antibodies were employed.

8.5.1 Purification of AChR

The conventional method for the purification of AChR involves affinity chromatography on α-neurotoxins or cholinergic ligands, linked to Sepharose or agarose. Recently the use of monoclonal anti-AChR antibodies for affinity purification of AChR was introduced (see also Chapter 13 by Schneider). Lennon *et al.* (1980) have conjugated to agarose a monoclonal antibody which binds to an antigenic determinant of AChR remote from the cholinergic-binding site. This affinity column was used to purify AChR from the electric organ of *Torpedo californica*. Elution at alkaline pH (10) resulted in a high yield of AChR which bound α-Bgt with a specific activity which did not differ significantly from that of AChR purified on α-neurotoxin agarose.

AChR from human muscle was also purified on an immunoadsorbent of anti-AChR monoclonal antibody conjugated to agarose (Momoi and Lennon, 1982). AChR was eluted with 1.2 M NaCl and further purified by chromatography on DEAE-Sephadex A-50 with NaCl gradient. Final recovery of AChR from the immunoadsorbent was 20–30% (higher than from α-toxin gel) and its specific binding of α-Bgt was 5–6 pmol/μg of protein. In polyacrylamide gel electrophoresis in SDS the purified AChR from human muscle revealed five predominant polypeptides with apparent molecular weights of 44 000, 53 000, 56 000, 61 000 and 66 000. When injected into rats it caused severe EAMG.

The use of an immunoadsorbent rather than α-toxin for purification of muscle AChR may have an additional advantage; monoclonal antibodies of appropriate specificity offer the potential for separating antigenically distinct subpopulations of muscle AChR, e.g. those of innervated and denervated muscle, or those of diseased muscle in pathological conditions.

8.5.2 Localization and orientation of AChR

AChR has been used as a model to answer some questions concerning integral membrane glycoproteins. It was of interest to find out whether the AChR spans the membrane, what parts of the receptor face the exterior and what parts face the cytoplasm, and whether all the subunits of the receptor penetrate the lipid bilayer. The advantage of antibodies over cholinergic ligands for elucidating such questions is that antibodies can serve as markers for various determinants of the receptor, including determinants which are remote from and not associated with the cholinergic site.

AChR was localized in a receptor-rich membrane preparation from the electric organ of *Torpedo californica* by applying an immunoferritin technique (Tarrab-Hazdai *et al.*, 1978). The membrane preparation was incubated with $F(ab')_2$ fragments derived from specific rabbit antibodies against the purified AChR and subsequently with ferritin-conjugated goat antiserum to rabbit immunoglobulin. It was demonstrated that more than 50% of the membrane vesicles were labeled with ferritin on both sides of the membrane. In closed vesicles the labeling was confined to the outer surface owing to the inability of the reagents to penetrate the membranes. These data indicated that antigenic sites of the receptor molecule are exposed on both sides of the excitable membrane, and that AChR is a transmembrane protein.

Localization of the AChR on *Torpedo* electroplax membrane vesicles using anti-AChR antibodies was also reported by Karlin *et al.* (1978). Since these authors have used a preparation of closed membrane vesicles they could not reach a conclusion concerning the presence of antigenic sites of AChR at the intracellular surface of the membrane.

Klymkowsky and Stroud (1979) used rabbit and goat anti-AChR antibodies coupled directly or indirectly to ferritin or colloidal gold spheres for labeling membrane-bound AChR of *Torpedo californica.* They have shown that the receptor corresponds to the 8.5 nm (85 Å) diameter rosette seen in membranes derived from the electroplax and have estimated the molecular weight of

AChR to be in the range 250 000 to 310 000. Electron micrographs of immunospecifically labeled receptor confirmed and extended their earlier conclusions based on X-ray diffraction analysis (Ross *et al.*, 1977) that the AChR molecule extends above the extracellular surface by 5.5 nm (55 Å) and 1.5 nm (15 Å) on the cytoplasmic side.

The use of monoclonal rather than polyclonal anti-AChR antibodies enables the identification of antibodies which react specifically with different subunits of AChR. Anderson *et al.* (1983) have applied monoclonal antibodies for a transmembrane orientation of an early biosynthetic form of the δ subunit of AChR. They have detected antibodies specific to the cytoplasmic domains of both the δ and β subunits and an antibody specific to the extracellular fragment of the β subunit.

8.5.3 Antigenic determinants of AChR in developing muscle

Two molecular forms of AChR are found in mammalian skeletal muscle. Junctional receptor (JR) is densely packed at the adult neuromuscular junctions, whereas extrajunctional receptor (EJR) is found in lower density over the entire surface of embryonic or denervated adult muscle fibers (Fambrough, 1979). In several studies sera of myasthenic patients were employed to follow antigenic differences between junctional and extrajunctional AChR. Antigenic differences between AChR extracted from normal and denervated rat muscle were indicated using sera from myasthenic patients (Almon and Appel, 1976). These sera reacted better with AChR from embryonic muscle and adult denervated rat muscle, than with AChR from normal muscle. By using highly purified preparations of rat JR and EJR, Weinberg and Hall (1979) have suggested that EJR has determinants that are not present on JR: these determinants are detected with myasthenic sera but not with several antisera raised in animals to AChR from various sources. Reiness and Hall (1981) have used a myasthenic serum to characterize the receptor in extracts of developing rat muscles and followed the antigenic changes which occur during the first 3 weeks after birth.

Dwyer *et al.* (1981a) have employed an assay in which α-Bgt binding to muscle AChR is inhibited by myasthenic sera. They reported that the inhibition assay reveals marked differences between the two types of receptor and that the binding of α-Bgt to EJR is more inhibited by the myasthenic sera. Anti-AChR antibody titers in myasthenic patients were found to be higher when tested with human denervated as compared to normal muscle AChR (Vincent and Newsom-Davis, 1982; Lefvert, 1982). Some myasthenic sera distinguished between receptors extracted from human leg muscle and human extraocular muscles (Vincent and Newsom-Davis, 1982).

Recently, Lotwick *et al.* (1983) have demonstrated that myasthenic serum bound to the same extent to AChR partially purified from fetal and adult human muscle. It thus appears that fetal human AChR does not bear antigenic determinants which are not present on normal adult junctional AChR.

We have applied *monoclonal* anti-*Torpedo* AChR antibodies rather than polyclonal antibodies from sera of myasthenic patients, in order to define whether the observed antigenic changes in AChR result from the existence of entirely different antigenic determinants in various stages of muscle development, or from different accessibility of similar determinants (Souroujon *et al.*, 1983b). Most of our monoclonal antibodies reacted preferentially though not exclusively with EJR from either newborn or denervated muscle than with normal adult muscle. These results with monospecific antibodies led us to conclude that there are antigenic determinants shared by JR and EJR. Such determinants may be differently exposed in these two receptor forms due to a different state of aggregation in the membrane.

It has been proposed that some biochemical differences between JR and EJR may result from post-translational modifications such as phosphorylation or glycosylation (see Fambrough 1979; Changeux, 1981). It has been demonstrated that JR in *Torpedo* (Saitoh and Changeux, 1981) and in muscle (Brockes and Hall, 1975) is more phosphorylated than EJR. We have reported recently on an anti-AChR monoclonal antibody which reacts specifically with a non-related highly phosphorylated protein, phosvitin (Pizzighella *et al.*, 1983). The binding of this antibody to phosvitin is inhibited by AChR. This particular antibody was also found to react preferentially with JR from rat and mouse muscle. It is possible that a phosphorylated amino acid residue may be part of the determinant on AChR which is recognized by this antibody and is unique to JR.

Dwyer *et al.* (1981a,b) have shown that EJR treated with glycosidases resembles immunologically untreated JR. These results suggest that carbohydrate residues may contribute to the differences between the two types of receptors.

8.5.4 Effect of anti-AChR antibodies on receptor metabolism

AChR was one of the first integral membrane proteins for which it was demonstrated that its aggregation and turnover can be modulated by specific reagents acting at the cell surface. The average life time of skeletal muscle AChR in cell culture and organ culture is shortened markedly by exposure of the muscle cells to sera from myasthenic patients (Appel *et al.*, 1977; Drachman *et al.*, 1978) or to anti-AChR antibodies. With polyclonal antibodies this phenomenon has been demonstrated in chick primary muscle cultures with rabbit anti-*Torpedo* AChR antibodies and the $F(ab')_2$ fragments derived from them (Prives *et al.*, 1979) and with mouse anti-*Torpedo* AChR antibodies (Souroujon *et al.*, 1983a) and in the BC3H cell line with rat anti-(*Torpedo* AChR) antibodies (Heineman *et al.*, 1977). In organ culture of rat diaphragm the degradation rate of both junctional and extrajunctional AChR was demonstrated to be accelerated by rat antiserum to *Torpedo* AChR, rat AChR and fetal calf AChR (Heinemann *et al.*, 1978) or by sera from myasthenic patients (Reiness *et al.*, 1978).

By using monoclonal antibodies it is possible to find out which antibody (ies) specificity is responsible for the accelerated degradation of AChR and it was observed that only particular monoclonal antibodies could mediate this effect (Conti-Tronconi *et al.*, 1981; Souroujon *et al.*, 1982, 1983a). We have tested the effect of our anti-AChR monoclonal antibodies on the degradation rate of AChR in primary chick muscle cell cultures. Of the antibodies tested, only two, mcAb 5.5 and mcAb 5.34, accelerated AChR degradation significantly. Binding of an antibody to the cultured muscle cells is not sufficient to affect the degradation rate and we had monoclonal antibodies (e.g. mcAb 5.14) which bind well to the cell culture but had no effect on the receptor degradation rate (Souroujon *et al.*, 1983a). The fact that monoclonal antibodies accelerate the degradation of AChR, as was also observed by Conti-Tronconi *et al.* (1981) in fetal calf muscle cell cultures, is interesting in itself, because of their mono-specificity. If one assumes an antigenic determinant which exists only once on each AChR molecule, the maximum number of AChR molecules that could be cross-linked is two. Another possibility is to assume that the antigenic determinant recognized by the monoclonal antibody appears more than once in every AChR molecule. This may be the case with antibodies directed against antigenic determinants on the two related chains bearing the ACh-binding site in muscle AChR and also with antibodies reacting with more than one subunit owing to sequence homology. The occurrence of an antigenic determinant more than once on an AChR molecule is in itself not sufficient to cause acceleration of its degradation, since there are anti-α subunit monoclonal antibodies, such as mcAb 5.14, which do not accelerate receptor degradation. Conti-Tronconi *et al.* (1981) have also reported on anti-α subunit monoclonal antibodies that do not cause accelerated degradation and suggested that such antibodies cross-link the α subunit within the receptor monomer.

8.5.5 Effect of anti-AChR antibodies on acetylcholine sensitivity and single channel properties

Electrophysiological studies of muscles obtained from myasthenic patients have shown a reduction in the amplitude of the miniature end-plate potential (mepp) and a reduced sensitivity to acetylcholine (ACh) (Elmquist *et al.*, 1964; Albuquerque *et al.*, 1976; Ito *et al.*, 1978; Cull-Candy *et al.*, 1978). The same was shown also in muscles derived from animals with EAMG (Green *et al.*, 1975; Bevan *et al.*, 1976; Lambert *et al.*, 1976; Lennon and Lambert, 1980; Alema *et al.*, 1981; Hohfeld *et al.*, 1981). Reduced mepps could result either from a reduction in the size of the ACh quanta released from the nerve terminals or from a reduced *post*-junctional response to ACh. The latter possibility was examined by measuring the effect of anti-AChR antibodies on muscle cells *in vitro*. It was found that anti-AChR antibodies reduced the ACh sensitivity of muscle cells in culture (Heinemann *et al.*, 1977; Bevan *et al.*, 1976, 1977) but had only a slight effect on the mepps when added for a prolonged time (4 h) to intercostal muscles of rabbits (Eldefrawi *et al.*, 1979).

Single-channel properties were unchanged in muscles obtained from myasthenic patients (Cull-Candy *et al.*, 1979) or immunized rats (Alema *et al.*, 1981). Peper *et al.* (1981) report that sera from myasthenic patients which were shown to bind to rat end-plates failed to exert any effect on the response of voltage-clamped end-plates to iontophoretically applied ACh. However, a small reduction in the mean single-channel conductance (γ) and open time (τ) in cultured rat muscle cells treated with rat anti-AChR antibodies was reported by Heinemann *et al.* (1977). There are several difficulties to the interpretation of the electrophysiological studies using polyclonal anti-AChR serum. One problem in adding antibodies to muscle cells and measuring the response to ACh is that the spectrum of antibodies from every human or immunized animal may be different and may account for various effects. This disadvantage can be overcome by using monoclonal antibodies each bearing a single specificity.

We have studied the effect of several anti-AChR monoclonal antibodies applied in acute experiments to cultured chick 'myoballs' during standard electrophysiological measurements *in vitro* (Goldberg *et al.*, 1983). Of the antibodies tested only mcAb 5.5, the one directed against the cholinergic-binding site of AChR, had an effect in this system. We found a time- and dose-dependent reduction in ACh sensitivity of the muscle cells. Moreover, noise analysis experiments indicated a decrease in the mean single-channel conductance and increase in the mean single-channel open time. We have suggested an interpretation for the observed modification of the channel properties which is based on the fact that the antibody behaves as a high-molecular-weight antagonist in the system (Goldberg *et al.*, 1983).

8.5.6 Effect of antibodies on the receptor ionophore

The ability of antibodies to affect ligand-induced sodium flux through the receptor channel is another measure of the effect of antibodies on receptor function. Most antibodies to AChR bind without affecting the ability of the receptor to bind ACh or regulate the opening of the cation channel. This has been demonstrated by the fact that channel open time and conductance are not detectably altered in muscles from myasthenic patients (Cull-Candy *et al.*, 1979).

We have studied the effect of several monoclonal antibodies which bind to chick muscle cultures on sodium transport in these cultures, by measuring the inhibitory effect of antibodies on carbamoylcholine-activated sodium uptake (Souroujon *et al.*, 1982, 1983a). Only mcAb 5.5 which binds to the receptor cholinergic site acted as an antagonist; it inhibited the initial rate of sodium uptake in a concentration-dependent manner. This antibody had no effect on the desensitization of AChR by carbamoylcholine. Polyclonal mouse anti-AChR serum had no effect on sodium uptake.

It would be interesting to have a monoclonal antibody which is not site-directed and can affect sodium uptake. Such an antibody could be directed against the lumen of the channel itself or against sites in the receptor molecule involved in mediating the ACh-induced conformation change. Lindstrom *et al.* (1981) reports on at least one monoclonal antibody which blocks sodium influx into vesicles containing reconstituted AChR whether added before or after reconstitution. This antibody does not inhibit carbamoylcholine binding. It is not known whether this antibody acts by interfering with some conformational change required for the opening of the channel or by actually occluding the lumen of the channel.

Polyclonal antibodies elicited in sheep and rabbits against *Torpedo* AChR (Desouki *et al.*, 1981) were shown to inhibit AChR-induced sodium flux in AChR-rich microsacs. These antibodies also inhibited the potentiation by carbamoylcholine of perhydrohistrionicotoxin binding to the ionic channel site in the membranes. It is suggested that these sera contain antibodies which inhibit AChR function by interfering with the agonist-induced conformational changes in the receptor channel complex. The assay of blocking agonist-induced sodium flux was also used to discriminate between the α-Bgt-binding site and the receptor ionophore in a rat sympathetic neuron cell line (PC12). Antibodies against eel AChR blocked the sodium flux into PC12 cells but failed to recognize the α-Bgt-binding component (Patrick and Stallcup, 1977).

8.5.7 Biosynthesis of AChR

In studies of the biosynthesis of AChR, assays involving indentification of new receptors by physicochemical properties and ligand binding have been employed. For such assays completed or nearly completed receptor molecules are required and unassembled subunits or incomplete polypeptide chains would not be detected. Anti-AChR antibodies and especially monoclonal antibodies provide powerful probes for precursor molecules of AChR. By using an anti-AChR serum prepared in rabbits against denatured receptor, Merlie and Sebbane (1981) were able to identify in the BC_3H cell line a precursor of AChR which does not bind α-Bgt. This result suggests that a post-translational step is involved in the assembly of functional AChR. Patrick *et al.* (1977) studied intracellular hidden pools of AChR and surface AChR molecules in BC_3H cells and found that they were immunologically indistinguishable in immunotitrations with rabbit anti-eel AChR serum.

Anti-AChR antibodies have also served for the identification and characterization of early biosynthetic forms of AChR produced in cell-free protein-synthesizing systems. Specific antibodies were shown to immunoprecipitate all four nascent chains of the newly synthesized AChR, translated by *Torpedo* mRNA *in vitro* (Mendez *et al.*, 1980; Anderson and Blobel, 1981; Soreq *et al.*, 1982) and *in ovo*, in microinjected *Xenopus* oocytes (Sumikawa *et al.*, 1981). Anderson and Blobel (1981) have studied the early events in AChR biosynthesis by using subunit-specific antisera to characterize the four *Torpedo* receptor polypeptide chains in a cell-free system. They report that each subunit is synthesized as an individual polypeptide, and all four chains are independently integrated into dog pancreas microsomes as a transmembrane protein. However, the *in vitro* synthesized material exhibits neither oligomeric assembly nor α-Bgt binding, and could thus not be identified by pharmacological markers of AChR. We used antibodies against the native and denatured AChR for immunoprecipitation of *in vitro* translated AChR. From the preferential immunoprecipitation with antibodies to denatured AChR we have concluded that the newly synthesized AChR subunits probably do not possess the final spatial conformation of the adult receptor, but rather appear to be closer in structure to the denatured receptor (Soreq *et al.*, 1982).

We are now applying monoclonal antibodies which cross-react with AChR in the central nervous system for immunoprecipitation of *in vitro* and *in ovo* translated rat brain AChR (Mochly-Rosen *et al.*, 1983). By using various monoclonal antibodies and α-Bgt it was indicated that translation *in vitro* in cell-free systems results in AChR polypeptides which are still devoid of the receptor native conformation. Such polypeptides could be immunoprecipitated only by an antibody specific for non-conformational determinants of AChR but not by α-Bgt or a monoclonal antibody specific for the receptor cholinergic site (mcAb 5.5). On the other hand, in injected oocytes, AChR was not only translated but also the correct post-translational modifications and maturation on the newly synthesized polypeptides took place, resulting in a native membrane-bound molecule, which probably possesses the receptor-binding site (Mochly-Rosen *et al.*, 1983, and unpublished data).

Anti-AChR antibodies have been recently used in genetic engineering systems for the immunoprecipitation of *in vitro* translation products of mRNA which is then used for cDNA cloning (see also Chapter 13 by Schneider). The resulting DNA is adsorbed to nitrocellulose filters and used to select mRNA from electric organ poly(A)$^+$ RNA by hybridization. The hybridization-selected mRNA species are then translated again and the *in vitro* translation products are specifically precipitated with anti-AChR sera and further analyzed (Ballivet *et al.*, 1982; Giraudat *et al.*, 1982).

8.5.8 Studies on AChR from non-muscle origins

The cross-reactivity of anti-(electric organ AChR) antibodies with the nicotinic AChR in the nervous system allowed several studies on the nature and development of this receptor in brain, PC12 cell line and chromaffin cells.

Immunological distinction between AChR and the α-Bgt-binding component in the nervous system was demonstrated by Patrick and Stallcup (1977). Using the PC12 cell line of rat sympathetic neurons they have shown that anti-electric eel AChR antibodies blocked the agonist-induced sodium flux, whereas α-Bgt did not, nor did anti-AChR sera immunoprecipitate α-Bgt-binding protein.

Tarrab-Hazdai and Edery (1980) applied anti-AChR antibodies to investigate the involvement of the brain nicotinic AChR in the control of motivated behaviour. Polyclonal immunoglobulins obtained from rabbits immunized with purified *Torpedo* AChR were injected every other day into the left or right cerebral ventricle of rabbits. A reduction in water and food intake was noticeable 7 days after the beginning of the experiment, culminating in total adipsia and aphagia.

Futerman *et al.* (1982) have reported on the *in vitro* effect of anti-muscle AChR antibodies on the presynaptic cholinergic modulation of dopamine release from striated nerve terminals. Immune serum and IgG from rabbit anti-rat muscle AChR were found to inhibit acetylcholine-evoked release of [^{3}H]dopamine up to a maximum of 50%. They have suggested that this effect points towards the presence in the system of an AChR which is structurally and immunologically related to the peripheral neuromuscular AChR.

Schwartz *et al.* (1980) have shown that anti-*Torpedo* AChR antibodies bind to the goldfish brain. The antibody binding was reduced in the denervated tectum following unilateral optic nerve crush or enucleation. Heilbronn *et al.* (1981) have shown that the IgG fraction from sera of myasthenic patients decreases K^+-evoked acetylcholine release from rat hippocampal nerve ending *in vitro*. However, they could not conclude whether the effects observed were linked to the nicotinic AChR in the brain.

We have employed monoclonal antibodies elicited against *Torpedo* AChR to identify the nicotinic AChR in the central nervous system and to study the putative receptor in the developing rat brain (Mochly-Rosen and Fuchs, 1983, and unpublished data). Several of the antibodies which cross-react with muscle AChR were shown to react specifically also with brain homogenates from rats of different ages. However, the antibody binding did not coincide with the α-Bgt binding at all ages tested (Fig. 8.5). Some differences in the binding patterns were also observed between the various monoclonal antibodies. We suggest that such differences may reflect antigenic changes in the AChR resulting from synaptogenesis. None of the monoclonal antibodies inhibited α-Bgt binding to brain, including mcAb 5.5 which is directed against the cholinergic-binding site of muscle AChR, neither did they immunoprecipitate the α-Bgt binding component from Triton extracts of rat brain homogenates. Our results suggest that the components in the rat brain which bind α-Bgt and monoclonal antibodies are not necessarily identical. Recently we have been using anti-*Torpedo* AChR monoclonal antibodies for specific immunopreci-

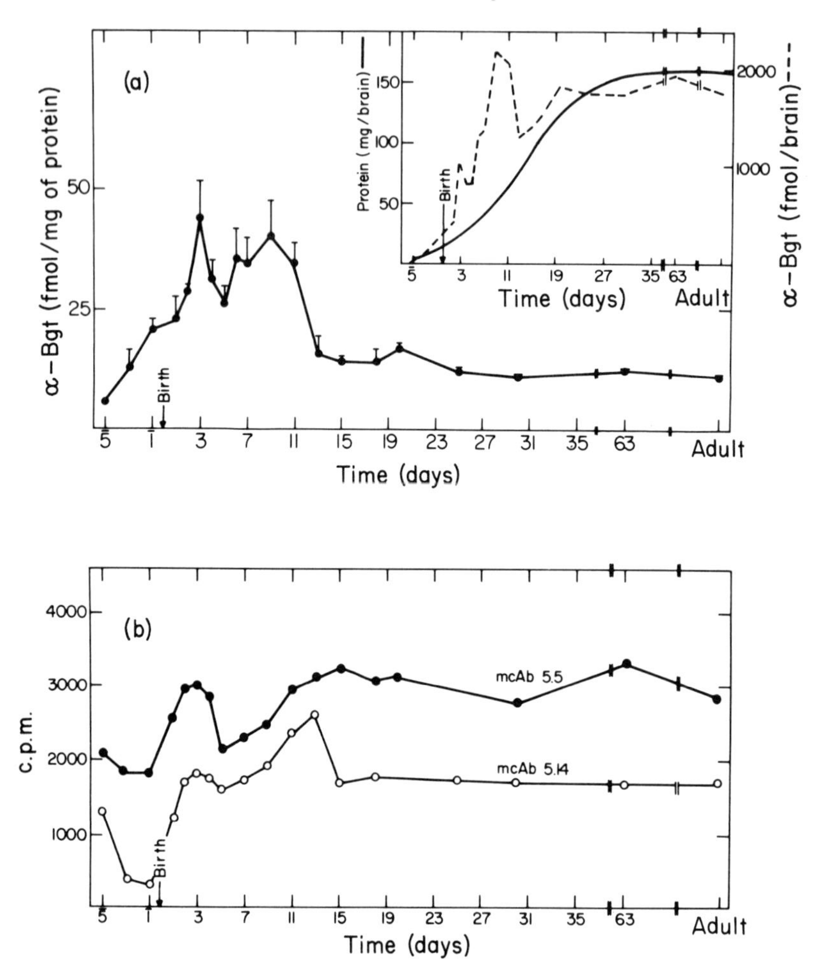

Fig. 8.5 Development of (a) α-Bgt binding, (b) monoclonal antibody binding to whole rat brain homogenate.

pitation of the putative rat brain AChR obtained following *in vitro* and *in ovo* translation of rat brain mRNA (Mochly-Rosen *et al.*, 1983; see Section 8.5.7).

Swanson *et al.* (1983) have applied anti-AChR monoclonal antibody for immunohistochemical studies in the midbrain of the chick. They have also observed that antibody binding did not parallel α-Bgt binding. Their observations suggest that there may be a cholinergic input to the lateral spiriform nucleus and that the projection fibers from it to the optic tectum may contain presynaptic AChR.

Both polyclonal and monoclonal anti-AChR antibodies were used to study the AChR in chromaffin cells of the adrenal medulla. These cells secrete catecholamines upon acetylcholine stimulation. Tarrab-Hazdai *et al.* (1982) have demonstrated that rabbit anti-*Torpedo* AChR antibodies bind to chromaffin cells and block more than 50% of the binding of α-Bgt. By using mcAb 5.5 we have observed an immunofluorescent staining of chromaffin cells, which is abolished by preincubation of the cells with acetylcholine (Fig. 8.6) or *d*-tubocuranine (Mochly-Rosen, Harish, Friedman, Rosenheck and Fuchs, unpublished data). McAb 5.5 had also a synergistic effect on the acetylcholine-induced catecholamine release of these cells, probably by interacting with the nicotinic AChR.

The association between the thymus and AChR is of special interest in view of the involvement of the thymus in myasthenia gravis. Using anti-AChR antibodies we have observed an immunological cross-reactivity, both humoral and cellular, between a thymic component and AChR (Aharonov *et al.*, 1975c). Studies from other laboratories have suggested that the cross-reactivity between AChR and the thymus could stem from myoid cells in the thymus

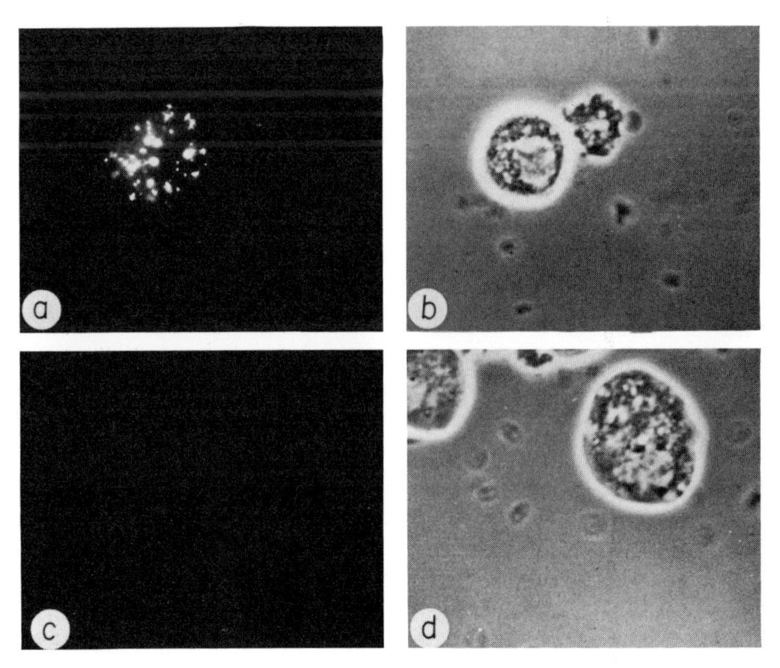

Fig. 8.6 Immunofluorescent staining of bovine adrenal medulla chromaffin cells with mcAb 5.5. (a),(b), Cells incubated with mcAb 5.5 followed by rhodamine–goat anti-mouse immunoglobulins; (c),(d), cells preincubated with acetylcholine (5×10^{-4} M) for 20 minutes at room temperature prior to the incubation with mcAb 5.5; (a),(c), fluorescence and (b),(d) phase photomicrographs of the same cells.

(Wekerle *et al.*, 1975; Kao and Drachman, 1977) and/or from epithelial cells of thymus (Engel *et al.*, 1977). Studies from our laboratory using immunofluorescence and radioimmunological techniques have demonstrated that thymic lymphocytes bear a surface antigen that binds specifically to antibodies against the nicotinic AChR (Fuchs *et al.*, 1980a).

8.5.9 Anti-idiotypes

Antibodies (anti-idiotypes) against anti-AChR antibodies can serve to study the idiotypic repertoire of the immune response to AChR and to assess the potential of anti-idiotypes for immune regulation of myasthenia. In previous studies from our laboratory (Schwartz *et al.*, 1978) we have analyzed the idiotypes of anti-AChR antibodies by anti-idiotypes raised in mice against syngeneic spleen cells incubated *in vitro* with AChR. These studies have demonstrated a broad cross-reactivity between idiotypes of anti-AChR antibodies, not only between different mouse strains but also among anti-AChR antibodies from other species. Other studies from our laboratory (Feingold and Fuchs, 1980; Fuchs *et al.*, 1980b, 1982) have indicated that rabbits preimmunized with rabbit anti-AChR antibodies produced anti-idiotypes which partially protected them against EAMG. Lefvert (1981) has analyzed sera of myasthenic patients and reported on a shared idiotype between anti-AChR antibodies from different patients (Lefvert, 1981; Lefvert *et al.*, 1982).

In order to analyze the fine specificity and repertoire of the idiotypes which compose the polyclonal anti-AChR antibody population it was necessary to raise anti-idiotypes to monoclonal anti-AChR antibodies and study their idiotypic specificities. Such anti-idiotypes were prepared by several laboratories (Lennon and Lambert, 1981; Dwyer *et al.*, 1981b; Souroujon *et al.*, 1983c). Lennon and Lambert (1981) demonstrated an idiotype common to four monoclonal antibodies which react with electric organ and mammalian muscle AChR. This particular idiotype was detectable in the sera of about half of the rats immunized with AChR of either *Torpedo*, eel or syngeneic muscle. However, rats with high titers of antibodies to this idiotype were not protected against the active induction of EAMG. We prepared in rabbits and mice anti-idiotypes against three well-characterized anti-AChR monoclonal antibodies which cross-react with muscle AChR as well as against polyclonal mouse anti-AChR antibodies. We have found that each of these antibodies bears 'private' idiotypes, each of which is represented in polyclonal serum. A marked suppression of the titer of both anti-*Torpedo* and anti-muscle AChR antibodies was observed in mice preimmunized with polyclonal anti-AChR antibodies and challenged later with AChR. Only partial suppression of the antibody titers was observed in mice producing anti-idiotypes against each of the monoclonal anti-AChR antibodies (Souroujon and Fuchs, unpublished data).

Recently, a novel approach to the production of anti-AChR antibodies in rabbits and mice based on an anti-idiotypic route was reported (Wasserman *et al.*, 1982; Cleveland *et al.*, 1983; see Chapter 2 by Strosberg and Schreiber). This procedure is based on the hypothesis that, regardless of functional differences, macromolecules of the same specificity might show homologies in their binding sites. Antibodies to the cholinergic agonist Bis Q mimicked the binding specificity of AChR. Some of the anti-idiotypes raised against these 'receptor-like' antibodies cross-reacted with determinant(s) on AChR preparations from *Torpedo californica, Electrophous* and rat muscle. By a similar approach we have raised anti-idiotypic antibodies against anti-α-Bgt antibodies; these anti-idiotypic antibodies partially mimicked α-Bgt and bound specifically to membrane-bound *Torpedo* AChR (Schreiber *et al.*, 1983).

REFERENCES

Aharonov, A., Kalderon, N., Silman, I. and Fuchs, S. (1975a), *Immunochemistry*, **12,** 765–771.

Aharonov, A., Abramsky, O., Tarrab-Hazdai, R. and Fuchs, S. (1975b), *Lancet*, **ii,** 340–342.

Aharonov, A., Tarrab-Hazdai, R., Abramsky, O. and Fuchs, S. (1975c), *Proc. Natl. Acad. Sci. U.S.A.*, **72,** 1456–1459.

Aharonov, A., Tarrab-Hazdai, R., Silman, I. and Fuchs, S. (1977), *Immunochemistry*, **14,** 129–137.

Albuquerque, E.X., Lebeda, J.F., Appel, S.H., Almon, R., Kaufman, F.C., Mayer, R.F., Narahashi, T. and Yeh, J.Z. (1976), *Ann. N.Y. Acad. Sci.*, **274,** 475–492.

Alema, S., Cull-Candy, S.G., Miledi, R. and Trautmann, A. (1981), *J. Physiol. (London)*, **311,** 251–266.

Almon, R. and Appel, S.H. (1976), *Biochemistry*, **15,** 3662–3667.

Almon, R.R., Andrew, C.G. and Appel, S.H. (1974), *Science*, **186,** 55–57.

Anderson, D.J. and Blobel, G. (1981), *Proc. Natl. Acad. Sci. U.S.A.*, **78,** 5598–5602.

Anderson, D.J., Blobel, G., Tzartos, S., Gullick, W. and Lindstrom, J. (1983), *J. Neurosci.*, **3,** 1773–1784.

Appel, S.H., Anwyl, R., McAdams, M.W. and Elias, S. (1977), *Proc. Natl. Acad. Sci. U.S.A.*, **74,** 2130–2134.

Ballivet, M., Patrick, J., Lee, J. and Heinemann, S. (1982), *Proc. Natl. Acad. Sci. U.S.A.*, **79,** 4466–4470.

Barkas, T., Gairns, J.M., Kerr, H.J., Coggins, J.R. and Simpson, J.A. (1982), *Eur. J. Immunol.*, **12,** 757–761.

Bartfeld, D. and Fuchs, S. (1977), *FEBS Lett.*, **77,** 214–218.

Bartfeld, D. and Fuchs, S. (1978), *Proc. Natl. Acad. Sci. U.S.A.*, **75,** 4006–4010.

Bartfeld, D. and Fuchs, S. (1979a), *Biochem. Biophys. Res. Commun.*, **89,** 512–519.

Bartfeld, D. and Fuchs, S. (1979b), *FEBS Lett.*, **105,** 303–306.

Berti, F., Clementi, F., Conti-Tronconi, B. and Folco, G.C. (1976), *Br. J. Pharmacol.*, **57,** 17–22.

Bevan, S. and Steinbach, J.H. (1977), *J. Physiol. (London)*, **267,** 195–213.
Bevan, S., Heinemann, S., Lennon, V.A. and Lindstrom, J. (1976), *Nature (London)*, **260,** 438–439.
Bevan, S., Kullberg, R.W. and Heinemann, S.F. (1977), *Nature (London)*, **267,** 263–265.
Brenner, T., Abramsky, O., Lisak, R.P., Zweiman, B., Tarrab-Hazdai, R. and Fuchs, S. (1978), *Isr. J. Med. Sci.*, **14,** 986–989.
Brockes, J.P. and Hall, Z.W. (1975), *Biochemistry*, **14,** 2092–2106.
Chang, H.W. and Bock, E. (1977), *Biochemistry*, **16,** 4513–4520.
Changeux, J.P. (1981), *The Harvey Lectures*, Series 75, 85–254.
Claudio, T. and Raftery, M.A. (1977), *Arch. Biochem. Biophys.*, **181,** 484–489.
Claudio, T. and Raftery, M.A. (1980), *J. Supramol. Struct.*, **14,** 267–279.
Claudio, T., Ballivet, M., Patrick, J. and Heinemann, S. (1983), *Proc. Natl. Acad.Sci. U.S.A.*, **80,** 1111–1115.
Cleveland, W.L., Wasserman, N.H., Sarangarajan, R., Penn, A.A. and Erlanger, B.F. (1983), *Nature (London)*, **305,** 56–57.
Cohen, J.B., Weber, M., Huchet, M. and Changeux, J.P. (1972), *FEBS Lett.*, **26,** 43–47.
Conti-Tronconi, B.M. and Raftery, M.A. (1982), *Annu. Rev. Biochem.*, **51,** 491–530.
Conti-Tronconi, B., Tzartos, S. and Lindstrom, J. (1981), *Biochemistry*, **20,** 2181–2191.
Cull-Candy, S.G., Miledi, R. and Trautmann, A. (1978), *Nature (London)*, **271,** 74–75.
Cull-Candy, S.G., Miledi, R. and Trautmann, A. (1979), *J. Physiol. (London)*, **287,** 247–265.
Damle, V.N., McLaughlin, M. and Karlin, A. (1978), *Biochem. Biophys. Res. Commun.*, **84,** 845–851.
Desouki, A., Eldefrawi, A.T. and Eldefrawi, M.E. (1981), *Exp. Neurol.*, **73,** 440–450.
Drachman, D.B. (1978), *N. Engl. J. Med.*, **298,** 136–142, 186–193.
Drachman, D.B., Angus, C.W., Adams, R.T., Michelson, J.D. and Hoffman, B.M. (1978), *N. Engl. J. Med.*, **298,** 1116–1122.
Dwyer, D.S., Bradley, R.J., Furner, R.L. and Kemp, G.E. (1981a), *Brain Res.*, **217,** 23–40.
Dwyer, D.S., Kearney, J.F., Bradley, R.J., Kemp, G.E. and Oh, S.J. (1981b), *Ann. N.Y. Acad. Sci.*, **377,** 143–157.
Einarson, B., Gullick, W., Conti-Tronconi, B., Ellisman, M. and Lindstrom, J. (1982), *Biochemistry*, **21,** 5295–5302.
Eldefrawi, M.E., Copio, D.S., Hudson, C.S., Rash, J.E., Mansour, N.A., Eldefrawi, A.T. and Albuquerque, E.X. (1979), *Exp. Neurol.*, **64,** 428–444.
Elmquist, D., Hoffmann, W.W., Kugelberg, J. and Quastel, D.M.J. (1964), *J. Physiol. (London)*, **174,** 417–434.
Engel, W.K., Trotter, J.L., McFarlin, D.E. and McIntosh, C.L. (1977), *Lancet*, **i,** 1310–1311.
Fambrough, D.M. (1979), *Physiol. Rev.*, **59,** 165–226.
Feingold, C. and Fuchs, S. (1980), *Isr. J. Med. Sci.*, **16,** 805.
Froehner, S.C. and Rafto, S. (1979), *Biochemistry*, **18,** 301–307.
Froehner, S.C., Reiness, C.G. and Hall, Z.W. (1977a), *J. Biol. Chem.*, **252,** 8589–8596.
Froehner, S.C., Karlin, A. and Hall, Z.W. (1977b), *Proc. Natl. Acad. Sci. U.S.A.*, **74,** 4685–4688.

Fuchs, S. (1979a), *Curr. Top. Microbiol. Immunol.*, **85,** 1–29.
Fuchs, S. (1979b), in *Plasmapheresis and the Immunobiology of Myasthenia Gravis* (P.C. Dau, ed.), Houghton Mifflin, Boston, pp. 20–31.
Fuchs, S. and Bartfeld, D. (1983), in *Cell Surface Receptors* (P.G. Strange, ed.), Ellis Horwood, Chister, U.K., pp. 126–141.
Fuchs, S., Nevo, D., Tarrab-Hazdai, R. and Yaar, I. (1976), *Nature (London),* **263,** 329–330.
Fuchs, S., Schmidt-Hopfeld, I., Tridente, G. and Tarrab-Hazdai, R. (1980a), *Nature (London),* **287,** 162–164.
Fuchs, S., Bartfeld, D., Eshhar, Z., Feingold, C., Mochly-Rosen, D., Novick, D., Schwartz, M. and Tarrab-Hazdai, R. (1980b), *J. Neurol. Neurosurg. Psychol.,***43,** 634–643.
Fuchs, S., Bartfeld, D. Mochly-Rosen, D., Souroujon, M. and Feingold, C. (1981a), *Ann. N.Y. Acad. Sci.,* **377,** 110–124.
Fuchs, S., Bartfeld, D., Mochly-Rosen, D., Schmidt-Hopfeld, I. and Tarrab-Hazdai, R. (1981b), in *Membrane Transport and Neuroreceptors* (D. Oxender, A. Blume, I. Diamond and C.F. Fox, eds), Alan R. Liss, New York, pp. 405–417.
Fuchs, S., Feingold, C., Bartfeld, D., Mochly-Rosen, D., Schmidt-Hopfeld, I. and Tarrab-Hazdai, R. (1982), in *Disorders of the Motor Unit* (D. Schotland, ed.), John Wiley & Sons, pp. 245–256.
Fulpius, B.W., Zurn, A.D., Granato, D.A. and Leder, R.M. (1976), *Ann. N.Y. Acad. Sci.,* **274,** 116–129.
Futerman, A.H., Harrison, R., Lunt, G.G. and Wonnacott, S. (1982), in *Presynaptic Receptors: Mechanism and Function* (J. De Belleroche, ed.), Ellis Horwood, Chichester, U.K., pp. 165–173.
Giraudat, J., DeVillers-Thiery, A., Auffray, C., Rongeon, F. and Changeux, J.P. (1982), *EMBO J.,* **1,** 713–717.
Goldberg, G., Mochly-Rosen, D., Fuchs, S. and Lass, Y. (1983), *J. Membr. Biol.,* **76,** 123–128.
Gomez, C.M., Richman, D.P., Berman, P.W., Burres, S.A., Arnason, B.G.W. and Fitch, F.W. (1979), *Biochem. Biophys. Res. Commun.,* **88,** 575–582.
Gordon, A.S., Davis, C.G. and Diamond, I. (1977), *Proc. Natl. Acad. Sci. U.S.A.,* **74,** 262–267.
Granato, D.A., Fulpius, B.W. and Moody, J.F. (1976), *Proc. Natl. Acad. Sci. U.S.A.,* **73,** 2872–2876.
Green, D.P.L., Miledi, R. and Vincent, A. (1975), *Proc. R. Soc. London Ser. B.,* **189,** 57–68.
Gullick, W.J. and Lindstrom, J.M. (1982), *Biochemistry,* **21,** 4563–4569.
Gullick, W.J. and Lindstrom, J.M. (1983), *Biochemistry,* **22,** 3801–3807.
Gullick, W.J., Tzartos, S. and Lindstrom, J. (1981), *Biochemistry,* **20,** 2173–2180.
Hamilton, S.L., McLaughlin, M. and Karlin, A. (1977), *Biochem. Biophys. Res. Commun.,* **79,** 692–699.
Hartzell, H.C. and Fambrough, D.M. (1972), *J. Gen. Physiol.,* **60,** 248–262.
Heilbronn, E. and Mattson, C.J. (1974), *J. Neurochem.,* **22,** 315–317.
Heilbronn, E., Haggblad, J. and Kubat, B. (1981), *Ann. N.Y. Acad. Sci.,* **377,** 198–207.
Heinemann, S., Bevan, S., Kullberg, R., Lindstrom, J. and Rice, J. (1977), *Proc. Natl. Acad. Sci. U.S.A.,* **74,** 3090–3094.

Heinemann, S., Merlie, J. and Lindstrom, J. (1978), *Nature (London)*, **274,** 65–67.

Hohfeld, R., Sterz, R., Kalias, I., Peper, K. and Wekerle, H. (1981), *Pfluegers Arch.*, **390,** 156–169.

Ito, Y., Miledi, R., Vincent, A. and Newsom–Davis, J. (1978), *Brain* **101,** 345–368.

James, R.W., Kato, A.C., Rey, M.J. and Fulpius, B.W. (1980), *FEBS Lett.*, **120,** 145–148.

Kao, I. and Drachman, D.B. (1977), *Science*, **195,** 74–75.

Karlin, A. (1980), in *The Cell Surface and Neuronal Function* (C.W. Cotman, G. Poste and G.L. Nicolson, eds), Elsevier, North-Holland Biomedical Press, pp. 191–260.

Karlin, A., Weill, C.H., McNamee, M.G. and Valderram, R. (1976), *Cold Spring Harbor Symp. Quant, Biol.*, **40,** 203–210.

Karlin, A., Holtzman, E., Valderrama, R., Damle, V., Hsu, K. and Reyes, F. (1978), *J. Cell Biol.*, **76,** 577–592.

Kemp, G., Morley, B., Dwyer, D. and Bradley, R.J. (1980), *Membr. Biochem.*, **3,** 229–257.

Klymkowsky, M.W. and Stroud, R.M. (1979), *J. Mol. Biol.*, **128,** 319–334.

Kohler, G. and Milstein, C. (1975), *Nature (London)*, **256,** 495–497.

Lambert, E.H., Lindstrom, J.M. and Lennon, V.A. (1976), *Ann. N.Y. Acad. Sci.*, **274,** 300–318.

Lefvert, A.K. (1981), *Scand. J. Immunol.*, **13,** 493–497.

Lefvert, A.K. (1982), *J. Neurol. Neurosurg. Psychiat.*, **45,** 70–74.

Lefvert, A.K., James, R.W., Alliod, C. and Fulpius, B.W. (1982), *Eur. J. Immunol.*, **12,** 790–792.

Lennon, V.A. and Lambert, E.H. (1980), *Nature (London)*, **285,** 238–240.

Lennon, V.A. and Lambert, E.H. (1981), *Ann. N.Y. Acad. Sci.*, **377,** 77–96.

Lennon, V.A., Lindstrom, J.M. and Seybold, M.E. (1975), *J. Exp. Med.*, **141,** 1365–1375.

Lennon, V.A., Thompson, M. and Chen, J. (1980), *J. Biol. Chem.*, **255,** 4395–4398.

Lindstrom, J. (1979), *Adv. Immunol.*, **27,** 1–50.

Lindstrom, J. and Dau, P. (1980), *Annu. Rev. Pharmacol. Toxicol.*, **20,** 337–362.

Lindstrom, J.M., Einarson, B.L., Lennon, V.A. and Seybold, M.E. (1976a), *J. Exp. Med.*, **144,** 726–738.

Lindstrom, J.M., Lennon, V.A., Seybold, M.E., and Whittingham, S. (1976b), *Ann. N.Y. Acad. Sci.*, **274,** 254–274.

Lindstrom, J., Einarson, B. and Merlie, J. (1978), *Proc. Natl. Acad. Sci. U.S.A.*, **75,** 769–773.

Lindstrom, J.M., Merlie, J. and Yogeeswaran, G. (1979), *Biochemistry*, **18,** 4465–4469.

Lindstrom, J., Tzartos, S. and Gullick, W. (1981), *Ann. N.Y. Acad. Sci.*, **377,** 1–19.

Lotwick, H., Harrison, R., Lunt, G. and Behan, P. (1983), *J. Neuroimmunol.*, (in press).

Lyddiatt, A., Sumikawa, K., Wolosin, J.M., Dolly, J.O. and Barnard, E.A. (1979), *FEBS Lett.*, **108,** 20–24.

Martinez, R.D., Tarrab-Hazdai, R., Aharonov, A. and Fuchs, S. (1977) *J. Immunol.*, **118,** 17–20.

Mehraban, F., Dolly, O. and Barnard, E.A. (1982), *FEBS Lett.*, **141,** 1–5.

Mendez, B., Valenzuela, P., Martial, J.A. and Baxter, J.D. (1980), *Science*, **209,** 695–697.

Merlie, J.P. and Sebbane, R. (1981), *J. Biol. Chem.*, **256,** 3605–3608.

Miledi, R. (1960), *J. Physiol (London),* **155,** 1–23.

Mochly-Rosen, D. (1983), Ph.D. thesis, The Weizmann Institute of Science.

Mochly-Rosen, D. and Fuchs, S. (1981), *Biochemistry,* **20,** 5920–5924.

Mochly-Rosen, D. and Fuchs, S. (1983), *Isr. J. Med. Sci.,* (in press).

Mochly-Rosen, D., Fuchs, S. and Eshhar, Z. (1979), *FEBS Lett.,* **106,** 389–392.

Mochly-Rosen, D., Souroujon, M.C., Eshhar, Z. and Fuchs, S. (1981), in *Monoclonal Antibodies to Neural Antigens* (eds. R. McKay, M.C. Raff and L.F. Reichardt, eds), Cold Spring Harbor Laboratory, New York, pp. 193–202.

Mochly-Rosen, D., Soreq, H. and Fuchs, S. (1983), *Isr. J. Med. Sci.,* (in press).

Momoi, M.Y. and Lennon, V.A. (1982), *J. Biol. Chem.,* **257,** 12757–12764.

Moore, H.H. and Raftery, M.A. (1979), *Biochemistry,* **18,** 1862–1867.

Nathanson, N.M. and Hall, Z.W. (1979), *Biochemistry,* **18,** 3392–3401.

Nathanson, N.M. and Hall, Z.W. (1980), *J. Biol. Chem.,* **255,** 1698–1703.

Nastuk, W.L., Niemi, W.D., Alexander, J.T., Chang, H.W. and Nastuk, M.A. (1979), *Am. J. Physiol.,* **236,** C53–C57.

Noda, M., Takahashi, H., Tanabe, T., Toyosato, M., Furutani, Y., Hirose, T., Asai, M., Inayama, S., Miyata, T. and Numa, S. (1982), *Nature (London),* **299,** 793–797.

Noda, M., Takahashi, H., Tanabe, T., Toyosato, M., Kikyotani, S., Hirose, T., Asai, M., Takashima, H., Inayama, S., Migata, T. and Numa, S. (1983), *Nature (London),* **301,** 251–255.

Patrick, J. and Lindstrom, J. (1973), *Science,* **180,** 871–872.

Patrick, J. and Stallcup, W.B. (1977), *Proc. Natl. Acad. Sci. U.S.A.,* **74,** 4689–4692.

Patrick, J., Lindstrom, J., Culp, B. and McMillan, J. (1973), *Proc. Natl. Acad. Sci. U.S.A.,* **70,** 3334–3338.

Patrick, J., McMillan, J., Wolfson, H. and O'Brien, J.C. (1977), *J. Biol. Chem.,* **252,** 2143–2153.

Penn, A.S., Chang, H.W., Lovelace, R.E., Niemi, W. and Miranda, A. (1976), *Ann. N.Y. Acad. Sci.,* **274,** 354–376.

Peper, K., Sterz, R. and Bradley, R.J. (1981), *Ann. N.Y. Acad. Sci.,* **377,** 519–543.

Pizzighella, G., Gordon, A.S., Souroujon, M.C., Mochly-Rosen, D., Sharp, A. and Fuchs, S. (1983), *FEBS Lett.,* **159,** 246–250.

Potter, L.T. (1973), *Drug Receptors* (H.P. Rang, ed.), Macmillan, London, pp. 295–312.

Prives, J., Hoffman, L., Tarrab-Hazdai, R., Fuchs, S. and Amsterdam, A. (1979), *Life Sci.,* **24,** 1713–1718.

Raftery, M.A., Hunkapiller, M.W., Strader, C.D. and Hood, L.E. (1980), *Science,* **208,** 1454–1457.

Reiness, C.G. and Hall, Z.W. (1981), *Dev. Biol.,* **81,** 324–331.

Reiness, C.G., Weinberg, C.B. and Hall, Z.W. (1978), *Nature (London),* **274,** 68–70.

Reynolds, J.A. and Karlin, A. (1978), *Biochemistry,* **17,** 2035–2038.

Ross, M.J., Klymkowsky, M.W., Agard, D.A. and Stround, R.M. (1977), *J. Mol. Biol.,* **116,** 635–659.

Saitoh, T. and Changeux, J.P. (1981), *Proc. Natl. Acad. Sci. U.S.A.,* **78,** 4430–4434.

Schreiber, M., Souroujon, M. and Fuchs, S. (1983), *Isr. J. Med. Sci.,* (in press).

Schwartz, M., Novick, D., Givol, D. and Fuchs, S. (1978), *Nature (London),* **273,** 543–545.

Schwartz, M., Lancet, D., Tarrab-Hazdai, R. and Fuchs, S. (1979), *Mol. Immunol.*, **16**, 483–487.

Schwartz, M., Axelrod, D., Feldman, E.L. and Agranoff, B.W. (1980), *Brain Res.* **194**, 171–180.

Shorr, R.G., Dolly, J.O. and Barnard, E.A. (1978), *Nature (London)*, **274**, 283–284.

Sobel, A. and Changeux, J.P. (1977), *Biochem. Soc. Trans.*, **5**, 511–514.

Soreq, H., Bartfeld, D., Parvari, R. and Fuchs, S. (1982), *FEBS Lett.*, **139**, 32–36.

Souroujon, M.C., Mochly-Rosen, D., Bartfeld, D. and Fuchs, S. (1982), in *Muscle Development: Molecular and Cellular Control* (H.F. Epstein and M.L. Pearson, eds), Cold Spring Harbor Laboratory, New York, pp. 527–534.

Souroujon, M.C., Mochly-Rosen, D., Gordon, A.S. and Fuchs, S. (1983a), *Muscle and Nerve*, **6**, 303–311.

Souroujon, M.C., Pizzighella, S., Mochly-Rosen, D. and Fuchs, S. (1983b) in *Molecular and Cellular Aspects of Myogenesis and Myofibrillogenesis* (H.M. Eppenberger and J.P. Perriard, eds), (in press).

Souroujon, M.C., Mochly-Rosen, D., Weiss, I. and Fuchs, S. (1983c) *Isr. J. Med. Sci.*, (in press).

Sugiyama, H., Benda, P., Meunier, J.C. and Changeux, J.P. (1973), *FEBS Lett.*, **35**, 124–128.

Sumikawa, K., Houghton, M., Emtage, J.S., Richards, B.M. and Barnard, E.A. (1981), *Nature (London)*, **292**, 862–864.

Sumikawa, K., Houghton, M., Smith, J.C., Bell, L., Richards, B.M. and Barnard, E.A. (1982), *Nucleic Acids Res.*, **10**, 5809–5822.

Swanson, L.W., Lindstrom, J., Tzartos, S., Schumed, L.C., O'Leary, D.D.M. and Cowan, W.M. (1983), *Proc. Natl. Acad. Sci. U.S.A.* (in press).

Tarrab-Hazdai, R. and Edery, H. (1980), *Exp. Neurol.*, **67**, 670–675.

Tarrab-Hazdai, R. (1981), *Ann. N.Y. Acad. Sci.*, **377**, 20–37.

Tarrab-Hazdai, R., Aharonov, A., Abramsky, O., Yaar, I. and Fuchs, S. (1975a), *J. Exp. Med.*, **142**, 785–789.

Tarrab-Hazdai, R., Aharonov, A., Silman, I., Fuchs, S. and Abramsky, O. (1975b), *Nature (London)*, **256**, 128–130.

Tarrab-Hazdai, R., Abramsky, O. and Fuchs, S. (1977), *J. Immunol.*, **119**, 702–706.

Tarrab-Hazdai, R., Geiger, B., Fuchs, S. and Amsterdam, A. (1978), *Proc. Natl. Acad. Sci. U.S.A.*, **75**, 2497–2501.

Tarrab-Hazdai, R., Schmidt-Sole, Y., Mochly-Rosen, D. and Fuchs, S. (1980), *FEBS Lett.*, **118**, 35–38.

Tarrab-Hazdai, R., Hernandez, M. and Martinez-Munoz, D. (1982), *Isr. J. Med. Sci.*, **18** (6), 20.

Tzartos, S.J. and Lindstrom, J.M. (1980), *Proc. Natl. Acad. Sci. U.S.A.*, **77**, 755–759.

Vandlen, R.L., Wu, C.S.W., Eisenach, J.C. and Raftery, M.A. (1979), *Biochemistry*, **18**, 1845–1854.

Vernet der Garabedian, B. and Morel, E. (1983), *Biochem. Biophys. Res. Commun.*, **113**, 1–9.

Vincent, A. (1980), *Physiol. Rev.* **60**, 756–824.

Vincent, A. and Newsom-Davis, J. (1982), *J. Clin. Exp. Immunol.*, **49**, 257–265.

Wassermann, N.H., Penn, A.S., Freimuth, P.I., Treptow, N., Wentzel, S. Cleveland, W.L. and Erlanger, B.F. (1982), *Proc. Natl. Acad. Sci. U.S.A.*, **79**, 4810–4814.

Watters, D. and Maelicke, A. (1983), *Biochemistry,* **22,** 1811–1819.
Weill, C.L., McNamee, M.G. and Karlin, A. (1974), *Biochem, Biophys. Res. Commun.*, **61,** 997–1003.
Weinberg, C.B. and Hall, Z.W. (1979), *Proc. Natl. Acad. Sci. U.S.A.*, **76,** 504.
Wekerle, H., Paterson, B., Ketelsen, U.P. and Feldman, M. (1975), *Nature (London),* **256,** 493–494.
Wonnacott, S., Harrison, R. and Lunt, G. (1982), *J. Neuroimmunol.*, **3,** 1–13.

9 Monoclonal Antibodies and the Thyrotropin Receptor

LEONARD D. KOHN, DONATELLA TOMBACCINI, MICHELE L. DE LUCA, MAURIZIO BIFULCO, EVELYN F. GROLLMAN and WILLIAM A. VALENTE

Glossary of terms and abbreviations

TSH	Thyrotropin (thyroid-stimulating hormone)
hCG	Human chorionic gonadotropin
LH	Luteinizing hormone
FSH	Follicle-stimulating hormone
LIS	Lithium diiodosalicylate
FRTL-5 cells	Functioning rat thyroid cells derived from normal Fisher rat thyroid glands
G_{M1}	Galactosyl-*N*-acetylgalactosaminyl-[*N*-acetylneuraminyl]galactosyl-glucosylceramide
G_{M2}	*N*-Acetylgalactosaminyl-[*N*-acetylneuraminyl]galactosylglucosylceramide
G_{M3}	*N*-Acetylneuraminylgalactosylglucosylceramide
G_{Dla}	*N*-Acetylneuraminylgalactosyl-*N*-acetylgalactosaminyl-[*N*-acetylneuraminyl] galactosylglucosylceramide
G_{Dlb}	Galactosyl-*N*-acetylgalactosaminyl-[*N*-acetylneuraminyl-*N*-acetylneuraminyl] galactosylglucosylceramide
G_{T1}	*N*-Acetylneuraminylgalactosyl-*N*-acetylgalactosaminyl-[*N*-acetylneuraminyl-*N*-acetylneuraminyl]galactosylglucosylceramide
T_3	Tri-iodothyronine
T_4	Thyroxine
CBA	Cytochemical bioassay
TSAb	Thyroid-stimulating antibody
FCCP	Carbonyl cyanide *p*-trifluoromethoxyphenylhydrazone

Monoclonal Antibodies to Receptors: Probes for Receptor Structure and Function
(*Receptors and Recognition*, Series B, Volume 17)
Edited by M. F. Greaves
Published in 1984 by Chapman and Hall, 11 New Fetter Lane, London EC4P 4EE

9.1 INTRODUCTION

Thyrotropin (TSH) is a pituitary glycoprotein hormone whose primary role is to regulate thyroid cell function (Dumont, 1971; Field, 1980; Robbins *et al.*, 1980; Kohn *et al.*, 1983). The interaction of TSH with a specific receptor on the thyroid cell surface induces changes in adenylate cylcase activity which result in the following tissue responses: enhanced iodide uptake, thyroglobulin biosynthesis, iodination of thyroglobulin, degradation of iodinated thyroglobulin to form thyroid hormone, and the release of thyroid hormone (T_3 and T_4) into the bloodstream. Although changes in all of the above activities have been and can be used to define receptor function, definition of the structure of the TSH receptor on a molecular level, as with all other receptors, has required the identification of specific membrane components using binding studies and ^{125}I-labeled hormone (Kohn, 1978; Kohn and Shifrin, 1982; Kohn *et al.*, 1983).

The sum of the binding studies identified a membrane ganglioside, as well as a membrane glycoprotein, as two potentially important components in the cell surface recognition event (Fig. 9.1). Studies of the *in vitro* properties of ^{125}I-TSH binding to each led to the following ideas concerning the process of recognition and the mechanism by which the hormone–receptor interaction induced the cell to a functional response. TSH binding to the glycoprotein

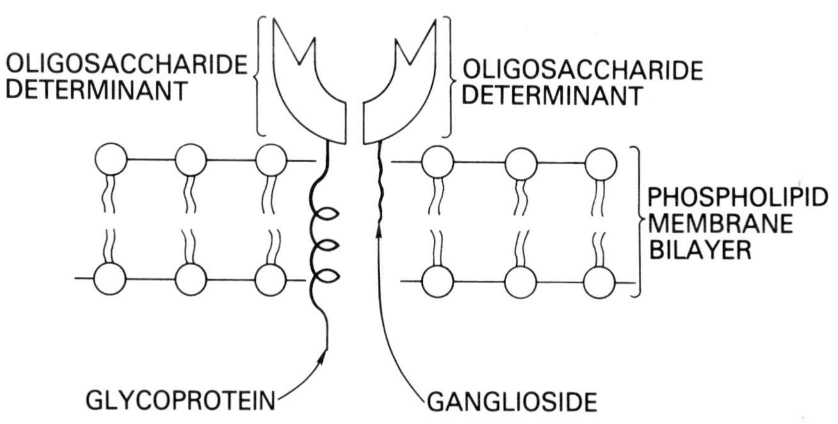

Fig. 9.1 Proposed model of TSH receptor composed of glycoprotein and ganglioside component. After the TSH β subunit interacts with the receptor, the hormone changes its conformation and the α subunit is brought into the bilayer where it interacts with other membrane components. The end result includes a change in organization of the membrane bilayer, a change in the transmembrane electrochemical ion gradient, changes in lipid turnover, and the expression of other receptors, in addition to the initiation of growth and cAMP signals. The β subunit of TSH is presumed to carry the primary determinants recognized by the glycoprotein receptor component, but in no way does the model exclude an α subunit contribution either direct or by conformational perturbations of β.

component of the membrane was proposed to be the initial high-affinity recognition event on the cell surface, i.e. the necessary first step in receptor recognition; however, a full functional response was postulated to require the ganglioside.

The ganglioside was suggested to contribute to the following receptor functions. It completed specificity by distinguishing among glycoprotein hormones and related ligands such as tetanus toxin, cholera toxin and interferon. It modulated the apparent affinity and capacity of the glycoprotein receptor component and induced a conformational change in the hormone believed necessary for subsequent message transmission. It allowed the ligand to perturb the phospholipid bilayer through alterations in lipid order and contributed to the ability of TSH to alter the ion flux across the membrane. Finally, the model proposed that after the glycoprotein receptor component trapped the TSH much as a sperm on the surface of the ovum, the ganglioside acted as an emulsifying agent to allow the hormone to interact with other membrane components within the hydrophobic environment of the lipid bilayer and thereby initiate the signal processes. The physical basis of the emulsification process was the formation of the anhydrous complex between the hormone and the oligosaccharide moiety of the ganglioside. By excluding water from the interface of the ligand–receptor complex, membrane penetration was facilitated, and interactions involving membrane components of the adenylate cyclase ensued. The evolution of these ideas can be traced in a series of reviews by these authors and their colleagues (Kohn, 1978; Kohn *et al.*, 1980, 1981, 1982, 1983; Kohn and Shifrin, 1982).

It is obvious that these ideas resemble many invoked in the study of cholera toxin–ganglioside interactions (Bennett and Cuatrecassas, 1977; Holmgren, 1978) and indeed they borrow heavily on these studies. They also borrow heavily on the presumptive role of the G_{M1} ganglioside to allow the α subunit of the toxin to intercalate within the bilayer structure where its ADP ribosyltransferase activity can catalyze the ADP ribosylation of the G regulatory protein of the adenylate cyclase complex using NAD as substrate (Gill, 1977; Moss and Vaughan, 1979). Thus, although TSH has no intrinsic ADP ribosylation activity (Moss and Vaughan, 1979), it has been shown to perturb a membrane-associated ADP ribosyltransferase which carried out this same reaction (De Wolf *et al.*, 1981; Vitti *et al.*, 1982). These reactions have been linked to TSH receptor-mediated regulation of adenylate cyclase activity (Kohn and Shifrin, 1982; Kohn *et al.*, 1982, 1983).

It is evident that if the model were valid, it would amplify our understanding of receptor structure and could provide us with clues as to the mechanism of receptor interactions for other ligands, which interact with gangliosides such as tetanus toxin and interferon (Kohn, 1978). The present report will summarize the results of two different methodologic approaches taken to test the model. The primary discussion will center on the use of monoclonal antibodies to identify receptor components and their link to message signals. The second, a reconstitution approach, will complement and amplify these conclusions.

The data will also provide major insight into the antigenic determinants recognized by antibodies in autoimmune thyroid disease which are believed etiologically important in their pathologic expression (Adams, 1981; Doniach, 1975; McKenzie and Zakarija, 1976; Pinchera *et al.*, 1982). In particular, it will be shown that thyroid-stimulating bodies (TSAbs) in Graves' patients are directed against gangliosides and are the autoimmune equivalents of cholera toxin.

Finally, the antibody data will confirm a hitherto less well accepted role of TSH in thyroid cell growth. These data will show that the same receptor components are used albeit with different emphasis or arrays and that an alternative signal system, perturbation of phospholipid metabolism, is involved.

9.2 THE MONOCLONAL ANTIBODY APPROACH

This approach is aimed at identifying monoclonal antibodies by assays of TSH receptor function and only then relating these to specific membrane molecules. The first step avoided a preconceived model wherein purified receptor components were injected, but rather involved the injection of crude solubilized thyroid membrane preparations into mice followed by spleen cell fusion with non-IgG-producing mouse myeloma cells; production of hybridomas secreting antibodies to the intact mouse preparation; functional identification of antibodies related to the TSH receptor structure; and, finally, characterization of the antigenic determinants of the antibodies with particular respect to the already identified TSH-binding components (Yavin *et al.*,1981a,b,c; Valente *et al.*, 1982b,c, 1984).

The second step encompassed attempts to directly relate TSH receptor structure to the autoantibodies in Graves' sera. This approach involved the following steps: fusion of lymphocytes from patients with active Graves' disease with a non-IgG-secreting mouse myeloma cell line; identification of heterohybridomas which secreted antibodies capable of stimulating thyroid function or blocking TSH binding; and characterization of the antigenic determinants of these antibodies with respect to the structural or functional components of the TSH receptor identified by the monoclonal antibodies derived from the first approach or postulated in the receptor model (Valente *et al.*, 1982c).

A two-stage screening procedure with thyroid membranes was utilized. In the first stage, hybridomas producing antibody reactive with thyroid membranes were identified by using ^{125}I-labeled staphylococcal protein A or a ^{125}I-labeled rabbit anti-mouse or anti-human IgG F(ab')$_2$ fragment-specific antibody. In the second stage, the assay was repeated with competing 1 μM TSH added during the initial incubation of membranes and hybridoma medium. Hybridoma antibodies reactive with thyroid membranes in the first screening assay, but blocked by unlabeled TSH in the second, were chosen as potential TSH receptor antibodies.

Table 9.1 TSH-specificity of antibody inhibition

Monoclonal antibody		^{125}I-Protein A or ^{125}I-anti-human IgG (Fab′)$_2$ bound* (c.p.m.)					
		11E8	22A6	59C9	52A8	129H8	307H6
Antibody alone		1470	2400	900	1600	1240	900
+ TSH	1×10^{-6} M	370	810	210	340	400	200
+ hCG	1×10^{-6} M	1390	2310	890	1590	1410	805
+ LH	1×10^{-6} M	1280	2290	905	1700	1390	795
+ Insulin	1×10^{-6} M	1408	2400	1005	1650	1210	910
+ Glucagon	1×10^{-6} M	1410	2510	985	1570	1260	870
+ ACTH	1×10^{-6} M	1385	2380	995	1625	1220	1080
+ Prolactin	1×10^{-6} M	1420	2500	1010	1540	1235	895
+ Albumin	1×10^{-6} M	1401	2450	995	1710	1245	945

* Binding of antibody was measured as described by Yavin *et al.* (1981a,b,c). These experiments used bovine thyroid membranes (11E8, 22A6) or human thyroid membranes (59C9, 52A8, 129H8, 307H6).

Thus far 15 separate fusions have been performed using solubilized human or bovine thyroid membrane preparations as the immunogen for classic hybridoma formation or using peripheral lymphocytes from patients with Graves' disease for heterohybridoma generation. The frequency of hybridoma-producing anti-thyroid membrane reactive antibodies was only a small fraction of the total number of antibody-producing clones (~ 10%); in turn, the majority of anti-thyroid membrane antibodies were unrelated to the TSH receptor (i.e. no effect of added TSH on antibody binding).* Nevertheless monoclonal antibodies were consistently found in each fusion which were inhibited by TSH from binding to membranes. Inhibition was specific as exemplified by the results in Table 9.1.

Having identified potential anti-TSH-receptor-producing clones, antibody preparations were evaluated in a series of assays measuring their ability (i) to competitively inhibit ^{125}I-TSH binding; (ii) to directly stimulate adenylate

*Editor's comment: Presumably this screening strategy would not identify monoclonal antibodies reactive with epitopes distinct from and some distance from the TSH-binding site of the receptor.

cyclase activity or to act as competitive antagonists or agonists of TSH-stimulated adenylate cyclase activity; (iii) to mimic the ability of TSH to release thyroid hormones (T_3/T_4) from the thyroid in an *in vivo* mouse bioassay; and (iv) to stimulate iodide uptake by thyroid cells in culture. As a first approximation the following criteria were used to identify a monoclonal antibody to the TSH receptor. (i) The antibody had to inhibit TSH binding to thyroid membranes or, conversely, be itself prevented from binding to thyroid membranes by TSH. (ii) Inhibition had to be reasonably *specific* and had to be *competitive* as opposed to non-competitive or uncompetitive. (iii) The antibody had to competitively inhibit TSH-stimulated functions or, conversely, had to mimic TSH activity and exhibit competitive agonism.

As will be noted, most antibodies to the TSH receptor, defined by these criteria, can be broadly grouped as 'inhibitors' or 'stimulators' on the basis of these assays. An 'inhibitor' has the following characteristics: competitive inhibition of ^{125}I-TSH binding; competitive inhibition of TSH-stimulated adenylate cyclase activity; inhibition of TSH-stimulatable thyroidal iodine release or uptake; and no direct stimulation of thyroid adenylate cyclase activity, T_3/T_4 release, or iodide uptake. A 'stimulator' (TSAb) in general had the following characteristics: significantly weaker inhibition of ^{125}I-TSH binding than the first group but equally potent competitive inhibition of TSH-responsive adenylate cyclase activity; direct TSH-like stimulatory action with respect to adenylate cyclase activity, as well as both iodide uptake and T_3/T_4 release by the thyroid; and competitive agonism when included with low concentrations of TSH in measuring adenylate cyclase activity.

An antibody with mixed activity is a stimulatory antibody (TSAb) which also has the ability to inhibit ^{125}I-TSH binding to a significant degree. The signifi-

Table 9.2 Representative monoclonal antibodies to the TSH receptor

Clone No.	TSH receptor source	'Activity' pattern
13D11	Bovine	Inhibitor
11E8	Bovine	Inhibitor
22A6	Bovine	Stimulator
59C9	Human	Inhibitor
60F5	Human	Inhibitor
52A8	Human	Mixed
122G3*	Human	Inhibitor
129H8*	Human	Inhibitor
206H3*	Human	Stimulator
208F7*	Human	Mixed
307H6*	Human	Stimulator

*Heterohybridomas.

cance of the existence of antibodies with 'mixed' properties will be discussed with respect to the organization of receptor components and their active site determinants. The antibodies discussed in this report, and their categorization as above, are summarized in Table 9.2.

9.3 CHARACTERIZATION OF ANTIBODIES TO THE GLYCOPROTEIN MEMBRANE COMPONENT WITH HIGH-AFFINITY TSH-BINDING PROPERTIES

Although all of the selected monoclonal antibodies were able to inhibit ^{125}I-TSH binding to thyroid membranes, significant differences existed in their relative potencies (Fig. 9.2). Antibodies which have been classified as inhibitors in Table 9.3 (11E8 and 13D11, bovine receptor; 59C9 and 60F5, human

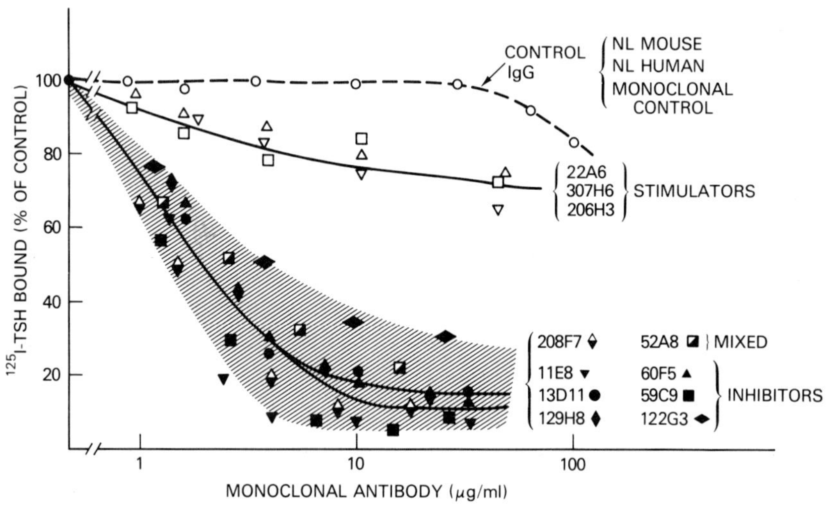

Fig. 9.2 Monoclonal antibody inhibition of ^{125}I-TSH binding to bovine thyroid membrane preparations. The ^{125}I-TSH binding assay was a solid-phase system (Yavin *et al.*, 1981a,b,c). Binding was measured in 10 μM Tris–chloride, pH 7.4, containing 50 mM NaCl and 0.5% bovine serum albumin. Binding was at 37 °C for 2 hours; specific binding was measured by including 1×10^{-6} M unlabeled TSH in control assays. The maximum binding of ^{125}I-TSH averaged 3–7% binding with 0.9% non-specific binding. Inhibition of ^{125}I-TSH binding could be measured either by including monoclonal IgG preparations from culture medium or ascites for 1 hour before ^{125}I-TSH was added (preincubation) or, alternatively, simultaneous with (competition) or after (displacement) the ^{125}I-TSH addition. In these experiments controls included ascites-derived monoclonal IgG preparations from hybridomas which reacted with thyroid membranes but did not inhibit TSH binding, normal mouse or human IgG, or 'naive' monoclonal IgG preparations with no reactivity with thyroid membranes.

Table 9.3 Ability of 'inhibitor' antibodies to prevent ^{125}I-TSH binding to liposomes containing the glycoprotein component of TSH receptor by comparison with 'stimulator' or 'mixed' antibodies

Antibody	^{125}I-TSH bound* (c.p.m.)				
	Low salt, pH 6.0, 0–4 °C			Low salt, pH 7.0, 22 °C	50 mM NaCl, pH 7.4, 37 °C
	Bovine	Human	Rat	Human	Human
Controls					
Monoclonal control†	18 200	14 800	29 200	12 500	3 900
N1 mouse IgG	17 940	14 700	29 500	12 000	4 000
N1 human IgG	18 400	14 800	31 200	12 900	3950
Inhibitors (blockers)					
13D11	4 100	3 800	5 400	2 750	700
11E8	2 100	400	2 050	450	300
59C9	6 200	1 200	5 900	1 300	200
60F5	4 100	1 050	4 200	850	300
129H8	1 400	980	2 800	1 200	200
122G3	6 400	4 200	9 200	3 200	700
Stimulators					
22A6	14 800	13 600	26 800	11 400	4 200
206H3	15 500	12 200	24 500	11 750	3 800
307H6	16 200	12 800	26 400	11 200	3 100
Mixed					
52A8	9 800	5 400	15 300	6 400	2 200
208F7	12 400	6 100	14 400	7 300	1 400

* Dipalmitoylphosphatidylcholine/cholesterol liposomes containing the portion of the TSH receptor were prepared as described (Aloj *et al.*, 1977, 1979). Liposomes were mixed with ^{125}I-TSH (50 000 c.p.m./50 μl) in the following buffers: Tris–acetate, 20 mM, pH 6.0 or 7.0; Tris-chloride, 20mM, pH 7.4, containing 50 mM NaCl. Liposomes were incubated at the noted temperatures for 1 hour. The reaction mixture was then filtered through EHWP-02500 Millipore filters, and liposome-bound ^{125}I-TSH (that trapped on the filter after washing with the incubation buffer) was counted. Control IgG or monoclonal IgG was added concomitantly with the ^{125}I-TSH. Values are ± 5% and are the average result of triplicate incubations.

† Monoclonal antibody from Table 9.1 which did not interact with thyroid membranes or interacted with thyroid membranes equally well in the presence or absence of TSH and was known not to perturb TSH binding to membranes.

receptor; 129H8 or 122G3, Graves' autoimmune) were inhibitors whether tested in preincubation, competition or displacement assays, i.e. independent of the order of addition of TSH or antibody. Inhibition was competitive with respect to TSH whether measured using ^{125}I-TSH and unlabeled antibody or unlabeled TSH and unlabeled antibody, binding of antibody being measured with ^{125}I-labeled protein A.

Table 9.4 Ability of monoclonal antibodies to stimulate or inhibit adenylate cyclase activity in human thyroid cells*, to release thyroid hormone from the thyroid into the blood in the mouse bioassay used to measure TSH bioactivity†, and to stimulate iodide uptake in rat FRTL-5 cells depleted of TSH for 5 days‡

	Adenylate cyclase activity in human thyroid cells* (cAMP level, pmol/μg of DNA)		Activity in a mouse bioassay measuring T_3/T_4 release (mouse blood ^{125}I; % of control)		Uptake of ^{125}I-labeled NaI (c.p.m.)	
Antibody added (0.1 mg/ml)	Direct effect on cells	Effect on 0.5 mU/ml TSH activity	Alone	+TSH	Alone	+TSH (2.5×10^{-10} M)
		±5%				
Controls						
No addition	1.2±0.2	8.8	120±15	520±18	700	7800
Normal human IgG	0.8±0.2	8.8	130±20	480±20	850	8100
Normal mouse IgG	0.8±0.2	8.8	120+10	480+40	720	8500
Monoclonal control	0.8±0.2	7.4	118±15	440±25	700	7900
Inhibitors						
13D11	0.8±0.2	3.8	112±12	170±15	700	2450
11E8	0.8±0.2	3.3	124±10	180±20	650	1410
59C9	0.8±0.2	4.1	110±14	200±15	730	2200
60F5	0.8±0.2	3.4	104±12	188±17	810	1100
129H8	0.8±0.2	3.9	123±20	210±15	740	1500
122G3	0.8±0.2	2.5	110±10	230±18	920	2980
Stimulators						
22A6	3.0±0.2	3.1	300±30	—	6840	—
206H3	1.5±0.2	2.9	340±30	—	8200	—
307H6	3.4±0.2	2.8	510±60	—	7400	—
Mixed						
5A28	2.8±0.2	4.2	340±25	—	7800	—
208F7	1.8±0.2	2.7	300±45	—	6400	—

* Human thyroid cell primary cultures were tested for the ability of bovine TSH and monoclonal antibodies to affect adenylate cyclase activity as follows. The medium was released with 0.2 ml of Krebs Ringer bicarbonate (KRB) buffer, pH 7.4, containing glucose (1.1 g/litre), bovine serum albumin (3.0 g/litre) and 3-isobutyl-1-methylxanthine (IMX) (0.6 mM). Direct stimulatory activity was measured at 37 °C in a 10% CO_2 environment for 120 minutes. Ability of monoclonal antibodies to inhibit TSH responsiveness was tested by incubating cells for 30 minutes with TSH, in the same conditions as above, but after a 120 minute preincubation in KRB buffer with the different monoclonal antibodies. Incubation was stopped by adding 0.2 ml of cold absolute ethanol and the plate was stored at −20 °C overnight. Thawed broken cells were detached by means of a rubber policeman, the suspension was centrifuged at 2000 *g* at 4 °C, and aliquots were tested in cAMP radioimmunoassay.

† The mice were prepared as described by McKenzie (1958). In all experiments, 0.1 ml of material was injected into each of six mice. The mean ±S.E.M are reported. Stimulatory activity was measured at 2 hours after injection. Antibody (0.2 mg/ml) and TSH (1 milliunit/ml) were mixed and injected together; assays were at 2 hours. Assays at different protein concentrations did reveal a dose–response effect for inhibition (+ inhibitors) in all cases (over the range 0.01–0.3 mg/ml). Similarly, stimulators also exhibited dose-related stimulation over the same range. Inhibitors were not stimulators when tested over the range 0.01–0.3 mg/ml.

‡ FRTL-5 rat cells are normally grown in medium containing TSH; cells are viable when TSH is removed for as long as 10 days, although cell growth stops. 5-Day TSH-depleted FRTL-5 cells were given TSH or antibody at the noted concentration for 60 hours, after which 75 000 c.p.m. of ^{125}I-labeled sodium iodide was added. After 40 minutes, cells were washed, recovered and the cell-associated iodide measured in the presence or absence of an ionophore (FCCP) which releases free but not bound or organified iodide. The difference is the free iodide uptake values presented above. Assays are in duplicate; standard errors are < ±5%. (See Weiss *et al.*, 1984 for further details.)

The antibodies selected from the potent 'inhibitory' group above were able to significantly block ^{125}I-TSH binding to human, bovine or rat thyroid membranes and to liposomes embedded with the high-affinity glycoprotein component of the thyroid membranes (Table 9.3). Inhibition was evident under many of the *in vitro* experimental conditions historically used to measure ^{125}I-TSH binding.

Their ability to interact with TSH receptors was not only evident in membrane preparations, but also whole cells (Yavin *et al.*, 1981a,b,c). Thus the antibodies from the inhibitory group were reactive with intact functioning rat thyroid cells, as well as thyroid membranes. Their interaction with the functioning thyroid cells was specifically inhibited by TSH but not hCG (human chorionic gonadotropin). Using these cells, the antibodies noted as inhibitors in Table 9.2 were all unable to mimic TSH as direct stimulators of adenylate cyclase activity (Fig. 9.3(A) and Table 9.4). They were, however, able to inhibit TSH-stimulated adenylate cyclase activity (Table 9.4); inhibition in

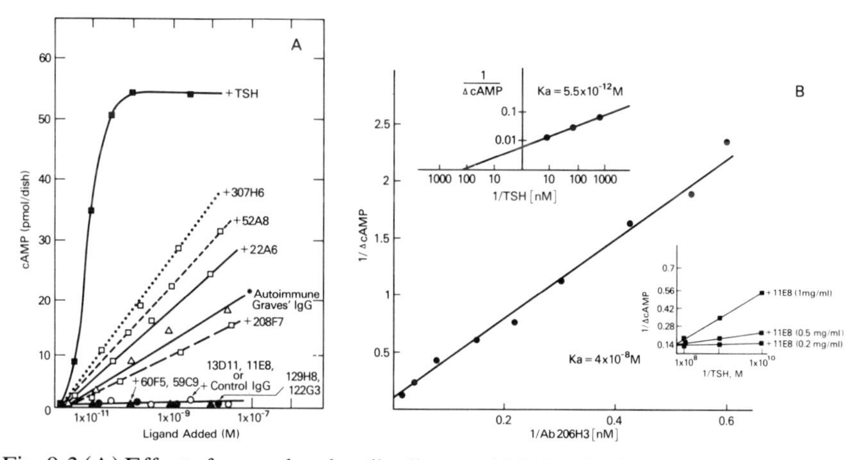

Fig. 9.3 (A) Effect of monoclonal antibodies on cAMP levels of FRTL-5 functioning rat thyroid cells in continuous culture. Antibodies 60F5 or 59C9 (▲), 13D11, 11E8 (O) and 129H8 and 122G3 (●) had no direct stimulatory action over the concentration range noted and were the same as a control IgG (O). Antibody molarity was calculated in this experiment using the total IgG protein content. Over the same concentration range, antibodies 307H6, 52A8, 22A6, 208F7 and 206H3 (similar to 208F7, see (B)) stimulated adenylate cyclase activity. The stimulatory action of a total IgG fraction from the sera of a Graves' patient is presented for comparison. The effect of the ligands on cAMP was measured using a technique similar to that in Vitti *et al.* (1983) and Valente *et al.* (1982a, 1983). (B) Double-reciprocal plot of the activity of 206H3 (body) by comparison with TSH (top insert) and double-reciprocal plot of the effect of 11E8 on the cAMP-stimulatory activity of TSH at the noted concentrations (bottom insert), with respect to 11C8 and TSH competitive inhibition is evident; the same results were evident for 129H8, 60F5 and 59C9. Concentration of antibody in this experiment was obtained by radioimmunoassay of IgG.

each case was competitive (Fig. 9.3(B), bottom insert). The antibodies also inhibited TSH-stimulated radioiodine uptake in thyroid cells (Table 9.4) and TSH-stimulated T_3/T_4 release in a mouse bioassay used to measure *in vivo* TSH activity (Table 9.4).

The ability of these antibodies to inhibit TSH binding and activity cannot be accounted for by the presence of a contaminating anti-TSH activity. This would be unlikely given (i) the competitive (as opposed to uncompetitive or mixed) inhibition data as well as (ii) data showing they have no anti-TSH activity in solid phase or soluble radioimmunoassay systems when directly tested using ^{125}I-labeled human or bovine TSH.

In summary, monoclonal antibodies classified as inhibitors in these studies, i.e. good competitive inhibitors of TSH binding and TSH-stimulated function, all appear to be antibodies directed at the membrane glycoprotein TSH-binding component. Further, these data established that, although the glycoprotein component is a portion of the physiologic TSH receptor on the surface of the thyroid cell, this component apparently is not the primary antigen against which autoimmune *stimulating* antibodies (TSAbs) are directed.

9.4 CHARACTERIZATION OF ANTIBODIES WHICH MIMIC TSH-STIMULATORY ACTION (TSABs) AS ANTIBODIES AGAINST THE GANGLIOSIDE RATHER THAN THE GLYCOPROTEIN MEMBRANE COMPONENT OF THE RECEPTOR

In evaluating the data above, it was noted that 22A6, 206H3 and 307H6, which were not significant inhibitors of TSH binding by comparison with other antibodies (Fig. 9.2 and Table 9.3), were, however, equipotent inhibitors of TSH-stimulated adenylate cyclase activity (Table 9.4). Resolution of this contradiction evolved with the observation that each of these antibodies was a potent direct stimulator of human (Table 9.4) and rat thyroid cell adenylate cyclase activity (Fig. 9.3(A)) and that their inhibitory action with respect to TSH was dependent on the TSH concentration. Thus, at low TSH levels, 22A6, 206H3, 307H6, 52A8 and 208F7 were more than additive competitive agonists, and at high TSH concentrations they were competitive antagonists (Valente *et al.*, 1982c). Also of interest, and in sharp contrast with the data in TSH-binding assays where the antibody exhibits a higher affinity than TSH (Davies *et al.*, 1977), the activation constant of the stimulatory IgG preparations, for example 206H3, was significantly poorer than TSH, 4×10^{-8} M versus 5.5×10^{-12} M, respectively (Figure 9.3(B)).

The antibodies were active in the mouse bioassay measuring thyroid hormone release (Table 9.4) and were able to enhance iodine uptake by thyroid cells in a manner identical with TSH (Table 9.4), i.e. under conditions where

Table 9.5 Ability of the stimulator antibodies to react with various ganglioside preparations by comparison with inhibitor and mixed antibodies*

Antibody	Ganglioside reactivity measured as ^{125}I-protein A or ^{125}I-anti-(human F(ab′)$_2$) binding to antibody–ganglioside complexes in a solid phase assay (c.p.m.)				
	No added lipid	Human thyroid ganglioside	Bovine thyroid ganglioside	Rat thyroid ganglioside	Mixed bovine brain ganglioside
Controls					
Monoclonal Control	150	160	140	155	155
Normal mouse IgG	128	185	160	125	135
Normal human IgG	149	144	138	160	145
Inhibitors					
13D11	141	235	210	175	215
11E8	156	206	310	210	204
59C9	121	195	215	195	198
60F5	110	210	240	210	200
129H8	101	285	180	256	170
122G3	146	268	195	235	155
Stimulators					
22A6	137	870	1101	982	410
206H3	156	2889	380	1210	128
307H6	129	4210	410	2450	240
Mixed					
52A8	135	1480	310	1210	205
208F7	136	920	260	1140	180

* Lipid extracts of human, bovine or rat thyroid membranes were prepared by chloroform/methanol extraction and Folch partitioning to generate ganglioside and neutral glycolipid fractions (Laccetti *et al.*, 1983, 1984). The binding system used gangliosides or neutral glycolipids adsorbed to microtiter wells (Laccetti *et al.*, 1983, 1984). After appropriate plate preparation, glycolipid in 50 μl of methanol was added to wells in a polyvinyl chloride 96-well plate, and evaporated; wells were washed with phosphate-buffered saline containing 1% bovine serum albumin. After 1 hour at room temperature, the wells were aspirated, and 50 μl of monoclonal antibody in the above buffer was added for overnight incubation at room temperature. The plate was centrifuged, washed twice and 20000 c.p.m. of ^{125}I-protein A or ^{125}I-labeled anti-(human IgG) [F(ab′)$_2$ fragment-specific] in 50 μl of bovine serum albumin buffer were added for 12 hours before terminal centrifugation and double washing. Bound radioactivity was counted in a gamma counter. Various specific gangliosides were used as the primary adsorbed antigen to assess the monoclonal antibody determinants. Controls included non-immune monoclonal antibodies as well as normal mouse and human IgG.

cAMP-dependent iodide uptake was the measured response (Weiss *et al.*, 1983).

The stimulating antibodies were all able to interact with ganglioside preparations of thyroid membranes (Table 9.5), whereas, with two exceptions, most were poorly reactive with the glycoprotein component (Table 9.3). The significance of the two exceptions, the 208F7 and 52A8 antibodies designated as mixed in Table 9.2 and which exhibit reactivity in both assays, will be discussed below.

The reactivity with the ganglioside preparations is relatively thyroid-specific when using total ganglioside extracts, i.e. when comparing bovine brain and bovine thyroid ganglioside preparations (Table 9.5). In addition, individual antibodies exhibit some species selectively for the thyroid ganglioside preparation, i.e. human better than bovine. In this respect, the poor reactivity of several Graves' autoimmune monoclonals with bovine by comparison with human ganglioside prepartions (206H3 and 307H6 for example) was coincident with a poorer adenylate cyclase-stimulatory activity in a bovine by comparison with a human thyroid system ($< 120\%$ as opposed to $> 250\%$ at comparable concentration). Species specificity of autoimmune stimulators has been a well-recognized phenomenon embodied in the definition of the Long acting thyroid stimulator (LATS) protector assay and the autoimmune stimulators (Adams, 1981; Doniach, 1975; Zakarija and McKenzie, 1980; Pinchera *et al.*, 1982).

The stimulating antibody reactivity with ganglioside preparations is lost if the glycolipid preparation is pretreated with neuraminidase (Table 9.6) and is

Table 9.6 Ability of monoclonal antibodies to react with modified or partially purified thyroid ganglioside preparations*

Ganglioside preparation	Ganglioside reactivity (c.p.m.)			
	22A6	307H6	52A8	Graves' IgG†
Control human thyroid ganglioside	930	3955	1525	2810
+ Neuraminidase	210	410	310	195
Control rat thyroid gangliosides	884	2290	1190	3100
+ Neuraminidase	195	290	206	301
Human neutral glycolipid‡	—	800	900	1100
monosialogangliosides‡	—	2000	1110	2100
disialogangliosides‡	—	10700	8050	14000
Rat FRTL-5 cell disialogangliosides	11400	22600	19400	13700

* Assayed as in Table 9.5.

† Graves' IgG from a patient with high stimulating activity. This is a 'polyclonal' preparation, i.e. a total IgG preparation with many potential individual antibodies including some which can inhibit TSH binding.

‡ Assayed as per Laccetti *et al.* (1983, 1984) using solid phase techniques.

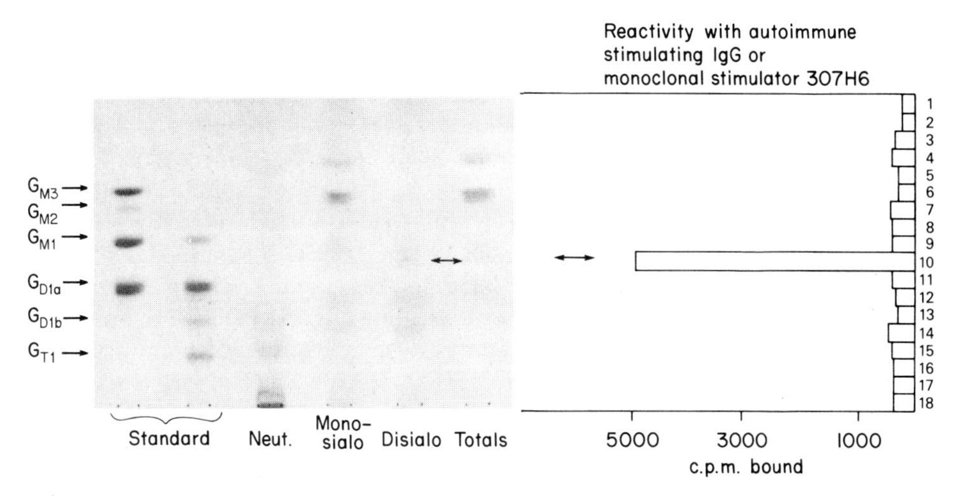

Fig. 9.4 Left. Thin-layer chromatograph of purified fractions, i.e. neutral glycolipids, monosialoglycolipids and disialoglycolipids, by comparison with the total human ganglioside starting preparation. Standards are in the first column; chromatography and purification was as per Laccetti *et al.* (1984). Right. Reactivity of 307H6 or a Graves' IgG preparation with individual gangliosides scraped from the thin-layer chromatograph eluted in methanol, and tested after drying and resolubilization in buffer in the solid phase assay (Laccetti *et al.*, 1983, 1984), wherein ganglioside is attached to the plate, incubated with one or the other antibody preparation, and then, after washing, reacted with ^{125}I-labeled protein A. As noted, the reactivity appears to coincide with a minor ganglioside component barely visible in the total human ganglioside extract but more apparent in the disialoganglioside fraction.

highest in disialoganglioside fractions obtained by column chromatographic techniques (Table 9.6). The highest reactivity is, however, evident with a single minor component of the disialoganglioside preparation (Table 9.6, Fig. 9.4).

In sum then the monoclonal data establish that a ganglioside is a component of the TSH receptor, that it is a minor component of the total ganglioside pool and that it appears to be vital in linking the recognition process to cAMP signal generation.

9.5 RELATIONSHIP BETWEEN THE TWO COMPONENTS OF THE TSH RECEPTOR DEFINED IN MONOCLONAL ANTIBODY STUDIES

Cytochemical assay techniques (Chayen *et al.*, 1973, 1976) have been used to study the activity of thyroid-stimulating autoantibodies (TSAbs) obtained from the sera of patients with Graves' disease, as well as TSH (Bitensky *et al.*,

1974; Ealey *et al.*, 1981). This assay was used to compare the activities of the antibodies and TSH when they were mixed together. The cytochemical bioassay (CBA) for thyroid stimulators is based upon quantification of changes in the staining of sections of guinea pig thyroids, utilizing leucine-2-naphthylamide as a substrate. In the CBA, 22A6, 307H6 and 208F7 are stimulators whose dose–response curves parallel these of a LATS-B standard, a Graves' serum TSAb (Fig. 9.5). In contrast to 22A6, 307H6 and 208F7, 11E8 is itself inactive as a stimulator in the CBA over a wide dose range (Fig. 9.5); it does, however, inhibit TSH stimulation in the CBA at a concentration where 22A6 is a stimulator (Fig. 9.6). In contrast to its effect on TSH, 11E8 shows relatively low potency (> 10 000-fold lower) when inhibiting stimulation of the thyroid-stimulating antibodies 22A6, 307H6 and LATS-B (Fig. 9.6), i.e. even 10^2 dilutions of antibody have no effect. As anticipated, the response to

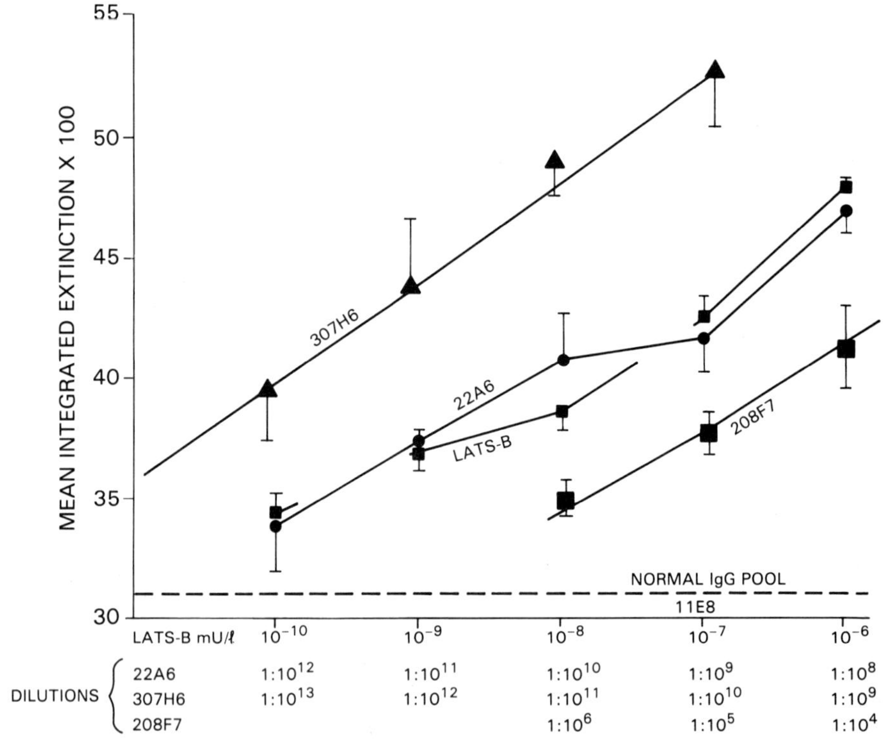

Fig. 9.5 Responses to increasing doses of monoclonal 22A6, 307H6, 208F7 and LATS-B reference preparation in the cytochemical bioassay for thyroid stimulators. The 12 μM sections of guinea pig thyroid tissue were exposed to the stimulators and the controls, normal IgG or 11E8, for 3 minutes. The results are the mean values of mean integrated extinction obtained from triplicate sections, ± S.D. Each preparation had a similar starting protein concentration, i.e. 15 to 25 ml of IgG protein/mg (Ealey *et al.*, 1984a,b).

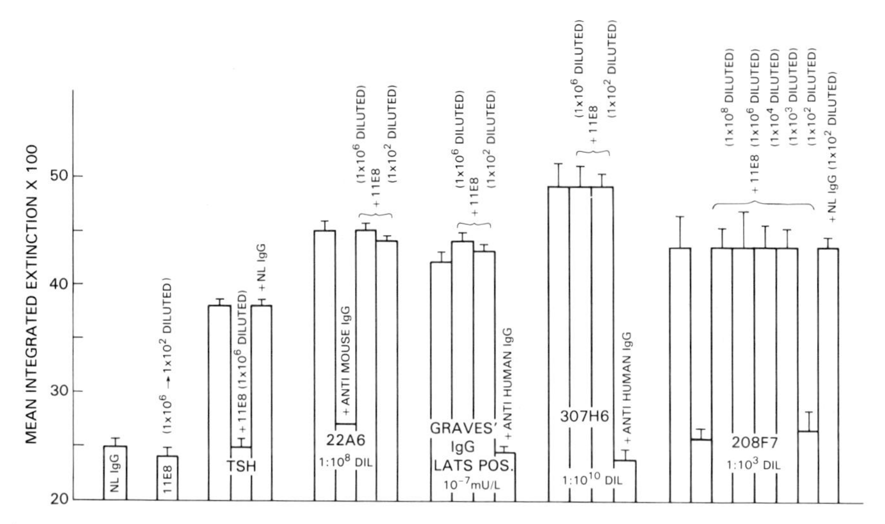

Fig. 9.6 Cytochemical bioassay showing the inhibitory effect of 11E8 on the activity of TSH as opposed to the lack of effect of 10 000-fold higher concentrations of 11E8 on TSAbs such as 22A6, 307H6 or the LATS-B standard. Inhibition by 11E8 can just be seen with 208F7 at a 10 000-fold higher 11E8 concentration. Stimulator activity is inhibited by an anti-(mouse IgG) or anti-(human IgG) as appropriate. Data obtained from Drs P. Ealey and N. Marshall, Middlesex Hospital, London. A preliminary account of these data was presented at the European Thyroid Meeting, 1982; manuscripts concerning these data have been submitted to *Endocrinology* and *J. Clin. Endocrinol. Metab.*

LATS-B and 307H6 is inhibited by anti-(human IgG), whereas that due to 22A6 is inhibited by anti-(mouse IgG) (Fig. 9.6).

Antibody 208F7 is inhibited by 10^2 dilutions of 11E8 (Fig. 9.6). This phenomenon, as well as its 'mixed' characteristics, is explored below.

Differences in the effect of 11E8 on TSH as opposed to 22A6, 307H6, 208F7 or LATS-B are not the result of a 10 000-fold difference in binding constants. This is intuitive from the observation that the concentration of 11E8 able to inhibit TSH activity was similar to the optimal concentration for 22A6 or 307H6 direct stimulatory action, i.e. both were active at similarly low concentrations. Direct measurements of the ability of 22A6, 307H6, 208F7 and 11E8 to bind to rat, human or guinea pig thyroid membranes using ^{125}I-labeled anti-(mouse IgG) and a radioimmunoassay detection technique (Yavin *et al.*, 1981a,b,c) confirmed this.

The ability of 11E8 as opposed to 22A6, 307H6 or 208F7 to stimulate or block selectively indicates that these two monoclonals are antibodies to different determinants on the TSH receptor. Since separate experiments indicate that the 11E8 monoclonal antibody interacts predominantly with a

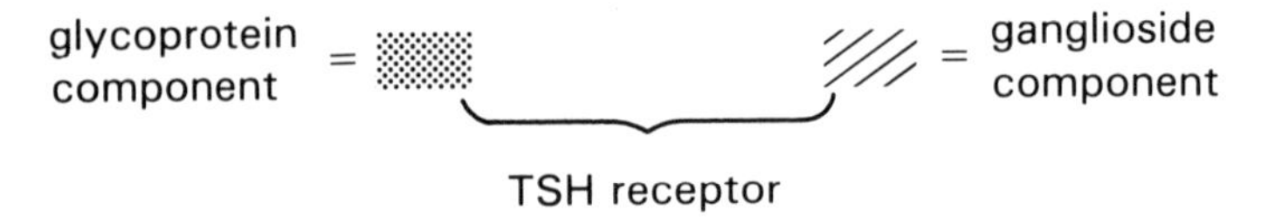

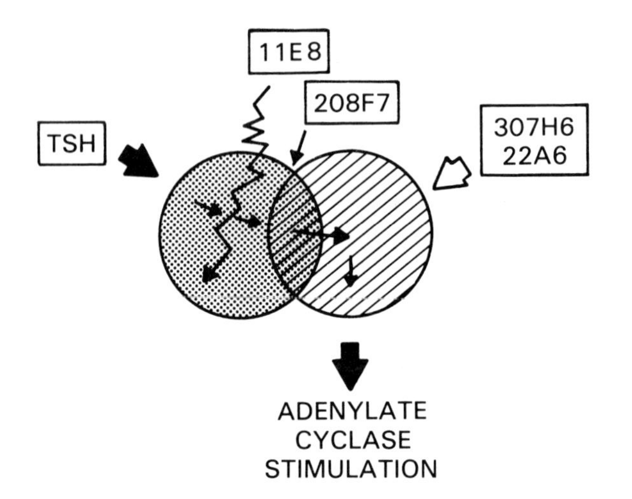

Fig. 9.7 Hypothetical model of TSH receptor as suggested by studies with monoclonal antibodies to the TSH receptor.

membrane glycoprotein, whereas the 22A6 and 307H6 monoclonal antibodies interact with a ganglioside (see above), the simplest way of reconciling the above observations is to apply the two-component receptor model suggested in receptor-binding studies (Figs 9.1 and 9.7). Thus, TSH can be envisaged to interact first with the glycoprotein component of the TSH receptor, which exhibits high-affinity binding properties. Its biological action, however, requires an additional or subsequent interaction with a ganglioside. In contrast, TSAbs represented by 307H6, LATS-B and 22A6 can be envisaged to bypass the glycoprotein receptor component and interact with the ganglioside to initiate the hormone-like signal. This model would be consistent with the observations that TSH action can be blocked by 11E8, an antibody binding to the glycoprotein receptor component. The three TSAbs, 307H6, LATS-B and 22A6, are in contrast minimally affected by 11E8 since they are directed at the next sequential step, the ganglioside, which is vital in ligand message transmission.

Antibody 208F7 raises another issue in this respect. It can interact with both ganglioside (Table 9.5) and glycoprotein receptor components (Table 9.3). It is a good inhibitor of TSH binding (Fig. 9.2) and a potent stimulator of adenylate cyclase activity (Fig. 9.3(A) and Table 9.4). It is also sensitive to 11E8 inhibition, albeit at very high concentrations by comparison with TSH (Fig. 9.6). These findings suggest that there is an 'overlap' or 'interaction' between the two receptor components and that 208F7 interacts at this common site (Fig. 9.7)*. It thus has a steric or spatial relationship between TSH or 11E8 and 307H6, 22A6, or LATS-B, as depicted in Fig. 9.7.

9.6 RECONSTITUTION STUDIES

The model in Fig. 9.7 has been supported by reconstitution studies. Thus, an important study implicating higher-order gangliosides as a potential component of the thyrotropin receptor showed that membrane preparations from the 1–8 rat thyroid tumor with a TSH receptor defect were coincidentally devoid of higher-order gangliosides (Meldolesi *et al.*, 1976). The TSH receptor defect was expressed as low TSH binding and no TSH-stimulated adenylate cyclase activity despite normal thyroid functional responses to prostaglandins or dibutyryl cyclic AMP (Mandato *et al.*, 1975; Meldolesi *et al.*, 1976). The original 1–8 tumor also had no cholera-toxin-stimulated adenylate cyclase response despite the ability of forskolin to activate adenylate cyclase activity (Table 9.7). During the course of 10 years of tumor passage, it was noted that a 1–8 variant evolved, 1–8N, which regained some TSH-binding activity and both TSH- and cholera-toxin-responsive adenylate cyclase activity (Table 9.7).

The defect in TSH and cholera-toxin-stimulated adenylate cyclase activity in the original 1–8 tumor line had been associated with a defect in the synthesis of higher-order gangliosides (Meldolesi *et al.*, 1976), i.e. only G_{M3} could be detected in its membrane fraction by thin-layer-chromatographic analysis. Analysis of the 1–8N tumor which had regained cholera toxin and TSH receptor sensitivity indicated that there were now small amounts of higher-order gangliosides chromatographically detectable in the G_{M1} region and that at least some of these were higher order as measured by their ability to interact with ^{125}I-labeled cholera toxin (Laccetti, *et al.*, 1983). Although these data were compatible with the conclusion that the return of cholera-toxin- and TSH-sensitive adenylate cyclase activity were coincident with the regained ability of the tumor cells to synthesize higher-order gangliosides, they did not exclude the requirement for a glycoprotein component to have full expression of TSH receptor activity. Thus trypsinized 1–8N membranes lost TSH binding and TSH-stimulated adenylate cyclase activity (Table 9.7) in the same manner as described earlier for trypsinized thyroid cell or membrane preparations

*Editor's comment: Presumably antibody 208F7 could recognize an oligosaccharide sequence present on both the glycoprotein and glycolipid of the postulated receptor structure.

Table 9.7 *In vivo* reconstitution of TSH receptor expression in 1–8 rat thyroid tumors

Thyroid membrane or cell preparation	^{125}I-TSH binding activity (c.p.m./20 μg of membrane protein)		Adenylate cyclase activity (pmol of cAMP/μg of membrane protein/20 min)					
					Cholera toxin		TSH	
	pH 7.4, 37 °C 50 mM NaCl	pH 6.0, 0 °C	Basal	Forskolin 1×10^{-5} M	1×10^{-9} M	1×10^{-8} M	1×10^{-9} M	1×10^{-8} M
Normal rat	1890	20 980	11	140	46	80	28	54
1–8 tumor (original)	320	4 050	10	135	11	8	10	12
1–8N tumor (variant)	1716	10 900	8	120	20	43	14	25
1–8N tumor after trypsin treatment	415	1 400	11	125	—	—	10	9

Table 9.8 Reconstitution of TSH receptor expression in 1–8 rat thyroid tumors

Thyroid membrane or cell preparation	Adenylate cyclase activity* (pmol of cAMP/μg of DNA)						
		Cholera toxin		TSH			
	Basal	1×10^{-9} M	1×10^{-8} M	1×10^{-9} M	1×10^{-8} M	307H6† (20 μg/ml)	Graves' IgG† (1 mg/ml)
FRTL-5 thyroid cell	0.5	6.4	9.5	18	17	5.4	9.0
1–8 tumor (original)	0.8	0.4	0.5	0.6	0.5	0.7	0.6
+1–8N thyroid tumor gangliosides	0.6	4.4	6.4	3.5	5.6	—	—
+FRTL-5 thyroid cell gangliosides	0.7	3.8	5.8	5.2	7.2	2.6	4.1
+NDase‡-treated 1–8N gangliosides	0.7	1.5	1.8	0.9	0.2	—	—
+NDase‡-treated FRTL-5 gangliosides	0.7	1.2	1.4	0.9	0.7	0.8	1.0
+Mixed brain gangliosides	0.8	3.1	4.8	1.1	1.3	0.8	0.9
+1–8 original gangliosides	0.5	0.7	0.7	0.5	0.6	—	—
+G_{M3}	0.8	0.7	0.6	0.7	0.8	0.8	0.6
+FRTL-5 thyroid cell gangliosides followed by trypsin treatment§	0.5	—	5.2	—	1.2	1.9	3.4

* Assays were performed in triplicate; results are the average of at least three separate experiments. In no case did the standard deviation of any value exceed ±10%.

† 307H6 is expressed as human IgG; the Graves' IgG is total IgG.

‡ NDase, neuraminidase.

§ Trypsin treatment of cell preparations used the procedures detailed in Winand and Kohn (1975).

(Winand and Kohn, 1975; Tate *et al.*, 1975; Pekonen and Weintraub, 1980). They also lost the ability to bind the 11E8 monoclonal antibody to the glycoprotein receptor component as measured by ^{125}I-protein A binding (1250±40 c.p.m. as opposed to 310±50 c.p.m. before and after trypsinization, respectively).

Incubation of a ganglioside extract from the 1–8N tumors with primary cultures of the original 1–8 tumor resulted in return of TSH-, TSAb-, and cholera-toxin-stimulated adenylate cyclase activity (Table 9.8). Reconstitution could also be effected by gangliosides from the cultured rat FRTL-5 thyroid cells which have a functional TSH receptor sensitive to the monoclonal anti-receptor stimulators which interact with gangliosides (see Fig. 9.3). No reconstitution of TSH receptor function occurred in control reconstitution incubations containing no added 1–8N ganglioside extract, a ganglioside extract from the original 1–8 tumor (G_{M3}), mixed brain gangliosides, or the 1–8N or FRTL-5 ganglioside extract treated with a mixture of neuraminidases capable of converting 87% to 92% of the sialic acid residues from a lipid-bound to free form.

Incorporation of the gangliosides from the 1–8N tumor variant or from FRTL-5 thyroid cell preparations into the 1–8 original tumor cells was monitored by reactivity with the 22A6 or 307H6 stimulating monoclonal anti-receptor antibodies; these antibodies have been shown above to react with thyroid ganglioside preparations (see above). Thus, for example, as measured by ^{125}I-protein A, 22A6 binding to the 'no-addition' cells, to cells incubated with G_{M3}, or to cells incubated with neuraminidase-treated 1–8N gangliosides was 170±50 c.p.m./mg of membrane protein. The 22A6 binding to cells incubated with 1–8N or FRTL-5 gangliosides was 480±50 and 610±40 c.p.m., respectively (p values < 0.01 compared with the controls above).

Again, the reconstitution data should in no way be construed to negate the importance of the glycoprotein component of the TSH receptor as noted from the trypsin-sensitivity data in Table 9.8. Thus 1–8 cells, reconstituted with FRTL-5 gangliosides, and then exposed to trypsin, lost their ability to respond to TSH. It is, however, notable that the TSAb- and cholera-toxin-stimulation activity were significantly less altered by the trypsin treatment (Table 9.8), i.e. TSAb behaves as an autoimmune equivalent of cholera toxin.

The FRTL-5 reconstitution data have relevance to a recent report (Beckner *et al.*, 1981) which argued that these cells did not have a ganglioside receptor component since they had few higher-order gangliosides, and TSH stimulation was not inhibited by mixed brain gangliosides. The data in Fig. 9.4 and Tables 9.6, 9.7 and 9.8 would suggest that simple identification of a bigger or smaller content of higher-order gangliosides in FRTL-5 thyroid cells or any other thyroid cell by thin-layer analysis of a total ganglioside extract does not establish that FRTL-5 gangliosides are or are not related to TSAb or TSH receptor function, particularly given the reconstitution data. Accordingly,

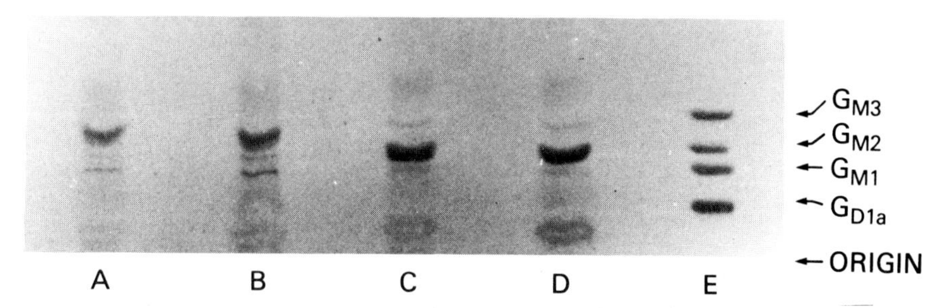

Fig. 9.8 Thin-layer chromatograms of ganglioside extracts from FRTL-5 thyroid cells evaluated by resorcinol staining. Lanes A and B: total FRTL-5 membrane gangliosides. Lanes C and D: FRTL-5 membrane disialogangliosides recovered from DEAE. Lane E: a mixture of pure G_{M3}, G_{M2}, G_{M1} and G_{D1a}.

both points were examined in greater detail (Laccetti *et al.*, 1984) to resolve this question.

A thin-layer chromatograph of total rat thyroid cell FRTL-5 gangliosides (lanes A and B) and the DEAE-purified disialoganglioside fraction of FRTL-5 thyroid cells (lanes C and D) is presented in Fig. 9.8; identification is by resorcinol staining. The appearance of the total FRTL-5 cell ganglioside pattern does not exactly resemble that presented by Beckner *et al.* (1981), who used radiolabeling techniques in their identification analysis; nevertheless, there is a relatively small proportion of ganglioside migrating in areas with high-order ganglioside standards. Despite this finding, neuraminidase digestion of the ganglioside(s) does result in a G_{M1} content of ~25% by quantitative scanning densitometry, i.e. at least 25% of the total ganglioside mixture are higher-order gangliosides (Laccetti *et al.*, 1984); the identification of G_{M1} and its increase after neuraminidase treatment was confirmed by overlaying thin-layer chromatograms with ^{125}I-labeled cholera toxin. The complex nature of the gangliosides in FRTL-5 cells is further suggested by experiments involving fucosidase digestion of the FRTL-5 mixture. Fucosidase treatment alone caused only minor changes in the total ganglioside pattern but totally altered the neuraminidase results. The data were not altered by assay incubation in Triton X-100 at concentrations between 0.1% and 5% or by ganglioside pretreatments with Pronase. Finally, lanes C and D show that one can isolate a significant amount of disialoganglioside reactive with 307H6, 22A6, 52A8 and Graves' IgG (Table 9.6). In sum, a portion of the FRTL-5 ganglioside mix is higher-order and reacts with stimulating autoantibodies to the TSH receptor.

The ability of mixed brain gangliosides and FRTL-5 gangliosides to alter 307H6- and TSH-stimulated cAMP activity of FRTL-5 cells is compared in Fig. 9.9. The significantly greater activity of the FRTL-5 by comparison with the mixed brain ganglioside preparation is consistent with the differences in Table 9.5 and with data concerning TSH–ganglioside interaction constants *in vitro*.

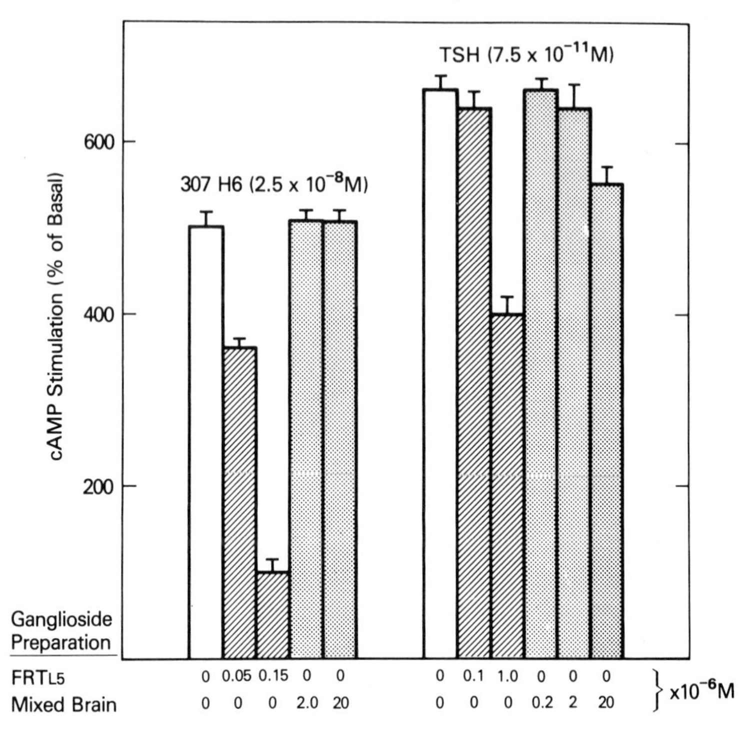

Fig. 9.9 Effect of gangliosides at the noted concentration on 307H6 and TSH stimulation of cAMP levels in FRTL-5 rat thyroid cells. Gangliosides were incubated with the ligands; assays were performed using cells in media with no TSH for 10 days; incubations were stopped and cAMP assays initiated 2 hours and 30 minutes, respectively, after the start of the 307H6 and TSH incubations.

9.7 DYNAMIC RECEPTOR MODEL: IMPLICATIONS FOR OTHER LIGANDS, RECEPTOR EXPRESSION AND MIXED ANTIBODIES

The monoclonal antibody and reconstitution data confirm the existence of both a glycoprotein and ganglioside component to the TSH receptor. The existence of mixed monoclonal antibodies would suggest (Fig. 9.7) that the receptor exists as a complex of the two glycoconjugates. Predominantly stimulating (cAMP) and predominantly inhibiting (TSH binding) could be taken to represent antibodies viewing the receptor complex from different orientations and/or antibodies which are directed at free glycoprotein or ganglioside (Fig. 9.7). In either case a model of a receptor complex of both components in equilibrium with free components is an intuitive extrapolation (Fig. 9.10, top). Further, the gangliosides can be viewed as a group or series of similar structures with greater

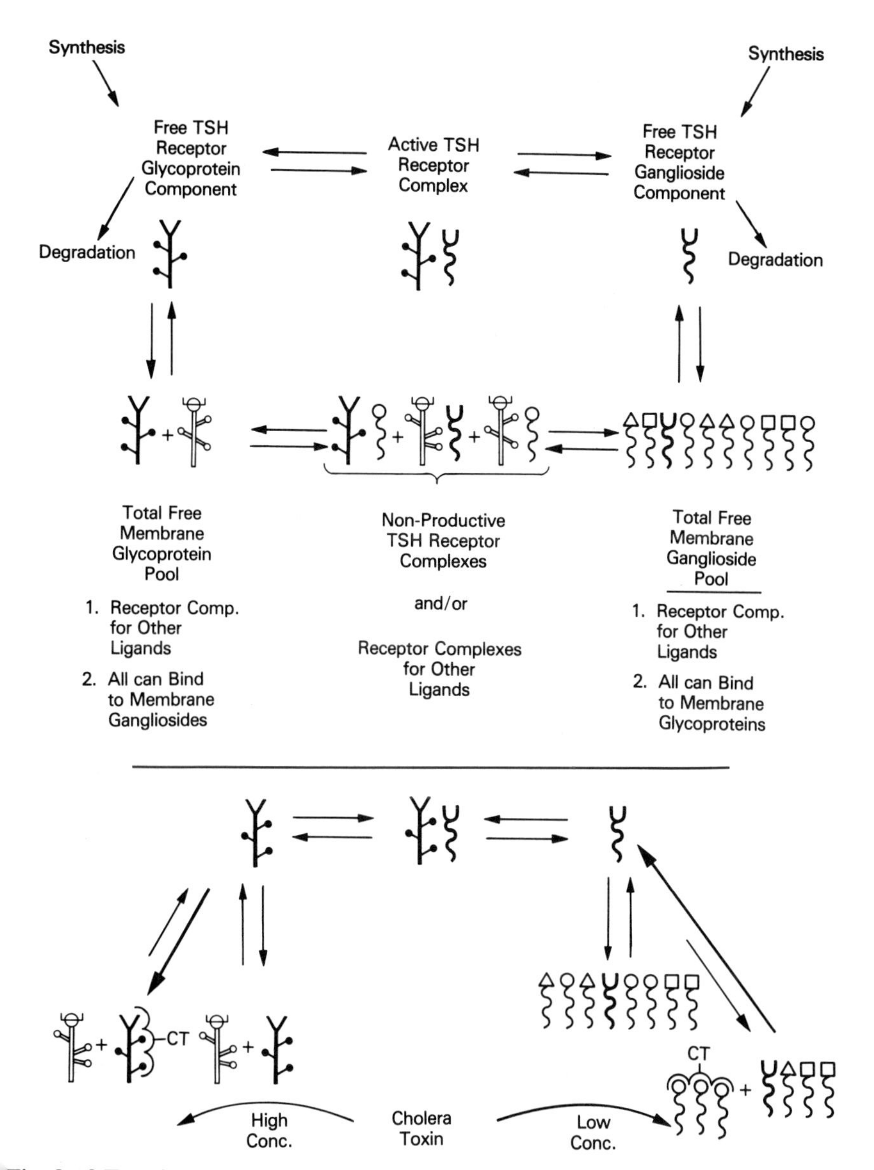

Fig. 9.10 Top, hypothetical model of active TSH–receptor complex in equilibrium with its free component parts which are in turn in equilibrium with other glycoproteins and gangliosides of the membrane. It is presumed the complex of both units exists since mixed antibodies exist, i.e. these are presumed to be directed against the complex as one possibility. The alternative, that TSH induces the complex to form, is not, however, excluded. Bottom, the presumption of the action of cholera toxin on this equilibrium system. These interactions could account for data concerned with the effect of cholera toxin on TSH binding and function (in the absence of exogenous NAD) when cholera toxin (CT) is added to membrane systems. This is not a unique model but is based on reasonable presumptions from the current available data.

or lesser affinities for a particular ligand but each with the potential for interacting with a glycoprotein – including the TSH receptor glycoprotein. The implications of such a model are far-reaching.

The TSH receptor has been related to receptor structures for cholera toxin, tetanus toxin and interferon (Kohn, 1978). Thus, each of these ligands has been associated with a receptor structure, involving gangliosides, albeit structures with different ganglioside specificities. Cholera toxin interacts with G_{M1} and, like the TSAb, bypasses any functional need for the glycoprotein receptor component despite the existence of glycoproteins capable of binding the cholera toxin (Grollman *et al.*, 1978). Similarly, interferon appears to interact with a glycoprotein and ganglioside receptor structure. All three, TSH, cholera and interferon, can influence the binding and bioactivity of each other (Kohn, 1978; Grollman *et al.*, 1978). Thus, cholera can first enhance TSH binding then inhibit. Similarly, interferon can enhance then inhibit cholera toxin binding and inhibit TSH binding. That these phenomena may reflect interactions of biologic consequence is exemplified by the fact that the anti-viral potency of mouse interferon using functioning, TSH-stimulated rat thyroid cells exceeds by 100- to 10000-fold the potency of interferon using non-functional, non-TSH-stimulatable rat thyroid cells and even by 10-fold the interferon activity in a sensitive mouse L cell strain (Friedman *et al.*, 1982). That these phenomena involve interactions with components of receptor structure is illustrated by the fact that the cholera toxin interaction results in enhanced surface exposure of higher-order gangliosides in the G_{M1} to G_{D1b} area, where the ganglioside with the highest affinity to TSH migrates (Kohn, 1978).

A possible simple explanation for these phenomena resides in the equilibrium model of Fig. 9.10 (top). Thus, cholera toxin at low concentrations (Fig. 9.10, bottom) interacts with G_{M1} and favors the interaction of the more relevant TSH receptor ganglioside with the TSH receptor glycoprotein, since the competing free ganglioside pool is smaller and its affinity is less. At high concentrations, cholera toxin would also interact with the TSH receptor glycoprotein and decrease its availability for forming the active TSH–receptor complex. A similar model might account for the interferon data.

9.8 THE ANTI-IDIOTYPE PROBLEM

Description of the theory of anti-idiotype and anti-anti-idiotype antibodies to receptors will be summarized elsewhere (see Chapter 2 by Strosberg and Schreiber). In essence, if one adapts these ideas, one can presume that the monoclonal TSAb which stimulates adenylate cyclase activity via the ganglioside component of the receptor is not only anti-receptor, but equivalent to hormone. Under this circumstance, the prediction should be that some antibody species in anti-TSH preparations should inhibit TSAb activity, i.e. the

Table 9.9 Ability of anti-(human TSH) to inhibit TSAb stimulatory activity

Ligands added	cAMP Level (pmol of cAMP/μg of DNA)*	Inhibition (%)
None	0.5	—
TSH (1×10^{-11} M)	24	—
+Anti-(TSH IgG)	0.4	100
+Anti-(cholera toxin	25	—
+Anti-(thyroglobulin IgG)	26.1	—
Normal IgG	0.4	—
Anti-(TSH IgG)	0.5	—
Graves' serum TSAb IgG	8.2	0
+Anti-(TSH IgG)	3.8	54
+NL IgG	8.1	0
+Anti-(thyroglobulin IgG)	8.4	0
208F7	3.2	0
+Anti-(TSH IgG)	1.5	54
+NL IgG	3.4	0
307H6	5.2	0
+Anti-(TSH IgG)	1.7	77
+NL IgG	5.1	0
+Anti-(thyroglobulin IgG)	5.4	0
52A8	4.1	0
+Anti-(TSH IgG)	2.8	32
+NL IgG	4.2	0
Cholera toxin	17	—
+Anti-(TSH IgG)	4	76
+NL IgG	18	—
+Anti-(thyroglobulin IgG)	17	—

*Values are the mean of triplicate assays; error values were in all cases less than ±4.5%.

determinants involved in the anti-receptor should be identical with those of TSH and some species of anti-TSH should interact with this. As noted in Table 9.9, this is true; further, it is true for a Graves' serum TSAb and cholera toxin as well.

These data support the conclusions of the two-component model of TSH receptor structure but raise issues concerning the cholera toxin model. Thus, reconstitution studies have established the importance of the G_{M1} ganglioside, but have neither eliminated the glycoprotein receptor component role, nor answered how cholera toxin avoids the glycoprotein 'forest' covering the glycolipid 'shrubs' on the cell surface. These data would suggest that cholera toxin does indeed interact with the glycoprotein receptor component and that the ganglioside is merely the kinetically detected step because of the unusual affinity and multivalent binding state of the β-subunit.

9.9 MONOCLONAL ANTIBODIES, THE TWO-COMPONENT RECEPTOR MODEL AND TSH-RECEPTOR-MEDIATED GROWTH

Recent studies have confirmed that TSH is an important stimulus of thyroid cell growth (Ambesi-Impiombato *et al.*, 1980; Valente *et al.*, 1982a,c, 1983a,b; Rotella *et al.*, 1984; Sluszkiewicz and Pawlikowski 1980) and that this action is not entirely mediated by a cAMP signal. In this respect one study (Valente *et al.*, 1982b) has shown that TSH-stimulated thymidine uptake in functioning rat thyroid cells is a value measure of TSH-dependent thyroid cell growth, is distinct from TSH effects on cAMP levels with respect to hormone concentration, thyroid cell pretreatments, or thyroid cell species and cannot be duplicated when dibutyryl cAMP replaces TSH. Thus TSH effects on thymidine uptake or growth appear to utilize a different receptor domain or transmission system.

The monoclonal antibodies above, as well as Graves' serum autoantibodies, have been evaluated as TSH-related 'growth' promoters in thymidine uptake assays in the functioning rat thyroid cells (Valente *et al.*, 1982a,c; Rotella *et al.*, 1984). The data indicate (Table 9.10) that some of the 'inhibitors' of the TSH-stimulated cAMP signal may be intrinsic stimulators in the growth assay in the absence of any effect on adenylate cyclase stimulation. Thus, monoclonal antibody 129H8, as well as monoclonal antibodies 22A6, 208F7, 52A8 and 206H3, were all found to be significant stimulators of [^{14}C]thymidine uptake. The greater role of the glycoprotein component was suggested by the

Table 9.10 Effect of monoclonal antibodies on [^{14}C]thymidine uptake in continuously cultured FRTL-5 cells*

Ligand added	[^{14}C]Thymidine uptake (c.p.m./μg of DNA)±S.D.
None	1890±197
Normal IgG	2287± 66
Buffer control	1990±200
TSH (10 mU/ml)	6070±350
TSH (1 mU/ml)	2880±418
TSH (0.1 mU/ml)	1650± 23
129H8	2780±157
122G3	2247± 85
122G3+TSH (10 mU/ml)	2860± 51
129H8+TSH (10 mU/ml)	9500±290

*Assays performed as detailed by Valente *et al.* (1982c). Values are the mean of quadruplicate samples in separate experiments. Antibodies were tested at 0.1 mg/ml.

fact that 129H8 was not a stimulator of adenylate cyclase (Table 9.4) activity nor significantly ganglioside reactive (Table 9.5) and the most potent stimulators on a concentration basis were 208F7 and 52A8, so-called mixed antibodies (Table 9.2). Antibody 122G3 was not a significant stimulator in this experiment (Table 9.10), but was an effective inhibitor of TSH-induced [^{14}C]thymidine uptake at the same concentration. This inhibitory effect has also been noted for bovine antibody 11E8 and human antibodies 59C9 and 60F5.

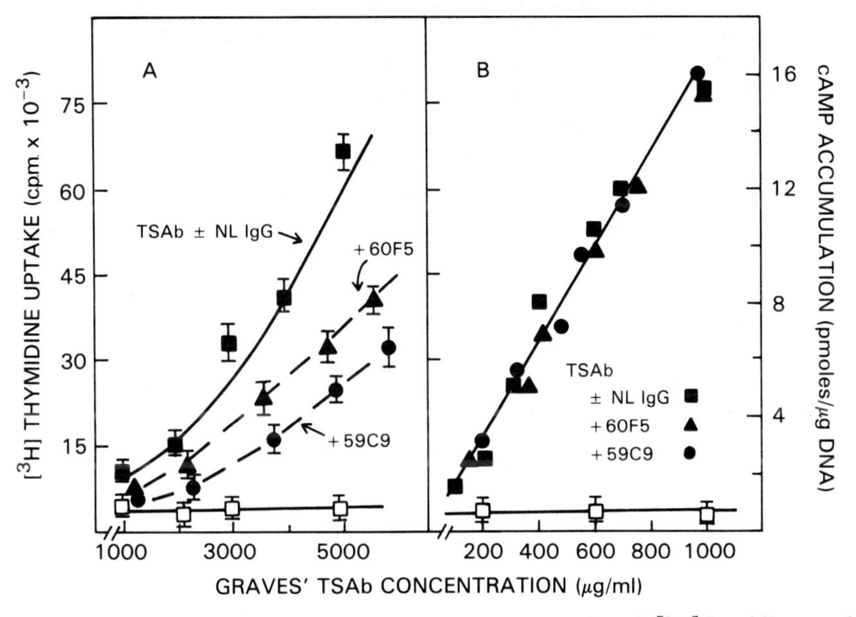

Fig. 9.11 Effect of an autoimmune Graves' TSAb preparation on [^{3}H]thymidine uptake (A) and cAMP levels (B) in the presence of a normal mouse IgG (■), 59C9 (●) or 60F5 (▲). The autoimmune TSAb IgG had the ability to increase thyroid cell growth as well as cAMP levels. The control normal human IgG (□) is compared with the TSAb. The 59C9 and 60F5 concentrations in A and B were 0.1 mg and 0.2 mg/ml respectively. In (B), increases of 60F5 and 59C9 to 1 mg/ml did not alter the data. Data with 122G3 are the same as 60F5 and 59C9 in A and B.

The ability of 122G3, 11E8, 60F5 or 59C9 to inhibit the stimulatory effect of TSH on thymidine uptake is paralelled by the inhibitory effect of each on TSH-stimulated adenylate cyclase activity. In contrast, although these four antibodies inhibit TSAb-stimulated thymidine uptake (Fig. 9.11(A)), they have no effect on TSAb-stimulated cAMP accumulation (Fig. 9.11(B)). All four antibodies are directed against the glycoprotein component of the TSH receptor. They competitively inhibit ^{125}I-TSH binding to thyroid membranes, competitively inhibit TSH-stimulated adenylate cyclase activity in thyroid cells, and have no apparent intrinsic adenylate cyclase-stimulatory activity. The results obtained with the four antibodies thus suggest that the thymidine stimulatory effect of the TSAbs is not only related to the same TSH receptor as characterized in the studies of monoclonal antibodies, but also indicates (i) that the locus of action within the TSH receptor, with respect to IgG stimulation of thymidine uptake, is distinct from that related to IgG-induced alterations in cAMP levels and (ii) that the glycoprotein receptor component has a more functional role with respect to growth than cAMP modulation.

It thus appears possible, on the basis of the monoclonal antibody studies, that thyroid stimulation in Graves' disease may result from a small goiter potently stimulated by a 307H6-type antibody to actively release thyroid hormones, from a larger goiter, caused by a 129H8 antibody, whose release of excess thyroid hormone reflects the existence of an excess of responsive tissue, or from a combination of both types of antibodies. In this regard, it is evident that the presence of an autoantibody with the characteristics of 129H8, i.e. an antibody to the glycoprotein component of the TSH receptor which is a growth stimulator, in addition to an antibody which is a cAMP stimulator such as

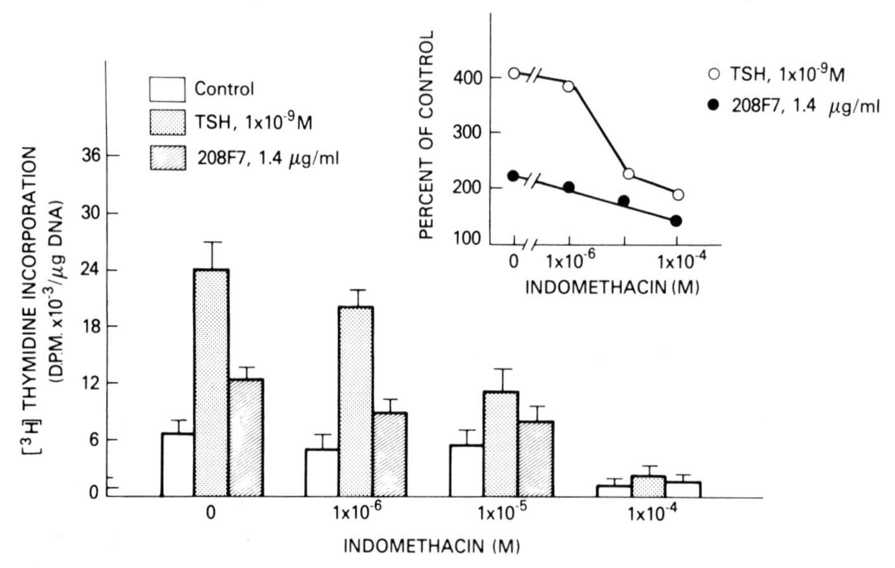

Fig. 9.12 Effect of TSH and a monoclonal 'mixed' autoantibody (208F7) on thyroid cell growth as measured by [^{3}H]thymidine incorporation in FRTL-5 thyroid cells maintained for 24 hours in a medium in the absence of TSH, an absolute growth requirement. [^{3}H]Thymidine uptake was compared (i) in cells with either no addition to the medium or with an equivalent concentration of normal IgG (control, open bars), (ii) in cells to whose media was added 1×10^{-9} M TSH, or (iii) in cells whose media contained 1.4 μg of TSAb/ml in addition to 0, 1×10^{-6} M, 1×10^{-5} M or 1×10^{-4} M indomethacin. [^{3}H]Thymidine incorporation was measured as detailed in Rotella *et al.* (1984). The inset presents the data in the body of the figure as a percentage of control, i.e. as a percentage of the value for the first bar in each series of three values at 0, 1×10^{-6} M, 1×10^{-5} M and 1×10^{-4} M indomethacin.

307H6, would result in two independent and potent stimulators being present simultaneously and, most likely, the most severe cases of thyroid stimulation. In this last respect, the 129H8 autoantibody capable of inhibiting TSH binding and function would have no inhibitory action on 307H6. The cAMP stimulatory antibody activity (TSAb) would dominate phenotypically since inhibition of TSAb activity requires a > 10000-fold higher activity coefficient than inhibition of TSH. The fundamental basis for this is their interaction with distinct sites of the TSH receptor. This means thyrotoxicosis would prevail in almost all cases.

In sum, studies of monoclonal antibodies have established that the same TSH receptor structure important in regulating adenylate cyclase activity is also important in regulating thyroid cell growth. The monoclonal antibody studies have further established that the signal pathway for growth involves intermediates of arachidonic acid since both the growth activity of these and TSH is inhibited by indomethacin (Fig. 9.12). Finally, the monoclonal antibodies suggest that there is a different relationship in the ganglioside–glycoprotein receptor components for growth as opposed to cAMP-stimulatory action.

In short, monoclonal antibodies have opened the door to studies at a molecular level of TSH-stimulated and TSH-receptor-mediated thyroid cell growth.

9.10 SUMMARY

The present report summarizes experiments with monoclonal antibodies to the TSH receptor. The data provide further insight into the TSH receptor structure and into the basis of autoimmune antibodies implicated in the pathogenesis of Graves' disease. They also establish an unequivocal role for the ganglioside in receptor structure which facilitates interpretation of *in vitro* experiments aimed at understanding the mechanism of ganglioside–ligand interactions. They offer some initial insight suggesting how the same receptor structure yields independent signals to the cell and how different ligands could utilize related membrane molecules to evolve a highly specific receptor structure.

REFERENCES

Adams, D.D. (1981), in *Vitam. Horm.*, **38,** 119–203.

Aloj, S.M., Kohn, L.D., Lee, G. and Meldolesi, M.F. (1977), *Biochem. Biophys. Res. Commun.*, **74,** 1053–1059.

Aloj, S.M., Lee, G., Grollman, E.F., Beguinot, F., Consiglio, E. and Kohn, L.D. (1979), *J. Biol. Chem.*, **254,** 9040–9049.

Ambesi-Impiombato, F.S., Park, L.A.M. and Coon, H.G. (1980), *Proc. Natl. Acad. Sci. U.S.A.*, **77,** 3455–3459.

Beckner, S.K., Brady, R.O. and Fishman, P.H. (1981), *Proc. Natl. Acad. Sci. U.S.A.*, **78,** 4848–4853.

Bennett, V. and Cuatrecasas, P. (1977), in *The Specificity and Action of Animal, Bacterial, and Plant Toxins* (Cuatrecasas, P. and Greaves, M.F, eds), Chapman and Hall, London, pp. 3–66.

Bitensky, L., Alaghband-Zadeh, J. and Chayen, J. (1974), *Clin. Endocrinol. (Oxf.)*, **3,** 363–374.

Chayen, J., Bitensky, L. and Butcher, R.G. (1973), *Practical Histochemistry,* Wiley, London.

Chayen, J., Daly, J.R., Loveridge, N. and Bitensky, L. (1976), *Recent Prog. Horm. Res.*, **32,** 33–79.

Davies, T.F., Yeo, P.P.B., Evered, D.C., Clark, F., Smith, R.R. and Hall, R. (1977), *Lancet,* **i,** 1181–1185.

De Wolf, M., Vitti, P., Ambesi-Impiombato, F.S. and Kohn, L.D. (1981), *J. Biol. Chem.*, **256,** 12287–12296.

Doniach, D. (1975), *Clin. Endocrinol. Metab.*, **4,** 267–285.

Dumont, J.E. (1971), *Vitam. Horm.*, **29,** 287–412.

Ealey, P.A., Marshall, N.J. and Ekins, R.P. (1981), *J. Clin. Endocrinol. Metab.*, **52,** 483–487.

Ealey, P.A., Kohn, L.D., Ekins, R.P. and Marshall, N.J. (1984a), *J. Clin. Endocrinol. Metab.*, (in press).

Ealey, P.A., Valente, W., Kohn, L.D., Ekins, R.P. and Marshall, N.J. (1984b), *Endocrinology,* (in press).

Field, J.B. (1980), in *The Thyroid* 4th ed.(Werner, S.C. and Ingbar, S.H., eds), Harper and Row, Hagerstown, Maryland, pp. 185–195.

Friedman, R.M., Lee, G., Shifrin, S., Ambesi-Impiombato, S., Epstein, D., Jacobsen, H. and Kohn, L.D. (1982), *J. Interferon Res.*, **2,** 387–400.

Gill, D.M. (1977), *Adv. Cyclic Nucleotide Res.*, **8,** 85–118.

Grollman, E.F., Lee, G., Ramos, S., Layo, P.S., Kaback, H.R., Friedman, R.M. and Kohn, L.D. (1978), *Cancer Res.*, **38,** 4172–4185.

Holmgren, J. (1978), in *Bacterial Toxin and Cell Membranes* (Jeljaczewicz, Z. and Wadstrom, E., eds), Academic Press, London, pp. 333–366.

Kohn, L.D. (1978), in *Receptors and Recognition* (Cuatrecasas, P. and Greaves, M.F., eds), Chapman and Hall, London, Vol. 5, Series A, pp. 133–212.

Kohn, L.D. and Shifrin, S. (1982), in *Horizons in Biochemistry and Biophysics, Hormone Receptors* (Kohn, L.D., ed.), J. Wiley and Sons, New York, Vol. 6, pp. 1–42.

Kohn, L.D., Consiglio, E., De Wolf, M.J.S., Grollman, E.F., Ledley, F.D., Lee, G. and Morris, N.P. (1980), in *Structure and Function of Gangliosides* (Svennerholm, L., Mandel, P., Dreyfus, H. and Urban, P.F., eds), Plenum Press, New York, pp. 487–504.

Kohn, L.D., Consiglio, E., Aloj, S.M., Beguinot, F., De Wolf, M.J.S., Yavin, E., Yavin, Z., Meldolesi, M.F., Shifrin, S., Gill, D.L., Vitti, P., Lee, G., Valente, W.A. and Grollman, E.F. (1981), in *International Cell Biology 1980–1981* (Schweiger, A.G., ed.), Lange and Springer, Berlin, pp. 696–706.

Kohn, L.D., Aloj, S.M., Beguinot, F., Vitti, P., Yavin, E., Yavin, Z., Laccetti, P., Grollman, E.F. and Valente, W.A. (1982a), in *Membranes and Genetic Diseases* (Shepard, J., ed.), Alan R. Liss, New York, Vol. 97, pp. 55–83.

Kohn, L.D., Valente, W.A., Laccetti, P., Cohen, J.L., Aloj, S.M. and Grollman, E.F. (1982b), *Life Sci.*, **32,** 15–30.

Kohn, L.D., Aloj, S.M., Shifrin, S., Valente, W.A., Weiss, S.J., Vitti, P., Laccetti, P., Cohen, J.L., Rotella, C.M. and Grollman, E.F. (1983), in *Receptors for Polypeptide Hormones* (Posner, B.T., ed.), Marcel Dekker, New York, (in press).

Laccetti, P. Grollman, E.F., Aloj, S.M. and Kohn, L.D. (1983), *Biochem. Biophys. Res. Commun.*, **110,** 772–778.

Laccetti, P., Vitti, P., Rotella, C.M., Grollman, E.F., Valente, W.A., Cohen, J.L., Aloj, S.M. and Kohn, L.D. (1984). *Endocrinology,* (in press).

Mandato, E.M., Meldolesi, F. and Macchia, V. (1975), *Cancer Res.*, **35,** 3089–3093.

KcKenzie, J.M. (1958), *Endocrinology,* **63,** 372–384.

McKenzie, J.M. and Zakarija, J. (1976), *J. Clin. Endocrinol. Metab.*, **42,** 778–781.

Meldolesi, M.F., Fishman, P.H., Aloj, S.M., Kohn, L.D. and Brady, R.O. (1976), *Proc. Natl. Acad. Sci. U.S.A.*, **73,** 4060–4064.

Moss, J. and Vaughan, M. (1979), *Annu. Rev. Biochem.*, **48,** 581–600.

Pekonen, F. and Weintraub, B.D. (1980), *J. Biol. Chem.*, **255,** 8121–8127.

Pinchera, A., Fenzi, G.F., Macchia, E., Vitti, P., Monzani, F. and Kohn, L.D. (1982), *Ann. Endocrinol. (Paris),* **43,** 520–533.

Robbins, J., Rall, J.E. and Gorden, P. (1980), in *Metabolic Control and Disease* (Bondy, P.K. and Rosenberg, L.E., eds), W.B. Saunders, Philadelphia, pp. 1325–1426.

Rotella, C.M., Tramantano, D., Kohn, L.D., Aloj, S.M., Ambesi-Impiombato, F.S. and Kohn, L.D. (1984), *Endocrinology,* (in press).

Sluszkiewicz, E. and Pawlikowski, M. (1980), *Endocrinol, Exp.*, **14,** 227–235.

Tate, R.L., Schwartz, H.I., Holmes, J.M., Kohn, L.D. and Winand, R.J. (1975), *J. Biol. Chem.*, **250,** 6509–6515.

Valente, W.A., Vitti, P., Kohn, L.D., Brandi, M.L., Rotella, C.M., Toccafondi, R., Tramontano, D., Aloj, S.M. and Ambesi-Impiombato, F.S. (1982a), *Endocrinology,* **112,** 71–79.

Valente, W.A., Yavin, Z., Yavin, E., Grollman, E.F., Schneider, M.D., Rotella, C., Zonefrati, R., Toccafondi, R.S. and Kohn, L.D. (1982b), *J. Endocrinol. Invest.*, **5,** 293–301.

Valente, W.A., Vitti, P., Yavin, Z., Yavin, E., Rotella, C.M., Grollman, E.F., Toccafondi, R.S. and Kohn, L.D. (1982c), *Proc. Natl. Acad. Sci. U.S.A.*, **79,** 6680–6684.

Valente, W.A., Vitti, P., Rotella, C.M., Vaughan, M.M., Aloj, S.M., Grollman, E.F., Ambesi-Impiombato, F.A. and Kohn, L.D. (1983), *N. Engl. J. Med.*, **309,** 1038–1044.

Valente, W.A., Yavin, E., Yavin, Z. and Kohn, L.D. (1984), *Endocrinology,* (in press).

Vitti, P., De Wolf, M.J.S., Acquaviva, A.M., Epstein, M. and Kohn, L.D. (1982), *Proc. Natl. Acad. Sci. U.S.A.*, **79,** 1525–1529.

Vitti, P., Rotella, C.M., Valente, W.A., Cohen, J.L., Aloj, S.M., Laccetti, P., Ambesi-Impiombato, F.S., Grollman, E.F., Pinchera, A., Toccafondi, R. and Kohn, L.D. (1983), *J. Clin. Endocrinol. Metab.*, **57,** 782–791.

Weiss, S., Philip,.N.J., Ambesi-Impiombato, F.S. and Grollman, E.F. (1984), *Endocriniology*, (in press).

Winand, R.J. and Kohn, L.D. (1975), *J. Biol. Chem.*, **250,** 6534–6540.

Yavin, E., Yavin, Z., Schneider, M.D. and Kohn, L.D. (1981a), *Proc. Natl. Acad. Sci. U.S.A.*, **78,** 3180–3184.

Yavin, E., Yavin, Z., Schneider, M.D. and Kohn, L.D. (1981b), in *Monoclonal Antibodies in Endocrine Research* (Fellows, R.E. and Eisenbarth, G., eds), Raven Press, New York, pp. 53–67.

Yavin, E., Yavin, Z., Schneider, M.D. and Kohn, L.D. (1981c), in *Monoclonal Antibodies to Neural Antigens* (McKay, R., Raff, M. and Reichardt, L., eds), Cold Spring Harbor Laboratories, New York, pp. 141–152.

Zakarija, J. and McKenzie, J.M. (1980), in *Autoimmune Aspects of Endocrine Disorders* (Pinchera, A., Doniach, D., Fenzi, G.F. and Baschieri, L., eds), Academic Press, New York, pp. 83–90.

10 Monoclonal Antibodies to Transferrin Receptors

IAN S. TROWBRIDGE and
ROLAND A. NEWMAN

Acknowledgments

We would like to acknowledge the contributions made by our colleagues to our own work cited in this review. These are Robert Hyman, Bishr Omary, Jayne Lesley, Derrick Domingo, Roberta Schulte, Catherine Mazauskas, Joseph Trotter and Frederick Lopez at the Salk Institute, and at Imperial Cancer Research Fund, Melvyn Greaves, Robert Sutherland and Claudio Schneider.

We would also like to thank Robert Hyman for his suggestions for improvement of the manuscript and Ami Koide for her patience and skill in typing this review.

This work was supported by Grants CA 34787 and CA 17733 from the National Cancer Institute and funds donated by the Armand Hammer Foundation, the Paul Stock Foundation, the Helen K. and Arthur E. Johnson Foundation, BankAmerica Foundation, Mr and Mrs Lyndon C. Whitaker Charitable Foundation, Frances and Charles G. Haynsworth and Constance and Edward L. Grund.

Monoclonal Antibodies to Receptors: Probes for Receptor Structure and Function
(*Receptors and Recognition*, Series B, Volume 17)
Edited by M. F. Greaves
Published in 1984 by Chapman and Hall, 11 New Fetter Lane, London EC4P 4EE

10.1 BACKGROUND

10.1.1 Early work on transferrin receptors

Iron plays an important role in cell growth and metabolism, and as many key reactions in energy metabolism and DNA synthesis are catalyzed by iron-containing enzymes, some biologists have considered it likely that iron was an obligatory requirement during the earliest phases of the evolution of life (Shapira, 1964; Neilands, 1972). Under most physiological conditions, however, the iron atom exists in its oxidized ferric (Fe^{3+}) state and at neutral pH ferric salts are hydrolyzed to insoluble ferric hydroxide. Thus ferric ions in excess of 2.5×10^{-18} M are insoluble. To combat this problem, organisms have developed various systems to maintain iron in a soluble form and transport it into the cell. Micro-organisms have solved this problem by producing and secreting various soluble iron-chelating molecules referred to as siderophores (Neilands, 1981) whereas vertebrates have developed a family of closely related iron-binding proteins collectively known as the transferrins (Aisen and Listowsky, 1980).

The cellular iron-transport system used by vertebrates involves the specific interaction of the iron-binding ligand with a cell surface receptor which then facilitates transport across the cell membrane. Early work on transferrin receptors was focused upon the maturing cells of the erythroid lineage which have a high-iron requirement for heme synthesis and the placental trophoblast which acts as a conduit through which the developing embryo can obtain iron from the maternal circulation. The existence of specific receptors for transferrin on erythroid cells was proposed by Jandl and Katz in 1963, but although many attempts were made to isolate and characterize the transferrin receptor from erythroid cells, the receptor was first purified and adequately characterized from human placenta (Seligman *et al.*, 1979; Wada *et al.*, 1979). Furthermore it was generally agreed that the transferrin receptor from either placenta or reticulocytes most likely had a subunit size of 90000–95000 molecular weight (Sullivan and Weintraub, 1978; Leibman and Aisen, 1977; Wada *et al.*, 1979).

Within the last 3 years, evidence that many other cell types also express transferrin receptors was obtained and concomitantly it was recognized that proliferating cells expressed much larger numbers of receptors than resting cells. Sussman's group and Faulk and his colleagues showed that a wide variety of cultured human tumor cell lines express large amounts of transferrin receptors (Hamilton *et al.*, 1979; Galbraith *et al.*, 1980), and Larrick and Cresswell (1979) elegantly demonstrated that, although transferrin receptors were not detectable on normal resting peripheral blood lymphocytes, after mitogenic stimulation with phytohemagglutin, large numbers of transferrin receptors were found on the proliferating blast cells. This was the existing state of knowledge about the transferrin receptor at the time monoclonal antibodies were identified against this structure as the result of a totally independent line of investigation. This is described in the following section.

10.1.2 Identification of monoclonal antibodies against the transferrin receptor of human cells

The identification of monoclonal antibodies against the transferrin receptor of human cells was mainly the result of serendipity. Omary *et al.*, (1980) obtained a murine monoclonal antibody designated B3/25 produced by a hybridoma derived from spleen cells of a mouse immunized with the myeloid/erythroid human leukemic cell line K562 (Lozzio and Lozzio, 1975). This antibody seemed interesting for several reasons. First, although it reacted with virtually all cultured human hematopoietic tumor cell lines, it did not bind to normal human lymphocytes. Second, under normal culture conditions, it was expressed upon the inducible human promyelocytic cell line, HL-60 (Collins *et al.*, 1978). However, after induction of HL-60 cells to differentiate *in vitro* along the myeloid pathway with dimethyl sulfoxide or butyric acid, the antigen was lost from the cell surface within 2–3 days and before morphological differentiation of the tumor cell line was evident. Third, characterization of the cell surface molecule with which this antibody reacted revealed that it was a cell surface glycoprotein with an apparent molecular weight of approximately 95 000 under reducing conditions and 200 000 under non-reducing conditions in SDS-polyacrylamide gel electrophoresis. Two years previously, Bramwell and Harris (1978, 1979) had described a cell surface glycoprotein with similar structural features that appeared to be rigorously associated with malignancy, and thus it seemed likely that B3/25 monoclonal antibody was directed against this malignancy-associated glycoprotein. However, upon further investigation, it became clear that the cell surface glycoprotein defined by B3/25 monoclonal antibody was a normal cell-surface component that was preferentially expressed upon proliferating cells (Omary *et al.*, 1980). A similar proliferation-associated antigen was independently identified by means of a conventional xenoantiserum by Judd *et al.* (1980). Further studies subsequently led to the realization that this cell-surface glycoprotein was the transferrin receptor. In one laboratory this came about from attempts to make a conventional rabbit antiserum against the purified glycoprotein (Trowbridge and Omary, 1981). The molecule was isolated from detergent lysates of the human T cell leukemic cell line CCRF-CEM (Foley *et al.*, 1965), by affinity chromatography on a B3/25 monoclonal antibody–Sepharose column. The antiserum obtained from a rabbit immunized with this purified glycoprotein preparation showed the same pattern of reactivity as B3/25 monoclonal antibody by immunofluorescence but immunoprecipitated not only a 95 000-molecular weight molecule from lysates of surface-iodinated CCRF-CEM cells but also an 80 000-

molecular weight component. This additional band was shown not to be a cell-surface component but rather a protein from the horse serum in the culture medium which was tightly bound to the cells. Careful inspection of radioautographs with immunoprecipitates prepared with B3/25 monoclonal antibody indicated that in some experiments small amounts of the 80 000-molecular weight component could also be found in these immunoprecipitates (Trowbridge and Omary, 1981). On the basis of these results, the hypothesis was formulated that the cell-surface glycoprotein defined by B3/25 monoclonal antibody was the receptor for the serum component. When the hypothesis was tested it was found to be correct. It was then shown that the serum component was transferrin and that B3/25 monoclonal antibody could immunoprecipitate transferrin bound to the cell-surface glycoprotein but not transferrin alone (Omary and Trowbridge, 1981). This provided formal proof that indeed the 95 000-molecular weight cell-surface glycoprotein was the transferrin receptor. At about the same time, by slightly different reasoning, two other groups (Sutherland *et al.*, 1981; Goding and Burns, 1981) showed that another monoclonal antibody, OKT9, which was originally thought to be a differentiation antigen of immature thymocytes (Reinherz *et al.*, 1980), was also directed against the transferrin receptor. Sutherland *et al.* (1981) showed that OKT9 monoclonal antibody could precipitate a 90 000–95 000-molecular weight cell-surface glycoprotein but in addition that if an unlabeled cell lysate was incubated with radiolabeled transferrin the antibody was able to precipitate the latter. If the cell lysate was omitted, OKT9 antibody did not precipitate transferrin directly. Furthermore, a rabbit antiserum against transferrin was able to indirectly precipitate the cell surface receptor as a complex with transferrin. Subsequently several other monoclonal antibodies against the human transferrin receptor have also been identified (Haynes *et al.*, 1981; Brown *et al.*, 1981; Lebman *et al.*, 1982). Such antibodies are frequently obtained from fusions of mouse spleen cells immunized with cultured cell lines as the receptor is an immunodominant surface component of human tumor cells.

10.2 STRUCTURE AND BIOSYNTHESIS OF TRANSFERRIN RECEPTORS

10.2.1 Structure

As mentioned earlier, the human transferrin receptor was first satisfactorily isolated from human placenta and shown to be a disulfide-bonded dimer consisting of two similar subunits of apparent molecular weight 95 000 (Seligman *et al.*, 1979). Subsequently the availability of monoclonal antibodies to the human transferrin receptor has allowed more extensive characterization

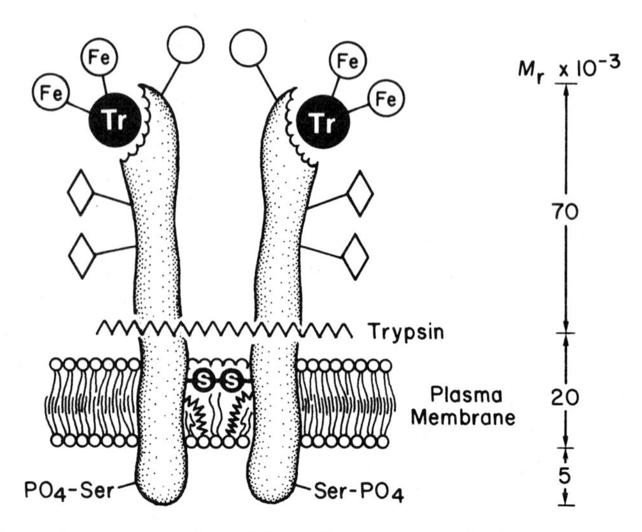

Fig. 10.1 Schematic representation of the cell-surface transferrin receptor. It should be noted that the definitive sites of fatty acid, phosphoserine and disulfide bonds have not been completely localized. Their likely locations in the tryptic fragment associated with the cell membrane are indicated and have been selected by analogy with what is known about other transmembrane glycoproteins. Also, the orientation of the transferrin receptor in the membrane and the number of times the receptor polypeptide spans the membrane is unknown. Key: ◊, high-mannose oligosaccharide; ○, complex-type oligosaccharide; ʌʌʌ, covalently bound fatty acid; Tr, transferrin.

of the molecule to be performed. The general structural features of the human transferrin receptor are illustrated schematically in Fig. 10.1. The molecule is a transmembrane glycoprotein that contains at least three *N*-asparagine-linked oligosaccharides. Both high-mannose and complex-type oligosaccharides are found on the mature glycoprotein (Omary and Trowbridge, 1981a,b; Schneider *et al.*, 1982). The bulk of the transferrin receptor molecule is exposed on the cell surface and it can be cleaved on intact cells with low concentrations of trypsin to give rise to a soluble fragment of apparent molecular weight 70000, which is still able to bind transferrin, leaving the region of the molecule containing the disulfide bond embedded in the membrane (Omary and Trowbridge, 1981a; Schneider *et al.*, 1982; Bleil and Bretscher, 1982). The portion of the molecule exposed on the cytoplasmic face of the plasma membrane has been estimated to be approximately 5000 molecular weight and, at least in some cultured cell lines and fetal liver cells, serine residues located in this region of the molecule are phosphorylated (Schneider *et al.*, 1982). As described in more detail below, the transferrin receptor is also modified post-translationally by the addition of fatty acid residues. Experiments using the cleavable cross-linking agent dithiobis(succinimidyl proprionate) and the non-cleavable reagent disuccinimidyl suberimidate indicate that each subunit of the dimeric transferrin receptor can bind a single molecule of transferrin (Schneider *et al.*, 1982). Measurements carried out on a variety of cell types, including placenta (Hamilton *et al.*, 1979), choriocarcinoma (Wada *et al.*, 1979), rat kidney (Fernandez-Pol and Klos, 1980), teratocarcinoma (Karin and Mintz, 1981) and hepatoma cells (Ciechanover *et al.*, 1983), have given a reasonably consistent value for the dissociation constant (K_d) of transferrin binding. Although there have been wide variations in K_d reported by some groups, all of the above authors have reported values in the range 2×10^{-9}–7×10^{-9} M. Although a monoclonal antibody has been identified that appears to preferentially react with transferrin receptors of human erythroid precursors (Lebman *et al.*, 1982), compared with non-hematopoietic cell lines, the evidence from immunological and peptide mapping studies (Stein and Sussman, 1983; Enns and Sussman, 1981) suggests that transferrin receptors of different cell types are closely related if not identical. The question of whether there is more than one type of transferrin receptor is likely to be resolved only by the isolation and characterization of the genes encoding transferrin receptors (see below). It is of interest, however, that it has recently been shown by Derks *et al.* (1983) that a previously described diallelic alloantigen designated TCA (van Leeuwen *et al.*, 1982) possibly represents a polymorphism of the human transferrin receptor.

10.2.2 Biosynthesis of transferrin receptors

The biosynthesis of transferrin receptors has been studied in several hematopoietic cell lines by Omary and Trowbridge (1981a,b) and Schneider *et al.* (1982) with concordant results. Newly synthesized receptor has an apparent molecular weight of 88000 on SDS/polyacrylamide gels and, within 4 hours, this species is totally converted into a mature form of a glycoprotein with an apparent molecular weight of 95000. The glycosylation of the transferrin receptor appears to follow the pathway previously described for other cell-surface and secreted glycoproteins (reviewed by Hubbard and Ivatt, 1981). Initially, the oligosaccharides on the immature form of the receptor are all high-mannose-type oligosaccharides and are susceptible to digestion by endoglycosidase H. At least one of these high mannose oligosaccharides is processed and converted to a complex oligosaccharide.

As mentioned previously, the transferrin receptor contains covalently bound lipid and can be metabolically labeled with [^{3}H]palmitate (Omary and Trowbridge, 1981a,b). Proteolysis experiments show that the lipid moiety is attached to the fragment of the transferrin receptor containing the membrane-associated region of the molecule. Pulse–chase experiments show that only the mature form of the transferrin receptor contains a lipid moiety and there is evidence from experiments employing tunicamycin that receptor molecules that have been synthesized more than 48 hours previously can be acylated. The significance of the covalently bound lipid is unknown, although several general suggestions have been put forward (Schmidt, 1982; Magee and Schlesinger, 1982). One suggestion is that it may play some role anchoring the receptor in the cell membrane. In the case of the Rous sarcoma virus-transforming protein, $pp60^{src}$, which also contains covalently bound lipid, there is evidence that acylation is important for association of this glycoprotein with the plasma membrane (Sefton *et al.*, 1982; Garber *et al.*, 1983). Another idea is derived from the fact that fatty acids are believed to facilitate membrane fusion and budding in certain viruses. Related to this is the process by which during the continuous internalization and recycling of transferrin receptors, a similar membrane fusion event would take place especially if the molecule was recycled via the Golgi. This may explain why presynthesized transferrin receptors may be acylated up to 48 hours later. Finally, acylation of the transferrin receptor may alter the polypeptide conformation slightly which could allow it to react with other proteins in the plane of the membrane or components of the cytoskeleton. One gap in our knowledge, because of the difficulty of obtaining sufficient palmitate-labeled receptors for analysis, is whether all or only a small proportion of transferrin receptors are acylated. Until this information is available it is difficult to rigorously assess the biological importance of this post-translational modification.

10.2.3 Turnover and recycling of transferrin receptors

There is now good evidence that during iron transport, transferrin receptors are endocytosed via coated pits and finally enter cells by means of an uncoated vesicle known as the endosome or receptosome (Bleil and Bretscher, 1982; Bretscher, 1983; FitzGerald *et al.*, 1983; Hughes and Trowbridge, 1983. After internalization, transferrin and transferrin receptors are both recycled to the cell surface. Measurements on HeLa cells (Bleil and Bretscher, 1982), teratocarcinoma cells (Karin and Mintz, 1981) and hepatoma cells (Ciechanover *et al.*, 1983) have shown that transferrin receptors enter the cell with a half-life of between 2 and 5 min. Bleil and Bretscher (1982) estimated that there are about three times as many transferrin receptors inside HeLa cells as on the surface and this means the transit through the cell to the cell surface takes approximately 21 min, assuming all receptors are on the same cycling pathway. The half-life of transferrin receptors has been shown to be of the order of 2–3 days (Omary and Trowbridge, 1981b) in CCRF–CEM T leukemic cells, although receptors in HeLa cells have been shown to have a half-life of 14 hours (Ward *et al.*, 1982) which may reflect different cellular metabolic rates. However, it is clear that very little degradation of the transferrin receptor occurs during each round of recycling. The details of the pathway that transferrin receptors follow inside the cell and how iron is released from the transferrin molecule before it returns to the cell surface still remain incomplete. However, Dautry-Varsat *et al.* (1983) and Klausner *et al.* 1983) have shown that although diferric transferrin binds to the transferrin receptor more strongly than apotransferrin at neutral pH, there was no difference in K_d between apotransferrin and the holoprotein at acidic pH. Both sets of authors have proposed an interesting mechanism of iron release and transferrin recycling based on differential dissociation constants of the ligands under acidic versus neutral conditions. It was postulated that diferric transferrin enters the cell and eventually an acidic endocytic vesicle distinct from the lysosomal compartment (van Renswoude *et al.*, 1982; see also Marsh *et al.*, 1983). At low pH, Fe^{3+} dissociates from the transferrin molecule but the dissociation rate of apotransferrin is low under these conditions so that the iron-free molecule remains bound to its receptor. After delivering its iron in this way, the apotransferrin–transferrin-receptor complex is returned to the cell surface by some ill-defined mechanism and at neutral pH, the rate of dissociation of the apotransferrin increases markedly ($\sim$ 30-fold). This allows dissociation of apotransferrin from the cell surface and the preferential binding of new iron-laden transferrin molecules from the external environment.

10.3 MONOCLONAL ANTIBODIES THAT BLOCK RECEPTOR FUNCTION

10.3.1 Monoclonal antibodies against the human transferrin receptor

The monoclonal antibodies originally obtained against the human transferrin receptor, such as B3/25, T56/14 (Trowbridge and Omary, 1981) 5E9 (Haynes *et al.*, 1981) and OKT9 (Reinherz *et al.*, 1980; Sutherland *et al.*, 1981), although binding to the transferrin receptor, do not interfere significantly with receptor function. In order to investigate what the consequences of blocking transferrin receptor function would be on cell growth, a deliberate attempt was made to obtain such antibodies (Trowbridge and Lopez, 1982). The strategy that was employed is as follows. Mice were immunized with purified human transferrin receptor glycoprotein. Supernatants of the hybridomas obtained from fusing spleen cells from the immunized mice with S194 myeloma cells were then tested sequentially for the presence of antibodies that (1) bound to CCRF–CEM cells, (2) immunoprecipitated labeled transferrin receptor from lysates of surface-iodinated cells and (3), when preincubated with CCRF–CEM cells, inhibited the binding of ^{125}I-labeled human transferrin. This approach led to the identification of a monoclonal antibody designated 42/6 that gave positive results in each of these tests. As shown in Fig. 10.2, this purified antibody blocked the binding of human transferrin to CCRF–CEM cells in a competitive manner. As a result of inhibiting transferrin binding to its receptor, the antibody not only had a profound effect upon the uptake of iron from transferrin by

CCRF–CEM cells, but also inhibited cell growth *in vitro* at concentrations of antibody as low as 2.5 μg/ml (Fig. 10.3). Although other antibodies have subsequently been obtained that also inhibit transferrin binding when tested in the radioimmune assay (Trowbridge, unpublished results), 42/6 monoclonal is still the most effective in terms of blocking of iron uptake and inhibiting cell growth. As shown previously (Trowbridge and Lopez, 1982), CCRF–CEM

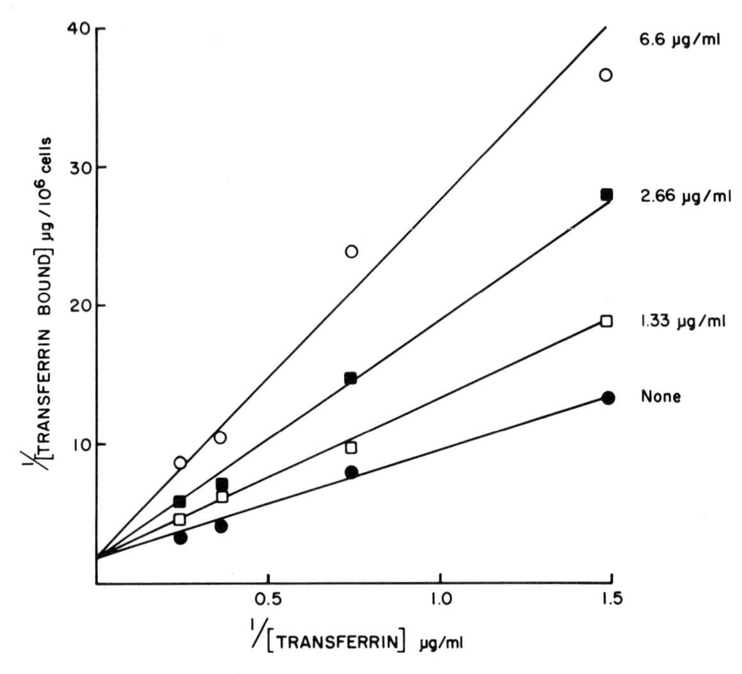

Fig. 10.2 Inhibition of transferrin binding to its receptor by 42/6 monoclonal antibody. Human transferrin was ^{125}I-labeled to a specific activity of 1 μCi per μg as described by Trowbridge and Lopez (1982). The initial rate of binding of transferrin to its receptor was measured by incubating CCRF-CEM cells with various concentrations of ^{125}I-labeled human transferrin. This was done by adding 1×10^6 CCRF-CEM cells in 50 μl of phosphate-buffered saline containing 0.1% bovine serum albumin (PBS–BSA) to duplicate microfuge tubes containing the appropriate amount of ^{125}I-labeled transferrin in 50 μl of the same buffer. After 2 min the cells were centrifuged for 10 seconds in a microcentrifuge and washed with 1 ml of PBS–BSA. The cell pellet was then counted in a gamma counter to determine the amount of transferrin bound to the cells. Under these conditions, binding of transferrin to the cells was linear with time for at least 2 min incubation period. The data shown are plotted as a double-reciprocal plot of transferrin bound as a function of transferrin concentration in the incubation medium. The inhibition of binding of transferrin by various concentrations of 42/6 monoclonal antibody is shown and the results indicate that 42/6 monoclonal antibody competitively inhibits the binding of transferrin to its cell surface receptor.

cells grown for 7 days in the presence of 42/6 monoclonal antibody accumulated in the S phase of the cell cycle. This suggests that, at least for this cell type, the major effect of iron deprivation is on some metabolic process associated with DNA synthesis. One candidate for this would be the enzyme ribonucleotide reductase which plays a key regulatory role in DNA synthesis and requires iron for the formation of an organic free radical necessary for the catalytic reduction of ribonucleotides (Reichard, 1978).

The effect of 42/6 monoclonal antibody on the growth of CCRF–CEM cells is very similar to the effect on cell growth of the metal-chelating agent, picolinic

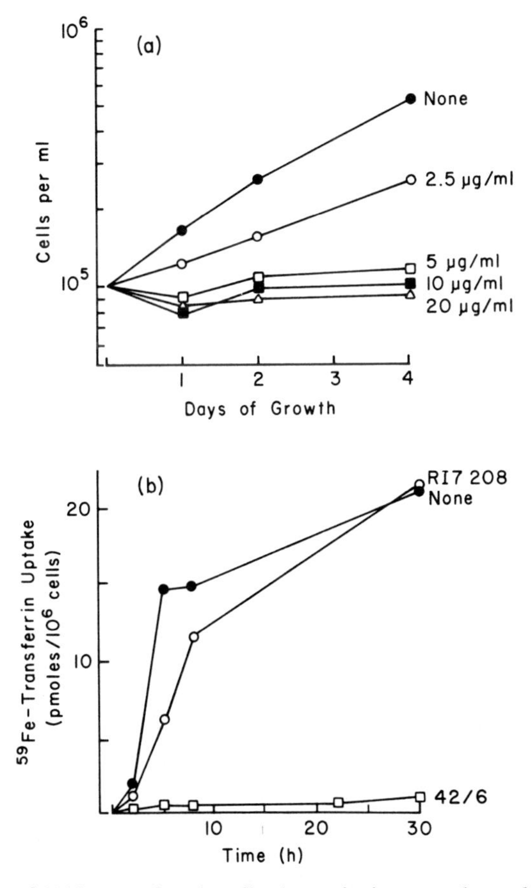

Fig. 10.3 Effects of 42/6 monoclonal antibody on the iron uptake and growth of CCRF-CEM cells. (a) shows the inhibition of growth of CCRF-CEM cells by 42/6 monoclonal antibody. Experimental details are given in Trowbridge and Lopez (1982). (b) shows the specific inhibition by 42/6 monoclonal antibody of transferrin-mediated iron uptake by CCRF-CEM cells. Experimental details of the assay are described in Trowbridge *et al.* (1982).

acid. This compound which both inhibits the incorporation of iron and can remove iron from cells (Fernandez-Pol, 1977b) arrests Chinese hamster ovary cells in S phase (Gurley and Jett, 1981). However, when other cell types are treated with picolinic acid they do not accumulate in S phase. For example, normal rat kidney cells are arrested in G_1 stage of the growth cycle whereas virally transformed rat kidney cells accumulated in G_1 and G_2 of the cell cycle (Fernandez-Pol *et al.*, 1977, 1978). The G_1 arrest induced in normal rat kidney cells by picolinic acid could be prevented by the addition of ferric, zinc or cobalt ions to the tissue culture medium (Fernandez-Pol, 1977a). Less extensive information is available about the growth inhibitory effects of antibodies to transferrin receptor; however, it is becoming clear that not all cells are arrested in S phase of the cell cycle after exposure to such antibodies. For instance, the mouse myeloma, S194, when treated with the monoclonal antibody R17.208 directed against the murine transferrin receptor (see Section 10.3.2), arrests in G_2 stage of the cell cycle (Trowbridge *et al.*, 1982). In contrast, when actively proliferating normal human lymphocytes are exposed to monoclonal antibody 42/6, cell growth is arrested but it appears that cells at all stages of the cell cycle are equally sensitive to the effects of the antibody (Mendelsohn *et al.*, 1983; see Section 10.4.1).

10.3.2 Monoclonal antibody against the murine transferrin receptor

After obtaining an antibody against the human transferrin receptor that blocked function, it became clear that if one wished to explore the possibility of using such antibodies *in vivo* to regulate cell growth it would be important to have an animal model system in which to carry out the initial studies. An animal system would also be convenient to determine the distribution of transferrin receptors on normal hematopoietic stem cells. The mouse is particularly useful for studying hematopoietic stem cells since pluripotent stem cells may be enumerated by means of the CFU-S assay (Till and McCulloch, 1961). A similar assay for the pluripotent stem cell is not available in man. Consequently, we searched a library of rat monoclonal antibodies against mouse lymphoid tumors for an antibody against the murine transferrin receptor (Trowbridge *et al.*, 1982). The approach we took to identify a rat monoclonal antibody against murine transferrin receptor was to first seek antibodies which precipitated a molecule of 95 000 molecular weight under reducing conditions and which was a disulfide-bonded dimer of 190 000 in its native state. Two monoclonal antibodies have been identified in this manner and have been designated R17 208 and R17 217 (Trowbridge *et al.*, 1982; Lesley *et al.*, 1984b). It was confirmed that these antibodies were indeed against the transferrin receptor by showing that both could precipitate ^{125}I-labeled transferrin complexed to the cell-surface molecule with which they reacted. The properties of the marine transferrin receptor are very similar to that of the human receptor, although the human and mouse molecules can be distinguished on SDS/polyacrylamide gel analysis by a small difference in their apparent molecular weight. This is probably due to differences in carbohydrate content as digestion of both molecules with endoglycosidase F abolishes this difference (R. Newman, unpublished results). When the two antibodies were tested for their effects on the function of the murine transferrin receptor, it was found that R17 208 behaved in a similar fashion to the anti-(human transferrin receptor) antibody 42/6 in that it specifically blocked the cellular uptake of iron in murine cells (Fig. 10.4) and inhibited the growth of a variety of murine hematopoietic cell lines (Trowbridge *et al.*, 1982). It is not clear at present whether R17 208 antibody inhibits iron uptake by interfering with transferrin binding or by affecting the ability of the receptor to be internalized or recycled. It may be significant that the 42/6 monoclonal antibody which blocks the function of human transferrin receptors is an IgA and that R17 208 monoclonal antibody is an IgM, whereas other antibodies that do not block receptor function are IgGs. It is possible that the bulk of the IgA and IgM antibodies are critical for the steric inhibition of transferrin binding.

Although R17 208 monoclonal antibody has profound effects upon the *in vitro* growth of some murine cell lines, other cell lines were not affected even though the iron transport in these cells is blocked as efficiently as in the sensitive cell lines. This is illustrated in Fig. 10.5 which shows that R17 208 reduces the uptake of iron into mouse L cells to about 20% of the normal rate.

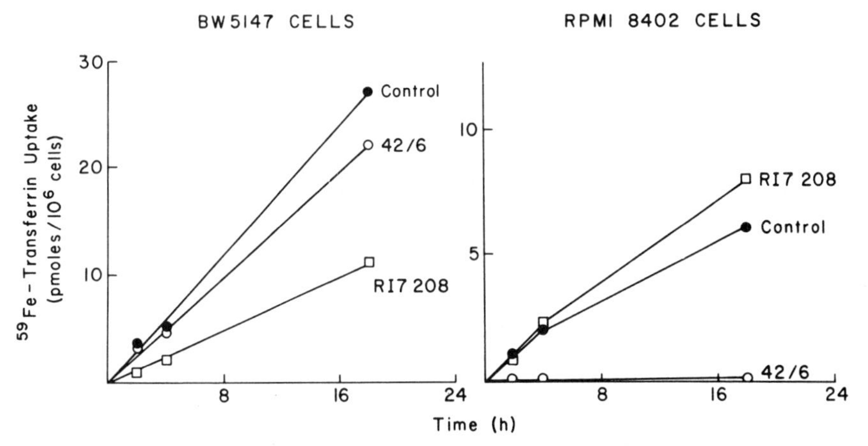

Fig. 10.4 Inhibition of transferrin-mediated iron uptake of RI7 208 monoclonal antibody in BW5147 cells. The figue shows the specific inhibition of iron uptake by a mouse T lymphoma cell line, BW5147, by R17 208 monoclonal antibody [anti-mouse transferrin receptor]. As a specificity control, the effect of R17 208 monoclonal antibody on iron uptake by a human RPMI-8402, was also measured as was the effect of 42/6 monoclonal antibody [anti-human transferrin receptor] on iron uptake in BW5147 cells. Experimental details are given in Trowbridge *et al.* (1982).

This is similar to the inhibition of iron uptake in cell lines that are sensitive to growth inhibition by the antibody, yet prolonged exposure of L cells to a high concentration of R17 208 monoclonal antibody has no effect on their growth rate (Fig. 10.5(b)). Similarly, monoclonal antibody 42/6 has little effect on the growth of many human carcinoma cell lines (I.S. Trowbridge, unpublished results). It is not clear at present why there should be differences in the growth

(a)

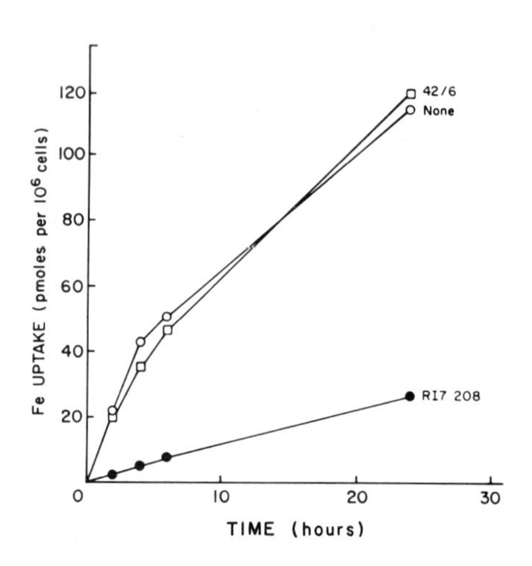

(b)

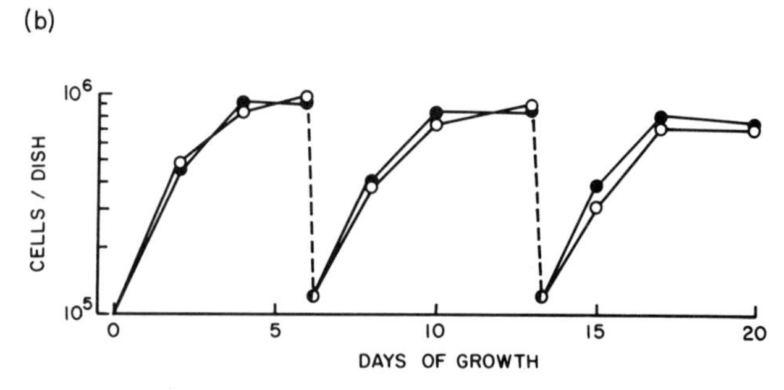

Fig. 10.5 Effect of RI7 208 monoclonal antibody on iron uptake and growth of mouse L cells. (a) shows the inhibition of transferrin-mediated iron uptake by RI7 208 monoclonal antibody into mouse L cells. Experimental details are as in Trowbridge *et al.* (1982). (b) shows the lack of effect of RI7 208 monoclonal antibody on growth of mouse L cells. The figure shows the growth of mouse L cells exposed continuously to 100 μg of RI7 208 monoclonal antibody/ml. The growth medium was changed twice weekly and cells transferred to new dishes when cultures became confluent. Note that the growth of L cells treated with 100 μg of RI7 208 monoclonal antibody/ml was identical with that of control cells grown in the absence of antibody for the 20-day period of the experiment.

inhibitory effect of anti-transferrin receptor monoclonal antibodies on various cell types. There are two major alternatives: (1) that the metabolic changes that occur in different cell types when deprived of iron are not the same, (2) that some differentiated cell types normally have the capacity to transport more iron into the cell than is required for cell growth, whereas the iron-transport system of other cells may only be sufficient to meet essential growth-related needs. If the second hypothesis were correct, then 80–90% inhibition of iron uptake by cells which normally take up more iron than required for immediate use may not block cell growth because the cells can respond to the deficit by channeling all incoming iron into the metabolic pathways necessary for continued cell growth at the expense of other non-essential uses and iron storage. On the other hand, cells whose requirements for iron can only just be satisfied if their iron-transport system is operating at full capacity will not be able to continue to grow. Ultimately, it is likely that the latter cells will suffer severe imbalances in other essential metabolic pathways such as energy metabolism that will lead to cell death. Relatively little is known about the regulation of iron metabolism *in vitro* and *in vivo* and progress in this area will be required before an understanding of the different cellular responses to iron deprivation is achieved. However, it may be possible to exploit the differential effect of partial iron deprivation on cell growth to develop an effective treatment of cancers that arise from sensitive cell types as they may be more sensitive to iron deprivation than most normal tissues including the actively proliferating component of those tissues (see Section 10.5.2).

10.4 EXPRESSION OF TRANSFERRIN RECEPTORS ON NORMAL AND MALIGNANT TISSUES

10.4.1 Studies in Man

The early work on the identification and distribution of transferrin receptors in human tissues was described earlier (see Section 10.1.1). The first hint that transferrin receptors may be selectively expressed on malignant tissues *in vivo* came when Faulk *et al.* (1980) reported that, whereas transferrin binding to normal breast was not detectable, a high proportion of breast carcinomas (16 of 22) showed extensive membrane staining with anti-transferrin antibodies. Subsequently, Shindelman *et al.* (1981) confirmed that transferrin receptors were expressed in greater amount on malignant breast tissue than on normal breast using a sensitive radioimmune assay for bound transferrin. The development of monoclonal antibodies against the human transferrin receptor has allowed more extensive studies comparing the distribution of transferrin receptors on normal and malignant tissues to be performed. Using several different monoclonal antibodies against the human transferrin receptor, Gatter *et al.* (1983) examined a wide variety of normal human tissues and neoplasms by anti-peroxidase immunoperoxidase staining of frozen tissue sections. They found that transferrin receptors could be detected only at a limited number of sites in normal human tissues. These included the basal layer of the epidermis, seminiferous tubules of the testes, and some cells of the Islets of Langerhans in the pancreas, the anterior pituitary and the liver. In contrast, a much more widespread distribution of transferrin receptors was found in a random sampling of human tumor biopsies. Of 87 samples tested 70 of these were classified as positive (Gatter *et al.*, 1983). Nevertheless, considerable heterogeneity was seen in the distribution of transferrin receptors in these tumors. A particularly striking example, was a basal cell carcinoma which contained two clearly different morphological regions of cell architecture (Mason and Gatter, 1983). In relatively undifferentiated areas of the tumour the cells stained strongly for the transferrin receptor and HLA-DR. In other regions which were organized into characteristic whorl-like structures, the cells were negative for transferrin receptor and HLA-DR but instead stained with antibodies for keratin and HLA-A, B, C. An interpretation of these data is that this tumour represents an actively differentiating cellular system and that the keratin-staining regions of the tumor contain cells that have differentiated and ceased to proliferate whereas the transferrin-receptor-positive regions of the tumor include immature cells that are still actively dividing.

With regard to the selective expression of transferrin receptors on tumor cells, it is of some interest that gallium-67 has been used as a marker for various malignancies (reviewed by Johnston, 1981). Although the localization of gallium to tumor cells is probably a complex phenomenon involving a number of different factors, it is of interest that gallium uptake into cells is mediated by transferrin (Harris and Sephton, 1977; Larson *et al.*, 1980; Sephton and Harris, 1974, 1981) and that gallium-transferrin inhibits cell growth in S phase of the cell cycle (Chitamber *et al.*, 1983).

The distribution of transferrin receptors on human hematopoietic cells has been analyzed in considerable detail using anti-receptor monoclonal antibodies. In early studies with the anti-transferrin receptor monoclonal antibody, OKT9, it was suggested that this antibody reacted preferentially with immature thymocytes and prothymocytes (Reinherz *et al.*, 1980; Reinherz and Schlossman, 1980), but it is now clear that this was an incomplete description of the distribution of transferrin receptors on hematopoietic cells. Sutherland *et al.* (1981), using the OKT9 antibody, showed that transferrin receptors can be detected in small numbers on 5–12% of thymocytes from children but that transferrin receptors are also found on 5–20% of cells in pediatric or adult bone marrow. Separation of these cells by fluorescence-activated cell sorting shows that the transferrin-receptor-positive cells were predominantly nucleated erythroid precursors and myelocytes (Sutherland *et al.*, 1981). More recently,

Greaves and his colleagues have analyzed the distribution of transferrin receptors on human hematopoietic stem cells enumerated by the appropriate *in vitro* colony assay (Sieff *et al.*, 1982). Cell-sorting experiments showed that, within the erythroid series, transferrin receptors are expressed in variable amounts on the early erythroid precursor cell, BFU-E, but are found in larger numbers on the more mature erythroid precursor CFU-E. During terminal erythroid differentiation, the numbers of transferrin receptors decline from as many as 4×10^5 to less than 2×10^4 molecules per cell as reticulocytes mature (Frazier *et al.*, 1982). Analysis of the expression of transferrin receptors on the myeloid precursor CFU-GM showed that the fraction of colony-forming cells expressing detectable receptors ranged from 6–29% (Sieff *et al.*, 1982).

The expression of transferrin receptors on human hematopoietic tumor cells seems to accurately reflect their proliferative status. Analysis of leukemic samples showed that whereas few cells from patients with chronic lymphocytic leukemia were stained with monoclonal anti-transferrin receptor antibodies, a variable proportion of cells from patients with common acute lymphocytic leukemia (c-ALL) (range usually 0–30%) were found to express detectable numbers of receptors (Omary *et al.*, 1980; Reinherz *et al.*, 1980; Sutherland *et al.*, 1981). In particular, a large proportion of malignant cells from patients with T-ALL had high amounts of transferrin receptors (Omary *et al.*, 1980; Reinherz *et al.*, 1980; Sutherland *et al.*, 1981). Furthermore, when T-ALL cells or ionomycin/TPA-stimulated pediatric thymocytes were sorted into transferrin-receptor-positive and transferrin-receptor-negative populations on the cell sorter, it was found that the receptor-positive populations were actively cycling whereas the transferrin-receptor-negative populations were all present in the G_1/G_0 stage of the cell cycle (Sutherland *et al.*, 1981).

At present, there have only been two reports on the expression of transferrin receptors on lymphomas. In these studies, binding of OKT9 monoclonal antibody was used to assess the proportion of malignant cells from patients with non-Hodgkins lymphoma expressing transferrin receptors in biopsy material from lymph nodes (Habeshaw *et al.*, 1983a,b). A significant correlation was found with the expression of transferrin receptors and with the histological class of tumor. Whereas high-grade lymphomas showed a mean transferrin-receptor-positive cell population of 22.5%, low-grade lymphomas showed a mean value of only 2.5% positive cells. It appears, therefore, that in non-Hodgkins lymphoma, the expression of transferrin receptors is also associated with the growth fraction of tumor cell population and the preliminary evidence indicates this may be of prognostic value. It remains to be established, however, that the assay of transferrin receptors has any advantage over earlier techniques of measuring the proliferative status of lymphomas such as labeling indices or flow cytometric measurements of DNA content. Furthermore, a prospective study is needed to determine whether the binding of anti-transferrin receptor antibodies to non-Hodgkins lymphomas can be used to predict relapse, histological transformation or long-term prognosis. The results of such a study may aid the timing of effective chemotherapy in those cases of low-grade lymphoma in which aggressive clinical behavior is demonstrated (Habeshaw *et al.*, 1983b).

Finally, a new area which is opening up is the study of transferrin receptors in relationship to cell activation within the hematopoietic system. Studies by Mendelsohn *et al.* (1983) have shown that the anti-human transferrin receptor antibody 42/6 which blocks the binding of transferrin to its receptor inhibits the proliferation of human lymphocytes stimulated by phytohemagglutinin. This demonstrates that the increased expression of transferrin receptors which occurs during lymphocyte activation is a necessary step in the stimulation of cell proliferation. If the antibody is added to cultures initially, lymphocytes are unable to traverse the S phase of the cell cycle, whereas if addition is delayed to a time when cells are already actively proliferating, they are sensitive to the inhibitory effects of the antibody throughout all phases of the cell cycle. Inhibition of cell growth by the antibody is reversed by removal of the antibody after up to 48 hours of exposure (Mendelsohn *et al.*, 1983). Flow cytometric analysis of human T cell activation showed that transferrin receptors appear within 24 hours of stimulation rising from about 30% positive cells to approximately 60% positive cells 3 days later (Cotner *et al.*, 1983). Neckers and Cossman (1983) showed that mitogen-induced T cell lymphocyte proliferation is dependent on the presence of both interleukin-2 (IL-2) and transferrin receptors even though resting lymphocytes express neither receptor. Using monoclonal antibodies to the IL-2 receptor and the transferrin receptors that block their function, these investigators showed that the requirement for the T cell growth factor, IL-2, and the expression of IL-2 receptors precedes that for transferrin receptors. Thus, antibody to the IL-2 receptor inhibited DNA synthesis in lymphocytes only if administered before the appearance of transferrin receptors on the cell surface.

10.4.2 Studies in the mouse

The requirement for transferrin as a growth factor for the *in vitro* growth of mouse hematopoietic cells was recognized very early (Vogt *et al.*, 1969). The distribution of transferrin receptors on murine hematopoietic cells has been determined using the rat monoclonal antibodies against the murine transferrin receptor described earlier (Trowbridge *et al.*, 1982) and the results are in general agreement with studies in Man. Flow cytometric analysis of the distribution of transferrin receptors on adult hematopoietic tissues showed that less than 1% of thymus or spleen cells expressed detectable receptors although a subpopulation of bone marrow cells were transferrin-receptor-positive. Although in fetal thymus few transferrin-receptor-positive cells could be detected, in fetal liver and in neonatal spleen which, in the mouse, is a hematopoetic tissue early in life, substantial numbers of transferrin-receptor-positive cells were found (Trowbridge *et al.*, 1982). In both tissues, when these receptor-positive populations were sorted by flow cytometry and stained histologically they were found to contain erythroid and myeloid precursor cells. Further studies with the RI7 208 monoclonal antibody confirmed earlier work by Hu *et al.* (1977), which showed that during the *in vitro* differentiation of Friend erythroid leukemia cells along the erythroid series induced by dimethyl sulfoxide, the number of transferrin receptors increased significantly (Trowbridge *et al.*, 1982). Using the monoclonal antibody, R17 217, which appears to have higher affinity for the murine transferrin receptor than R17 208 antibody, Lesley *et al.* (1984b) have characterized the erythroid (CFU-E and BFU-E) and myeloid (CFU-C) precursors in mouse bone marrow in terms of transferrin receptor expression. With this antibody, approximately 20–30% of the total bone marrow cell population is transferrin-receptor-positive. Although CFU-E represent only 1–4% of the transferrin-receptor-positive cells, virtually all CFU-E are found in the transferrin-receptor-positive fraction isolated by cell sorting. Thus, although there is selective loss of CFU-E during sorting procedures, and while 'modulation' of the receptor was found to occur during handling, the results argue strongly that most or all CFU-E are transferrin-receptor-positive. In contrast, the bulk of BFU-E sort into the transferrin-receptor-negative fraction (Lesley *et al.*, 1984b). In the case of myeloid precursors approximately 20% of the CFU-C were found in the transferrin-receptor-positive population. These results are in general agreement with those found in studies of human hematopoietic stem cells and confirm the correlation between transferrin receptor expression and the proliferative status of these different stem cell populations. Furthermore, it has not been possible to demonstrate the presence of transferrin receptors on murine CFU-S in the mouse (Lesley *et al.*, 1984a) consistent with earlier work showing that the bulk of the pluripotential stem cell population is not normally actively cycling (Becker *et al.*, 1965; Necas and Neuwirt, 1976).

10.5 MONOCLONAL ANTIBODIES AGAINST TRANSFERRIN RECEPTORS AS THERAPEUTIC AGENTS IN THE TREATMENT OF CANCER

10.5.1 Monoclonal antibody–toxin conjugates

One use of monoclonal antibodies in the treatment of cancer that is being actively pursued currently is to selectively target cytotoxic agents such as conventional chemotherapeutic drugs, radioisotopes, or toxins to tumor cells. In principle, conjugates prepared with monoclonal antibodies and the A chain of various plant and bacterial toxins such as ricin, abrin or diphtheria toxin are particularly attractive (for a review see *Cancer Surveys*, Vol. 1, No. 3, 1982). The specificity of such conjugates is determined solely by the antibody moiety of the conjugate while the A chain of the toxin upon entry to the cell acts catalytically to inhibit protein synthesis. As antibodies against the transferrin receptor show some selectivity for tumor cells, it offers a possible target for this so-called 'magic bullet' therapy in a practical sense. However, because the transferrin receptor is well characterized and its function is known, it offers a favorable model system with which to study the basic biology of drug targeting. Experiments were first carried out with the mouse monoclonal antibody against the human transferrin receptor designated B3/25 (Trowbridge and Domingo, 1981). Antibody was conjugated to ricin A chain using the bifunctional cross-linking reagent succinimidyl-3-(2-pyridyldithioproprionate) (Carlsson *et al.*, 1978). Using this reagent a conjugate was prepared that on average contained one to two ricin A chain subunits per IgG molecule. This antibody ricin A chain conjugate gave potent and specific killing *in vitro*. The concentration of conjugate required for 50% inhibition of protein synthesis in the human T leukemic cell line, CCRF-CEM, was approximately 10^{-10} M, whereas the intact ricin toxin itself gave 50% killing at a concentration of 3×10^{-11} M (Trowbridge and Domingo, 1981). Ricin A chain alone was only toxic at levels greater than 10^{-8} M. Further studies with monoclonal antibodies against the murine transferrin receptor showed that ricin A conjugates prepared with these antibodies were also highly toxic *in vitro* (Trowbridge and Domingo, 1982). Thus, in contrast to a variety of other cell surface glycoproteins, the transferrin receptor seems to be a particularly efficient target antigen for 'magic bullet' therapy. It is possible that receptors such as the transferrin receptor that are internalized while performing their normal physiological function are particularly efficient at delivering the toxic subunits of bound conjugates in an undegraded form to the cell cytoplasm. Cytotoxic conjugates of antibodies specific for transferrin receptors have also been prepared with *Pseudomonas* exotoxin (FitzGerald *et al.*, 1983). In these studies it was shown that the antibody conjugates were taken up in coated pits and transferred to endosomes (receptosomes). Toxicity of these antibody–toxin conjugates due to entry via the transferrin receptor was enhanced 100–300 fold in the presence of adenovirus (FitzGerald *et al.*, 1983). It is suggested that the enhanced toxicity resulted when adenovirus and the toxin conjugates were internalized into the same vesicles. During the process of infection, adenovirus enters cells and brings about the virus-mediated disruption of endosomes; and it is believed that this disruption may liberate many more toxin molecules into the cytoplasm than is possible in the absence of virus.

Although conjugates of ricin A chain plus antibody to transferrin receptor were extremely potent *in vitro*, they were much less effective *in vivo*. Although conjugates prepared with B3/25 monoclonal antibody inhibited the growth of a human melanoma cell line, M21, implanted subcutaneously in athymic nude mice, antibody alone was equally effective (Trowbridge and Domingo, 1981). It is thought that the reason such conjugates are not effective *in vivo* is that they are rapidly inactivated either by cleavage of the antibody from the toxin A chain or by rapid clearance of the antibody conjugates from the bloodstream by the reticuloendothelial system. However, a number of approaches can be taken to minimize these problems (Thorpe and Ross, 1982; Thorpe *et al.*, 1983), and future progress will depend upon the outcome of these manipulations.

10.5.2 Therapeutic potential of antibodies to transferrin receptors that block biological function

The availability of antibodies to transferrin receptors that block the function of the receptor (see Section 10.3.1) suggests another general approach to using monoclonal antibodies in cancer therapy, namely, employing monoclonal antibodies as pharmacological agents to directly block biological functions essential for cell proliferation. As the rat monoclonal antibody RI7 208 blocks the function of the murine transferrin receptor, it has been possible to test the *in vivo* anti-tumor activity of an antibody that blocks iron uptake into cells in a mouse model system (Trowbridge, 1983). Initial experiments have been carried out using the transplantable AKR mouse T cell leukemia SL-2. This is the cell line previously used by Bernstein and his colleagues to investigate requirements for effective serotherapy with anti-Thy-1 monoclonal antibodies (Bernstein *et al.*, 1980; Bernstein and Nowinski, 1982). These earlier studies provide a standard against which one can compare the anti-tumor activity of anti-transferrin receptor antibodies. The murine model is also attractive in that if therapeutic effects can be obtained with the transplantable T cell leukemia, studies can then be extended to the spontaneous T cell leukemias that arise in aged AKR mice. Furthermore, it is known that human T cell leukemias frequently express high numbers of transferrin receptors (see Section 10.4.1) and thus present a favorable opportunity for attempting immunotherapy with receptor antibodies to these receptors in man. Several trials of immunotherapy with RI7 208 monoclonal antibody using the SL-2 leukemia model have been completed, as described in detail elsewhere (Trowbridge, 1983). Intravenous or intraperitoneal injection of the monoclonal antibody had a significant effect upon the growth of SL-2 cells inoculated at a subcutaneous site. This was manifested not only as prolonged survival of tumor-bearing mice but also in a marked inhibition of growth of the tumor at the primary site. Of equal importance, there appeared to be no acute toxic side effects of administering the antibody systemically. Further studies are required to determine the *in vivo* binding pattern of antibodies in normal and tumor-bearing mice and to look for possible deleterious effects caused by chronic antibody administration. Clearly, this approach is not limited to transferrin receptors and can be applied to other receptors that are essential to tumor cell growth.

It should be pointed out that there are a number of problems associated with monoclonal antibody therapy including antibodies to transferrin receptors that have still to be investigated. One of these is the problem of tumor-escape mechanisms. It is well established that the tumor cells can avoid the deleterious effects of antibodies directed against the cell surface antigens by either genetic loss of the antigen (Hyman and Stallings, 1972; Hyman *et al.*, 1980), or by antigenic modulation (Old *et al.*, 1968). This problem may be less severe for receptors required for cell growth than for non-essential surface molecules because loss mutations would be expected to be lethal. However, there is already evidence that variants can be obtained from mouse hematopoietic tumor cell lines that are no longer sensitive to the growth inhibitory activity of R17 208 monoclonal antibody (J. Lesley and R. Schulte, unpublished results). There are a number of strategies for minimizing this difficulty as for other types of therapy; however, its importance in potentially limiting monoclonal antibody therapy should not be discounted.

10.6 SELECTION AND PROPERTIES OF MOUSE L CELL TRANSFORMANTS EXPRESSING THE HUMAN TRANSFERRIN RECEPTOR

It is likely that much of the progress to be made in the next few years in understanding the structure and function of membrane molecules will come from the application of molecular biological techniques (see Chapter 13 by Schneider). In the case of the transferrin receptor, cloning the gene would not only contribute to an understanding of its primary structure but also provide information germane to several pertinent genetic questions. Recent chromosomal analysis has shown that structural genes for the transferrin receptor, probably transferrin itself, and p97, a melanoma tumor-associated antigen that exhibits primary sequence homology with transferrin and can bind ferric ions, each map in Man to chromosome 3 (Goodfellow *et al.*, 1982; Enns *et al.*, 1982; Miller *et al.*, 1982; Plowman *et al.*, 1983). On this basis it has been suggested that there may be a region within chromosome 3 containing genes involved in iron transport and that in some circumstances rearrangements in this region may be associated with malignant transformation (Plowman *et al.*, 1983). This question together with the problem of how the expression of transferrin receptors is co-ordinately regulated with cell proliferation and the continuing debate as to whether all cell types express structurally identical transferrin receptors require study at the genetic level.

As an approach to investigating some of these problems, our laboratory has obtained mouse L cell transformants expressing the human transferrin receptor. The method we employed was to co-transform mouse L (tk^-) cells with total high-molecular weight DNA from a human leukemic cell line CCRF-CEM and a plasmid containing the *Herpes simplex* thymidine kinase (*tk*) gene using the calcium phosphate method described by Graham and Van der Eb (1973). Transformants containing the *tk* gene were selected by growth in Dulbecco's modified Eagles medium containing 10% horse serum and hypoxanthine/aminopterin/thymidine (HAT medium). In independent trials 2×10^3–4×10^3 HAT-resistant colonies were pooled, stained with a mixture of mouse monoclonal antibodies against the human transferrin receptor followed by fluorescein isothiocyanate-conjugated goat-anti-mouse immunoglobulin, and then sorted by flow cytometry. The brightest 1–2% of the viable fluorescent cell population was collected, grown up and resorted (Newman *et al.* 1983). Two L cell transformants were obtained which expressed the human transferrin receptor on their cell surface. One of these transformants, J4, expressed approximately equal amounts of the human and murine transferrin receptors when analyzed by fluorescence-activated cell analysis (Newman *et al.*, 1983). SDS-polyacrylamide gel analysis of immunoprecipitates prepared from detergent lysates of these cells after labeling by lactoperoxidase-catalyzed iodination confirmed that the transformed cells express human transferrin receptors. The human and murine transferrin receptors on J4 cells can be distinguished by their apparent molecular weight in SDS-polyacrylamide gel electrophoresis; the mouse receptor migrates significantly slower than the human receptor. There is evidence that the mouse and human polypeptide chains can also form heterodimers on the surface of the transformed cells (Newman *et al.*, 1983). Antibodies are available that specifically block the uptake of transferrin-bound iron mediated by the human and mouse transferrin receptors respectively, and thus it has been possible to show that the product of the transfected human gene was functional in the transformants (Newman *et al.*, 1983). The expression of the human transferrin receptor in J4 cells appears to be regulated co-ordinately with cell growth in the same way as the native receptor and clones of this transformant have been isolated that stably express the human molecule. While the long-term goal of this approach is to isolate the gene for the human transferrin receptor, the availability of mouse L cell transformants expressing the human structure on its surface provides the opportunity to study various aspects of its genetic regulation and function.

10.7 CONCLUDING REMARKS

The study of transferrin receptors has become a very active area of research in the past few years. This in part has been due to the availability of monoclonal antibodies against the receptor. Continuing progress in the future seems likely to be in three major areas: (1) in understanding in more detail the molecular basis of transferrin receptor function, including how transferrin is bound and the internalization and recycling of transferrin and its receptor to and from the cell surface. Another facet of transferrin receptor function is the intriguing possibility that it may be a major target for natural killer cells (Vodinelich *et al.*, 1983). (2) The second major area will be in isolating the gene for the human transferrin receptor. Its isolation should provide invaluable structural information about the transferrin receptor and lead to an understanding of how the expression of the transferrin receptor is co-ordinately regulated with growth. (3) The third major area will be in the clinical applications of monoclonal antibodies. At this stage, it is not clear whether monoclonal antibodies against the transferrin receptor will ultimately be of clinical value either as therapeutic agents or as diagnostic tools. However, it is our view that in comparison with other monoclonal antibodies against the cell surface antigens expressed on tumor cells, monoclonal antibodies against transferrin receptors have a reasonable chance of being truly useful in the clinic.

REFERENCES

Aisen, P. and Listowsky, I. (1980), *Annu. Rev. Biochem.*, **49,** 357–393.

Becker, A.J., McCulloch, E.A., Siminovitch, L. and Till, J.E. (1965), *Blood,* **26,** 296–308.

Bernstein, I.D. and Nowinski, R.C. (1982), in *Hybridomas in Cancer Diagnosis and Treatment* (M.S. Mitchell and H.F. Oettgen, eds), Raven Press, New York, pp. 97–112.

Bernstein, I.D., Tam, M.R. and Nowinski, R.C. (1980), *Science,* **207,** 68–71.

Bleil, J.D. and Bretscher, M.S. (1982), *Eur. Mol. Biol. Organ. J.*, **1,** 351–355.

Bramwell, M.E. and Harris, H. (1978), *Proc. R. Soc., London, Ser. A,* **201,** 87–106.

Bramwell, M.E. and Harris, H. (1979), *Proc. R. Soc. London, Ser. A,* **203,** 93–96.

Bretscher, M.S. (1983), *Proc. Natl. Acad. Sci. U.S.A.*, **80,** 454–458.

Brown, G., Kourilsky, F.M., Fisher, A.G., *et al.* (1981), *Hum. Lymph. Diff.*, **1,** 167–192.

Carlsson, J., Drevin, H. and Axen, R. (1978), *Biochem. J.*, **173,** 723–737.

Chitamber, C.R., Massey, E.J. and Seligmann, P.A. (1983), *J. Clin, Invest.*, **72,** 1314–1325.

Ciechanover, A., Schwartz, A.L. and Lodish, H.F. (1983), *Cell,* **32,** 267–275.

Collins, S.J., Ruscetti, F.W., Gallagher, R.E. and Gallo, R.C. (1978), *Proc. Natl. Acad. Sci. U.S.A.*, **75,** 2458–2462.

Cotner, T., Williams, J.M., Christenson, L., Shapiro, H.M., Strom, T.B. and Strominger, J. (1983), *J. Exp. Med.*, **157,** 461–472.

Dautry-Varsat, A., Ciechanover, A. and Lodish, H.F. (1983), *Proc. Natl. Acad. Sci. U.S.A.*, **80,** 2258–2262.

Derks, J.P.A., Hofmans, L., Bruning, H.W. and van Rood, J.J. (1983), *Cancer Res.*, **43,** 1914–1920.

Enns, C.A. and Sussman, H.H. (1981), *J. Biol. Chem.*, **256,** 12 620–12 623.

Enns, C.A., Suomaliainen, H.A., Gebhardt, J.E., Schroder, J. and Sussman, H.H. (1982), *Proc. Natl. Acad. Sci. U.S.A.*, **79,** 3241–3245.

Faulk, W.P., Hsi, B.-L. and Stevens, P.L. (1980), *Lancet*, **ii**, 390–392.
Fernandez-Pol, J.A. (1977a), *Biochem. Biophys. Res. Commun.*, **76**, 413–419.
Fernandez-Pol, J.A. (1977b), *Biochem. Biophys. Res. Commun.*, **78**, 136–143.
Fernandez-Pol, J.A. and Klos, D.J. (1980), *Biochemistry*, **19**, 3904–3912.
Fernandez-Pol, J.A., Bono, Jr., V.H. and Johnson, G.S. (1977), *Proc. Natl. Acad. Sci. U.S.A.*, **74**, 2889–2893.
Fernandez-Pol, J.A., Klos, D. and Donati, R.M. (1978), *Cell Biol. Int. Rep.*, **2**, 433–439.
FitzGerald, D.J.P., Trowbridge, I.S., Pastan, I. and Willingham, M.C. (1983), *Proc. Natl. Acad. Sci. U.S.A.*, **80**, 4134–4138.
Foley, G.E., Lazarus, H., Farber, S., Uzman, B.G., Boone, B.A. and McCarthy, R.E. (1965), *Cancer*, **18**, 522–529.
Frazier, J.L., Caskey, J.H., Yoffe, M. and Seligmann, P.A. (1982), *J. Clin. Invest.*, **69**, 853–865.
Galbraith, G.M.P., Galbraith, R.M. and Faulk, W.P. (1980), *Cell Immunol.*, **49**, 215–222.
Garber, E.A., Krueger, J.G., Hanafusa, H. and Goldberg, A.R. (1983), *Nature (London)*, **302**, 161–163.
Gatter, K.C., Brown, G., Trowbridge, I.S., Woolston, R.-E. and Mason, D.Y. (1983), *J. Clin. Pathol.*, **36**, 539–545.
Goding, J.W. and Burns, G.F. (1981), *J. Immunol.*, **127**, 1256–1258.
Goodfellow, P.M., Banting, G., Sutherland, R., Greaves, M., Solomon, E. and Povey, S. (1982), *Som. Cell Genet.*, **8**, 197–206.
Graham, F.L. and Van der Eb, A.J. (1973), *Virology*, **52**, 456–467.
Gurley, L.R. and Jett, J.H. (1981), *Cell Tissue Kinet.*, **14**, 269–283.
Habeshaw, J.A., Bailey, D., Stansfeld, A.G. and Greaves, M.F. (1983a), *Br. J. Cancer*, **47**, 327–351.
Habeshaw, J.A., Lister, T.A., Stansfeld, A.G. and Greaves, M.F. (1983b), *Lancet*, **i**, 498–500.
Hamilton, T.A., Wada, H.G. and Sussman, H.H. (1979), *Proc. Natl. Acad. Sci. U.S.A.*, **75**, 6406–6410.
Harris, A.W. and Sephton, R.G. (1977), *Cancer Res.*, **37**, 3634–3638.
Haynes, B.F., Hemler, M., Cotner, T., Mann, D.L., Eisenbarth, G.S., Strominger, J.L. and Fauci, A.S. (1981), *J. Immunol.*, **127**, 347–351.
Hu, H.-Y.Y., Gardner, J. and Aisen, P. (1977), *Science*, **197**, 559–561.
Hubbard, S.C. and Ivatt, R.J. (1981), *Annu. Rev. Biochem.*, **50**, 555–583.
Hughes, C.R. and Trowbridge, I.S. (1983), *J. Cell Biol.*, **97**, 508–521.
Hyman, R. and Stallings, V. (1974), *J. Natl. Cancer Inst.*, **52**, 429–436.
Hyman, R., Cunningham, K. and Stallings, V. (1980), *Immunogenetics*, **10**, 261–271.
Jandl, J.H. and Katz, J.H. (1963), *J. Clin. Invest.*, **42**, 314–326.
Johnston, G.S. (1981), *Int. J. Nuclear Med. Biol.*, **8**, 249–255.
Judd, W., Poodry, C.A. and Strominger, J.L. (1980), *J. Exp. Med.*, **152**, 1430–1435.
Karin, M. and Mintz, B. (1981), *J. Biol. Chem.*, **256**, 3245–3252.
Klausner, R.D., Ashwell, G., van Renswoude, J., Harford, J.B. and Bridges, K.R. (1983), *Proc. Natl. Acad. Sci. U.S.A.*, **80**, 2263–2266.
Larrick, J.W. and Cresswell, P. (1979), *J. Supramol. Struct.*, **II**, 579–586.
Larson, S.M., Rasey, J.S., Allen, D.R., Nelson, N.J., Grunbaum, Z., Harp, G.D. and Williams, D.L. (1980), *J. Natl. Cancer Inst.*, **64**, 41–52.

Lebman, D., Trucco, M., Bottero, L., Lange, B., Pessano, S. and Rovera, G. (1982), *Blood*, **59**, 671–678.

Leibman, A. and Aisen, P. (1977), *Biochemistry*, **16**, 1268–1272.

Lesley, J., Domingo, D.L., Schulte, R. and Trowbridge, I.S. (1984a), *Exp. Cell. Res.*, **150**, 400–407.

Lesley, J., Hyman, R., Schulte, R. and Trotter, J. (1984b), *Cell. Immunol.*, **83**, 14–25.

Lozzio, C.B. and Lozzio, B.B. (1975), *Blood*, **45**, 321–334.

Magee, A.I. and Schlesinger, M.J. (1982), *Biochim. Biophys. Acta*, **694**, 279–289.

Marsh, M., Bolzau, E., White, J. and Helenius, A. (1983), *J. Cell. Biol.*, **96**, 455–461.

Mason, D.Y. and Gatter, K.C. (1983), in *Monoclonal Antibodies in Cancer* (B.D. Boss, R. Langman, I.S. Trowbridge, and R. Dulbecco, eds), pp. 187–196.

Mendelsohn, J., Trowbridge, I.S. and Castagnola, J. (1983), *Blood*, **62**, 821–826.

Miller, Y., Jones, C., Scoggin, C., Morse, H. and Seligman, P. (1983), *Am. J. Hum. Genet.*, **35**, 573–583.

Necas, E. and Neuwirt, J. (1976), *Cell Tissue Kinet.*, **9**, 479–487.

Neckers, L.M. and Cossman, J. (1983), *Proc. Natl. Acad. Sci. U.S.A.*, **80**, 3494–3498.

Neilands, J.B. (1981), *Ann. Rev. Biochem.*, **50**, 715–731.

Neilands, J.B. (1972), *Struct. Bonding (Berlin)*, **2**, 145–170.

Newman, R., Domingo, D.L., Trotter, J. and Trowbridge, I.S. (1983), *Nature*, **304**, 643–645.

Old, L., Stockert, E., Boyse, E. and Kim, J.H. (1968), *J. Exp. Med.*, **127**, 523–539.

Omary, M.B. and Trowbridge, I.S. (1981a), *J. Biol. Chem.*, **256**, 4715–4718.

Omary, M.B. and Trowbridge, I.S. (1981b), *J. Biol. Chem.*, **256**, 12 888–12 892.

Omary, M.B., Trowbridge, I.S. and Minowada, J. (1980), *Nature (London)*, **286**, 888–891.

Plowman, G.D., Brown, J.P., Enns, C.A., Schroder, J., Nikinmaa, B., Sussman, H.H., Hellstrom, K.E. and Hellstrom, I. (1983), *Nature (London)*, **303**, 70–72.

Reichard, P. (1978), *Fed. Proc.*, **37**, 9–14.

Reinherz, E.L. and Schlossman, S.F. (1980), *Cell*, **19**, 821–827.

Reinherz, E.L., Kung, P.C., Goldstein, G., Levey, R.H. and Schlossman, S.F. (1980), *Proc. Natl. Acad. Sci. U.S.A.*, **77**, 1588–1592.

Schmidt, M.F.G. (1982), *Trends Biochem. Sci.*, **7**, 322–324.

Schneider, C., Sutherland, R., Newman, R.A. and Greaves, M.F. (1982), *J. Biol. Chem.*, **257**, 8516–8522.

Sefton, B.M., Trowbridge, I.S. and Cooper, J.A. (1982), *Cell*, **31**, 465–474.

Seligman, P.A., Schleicher, R.B. and Allen, R.H. (1979), *J. Biol. Chem.*, **254**, 9943–9946.

Sephton, R.G. and Harris, A.W. (1974), *J. Natl. Cancer Inst.*, **54**, 1263–1266.

Sephton, R.G. and Harris, A.W. (1981), *Int. J. Nucl. Med. Biol.*, 333–339.

Shapira, G. (1964), in *Iron Metabolism* (F. Gross, ed.), Springer-Verlag, Berlin, pp. 1–8.

Shindelman, J.E., Ortmeyer, A.E. and Sussman, H.H. (1981), *Int. J. Cancer*, **27**, 329–334.

Sieff, C., Bicknell, D., Caine, G., Robinson, J., Lam, G. and Greaves, M.F. (1982), *Blood*, **60**, 703–713.

Stein, B. and Sussman, H.H. (1983), *J. Biol. Chem.*, **258**, 2668–2673.

Sullivan, A.L. and Weintraub, L.R. (1978), *Blood*, **52**, 436–446.

Sutherland, R., Delia, D., Schneider, C., Newman, R., Kemshead, J. and Greaves, M. (1981), *Proc. Natl. Acad. Sci. U.S.A.*, **78,** 4515–4519.

Thorpe, P.E. and Ross, W.C.J. (1982), *Immunol. Rev.*, **62,** 119–158.

Thorpe, P.E., Brown, A., Foxwell, B., Myers, C., Ross, W., Cumber, A. and Forrester, T. (1983), in *Monoclonal Antibodies in Cancer* (B.D. Boss, R. Langman, I.S. Trowbridge and R. Dulbecco, eds), Academic Press, New York, pp. 117–124.

Till, J. and McCulloch, E. (1961), *Radiat. Res.*, **14,** 213–222.

Trowbridge, I.S. (1983), in *Monoclonal Antibodies in Cancer* (B.D. Boss, R. Langman, I.S. Trowbridge and R. Dulbecco, eds), Academic Press, New York, pp. 53–61.

Trowbridge, I.S. and Domingo, D. (1981), *Nature (London)*, **294,** 171–173.

Trowbridge, I.S. and Domingo, D.L. (1982), *Cancer Surveys*, **1,** 543–556.

Trowbridge, I.S. and Lopez, F. (1982), *Proc. Natl. Acad. Sci. U.S.A.*, **79,** 1175–1179.

Trowbridge, I.S. and Omary, M.B. (1981), *Proc. Natl. Acad. Sci. U.S.A.*, **78,** 3039–3043.

Trowbridge, I.S., Lesley, J. and Schulte, R. (1982), *J. Cell. Physiol.*, **112,** 403–410.

van Leeuwen, A., Festenstein, H. and van Rood, J.J. (1982), *Hum. Immunol.*, **4,** 109–121.

van Renswoude, J., Bridges, K.R., Hartford, J.B. and Klausner, R.D. (1982), *Proc. Natl. Acad. Sci. U.S.A.*, **79,** 6186–6190.

Vodinelich, L., Sutherland, R., Schneider, C., Newman, R. and Greaves, M. (1983), *Proc. Natl. Acad. Sci. U.S.A.*, **80,** 835–839.

Vogt, A., Mishell, R.I. and Dutton, R.W. (1969), *Exp. Cell Res.*, **54,** 195–200.

Wada, H.G., Hass, P.E. and Sussman, H.H. (1979), *J. Biol. Chem.*, **254,** 12629–12635.

Ward, J.H., Kushner, J.P. and Kaplan, J. (1982), *J. Biol. Chem.*, **257,** 10317–10323.

11 Monoclonal Antibodies as Probes for Insulin and Insulin-like Growth Factor-I Receptors

STEVEN JACOBS, FREDERICK C. KULL, Jr
and PEDRO CUATRECASAS

Glossary of terms and abbreviations

Insulin-like growth factor-I	Also called somatomedin-C
αIR-1, αIR-2, αIR-3	Three monoclonal antibodies against the insulin receptor

Monoclonal Antibodies to Receptors: Probes for Receptor Structure and Function
(*Receptors and Recognition*, Series B, Volume 17)
Edited by M. F. Greaves
Published in 1984 by Chapman and Hall, 11 New Fetter Lane, London EC4P 4EE

11.1 INTRODUCTION

Receptors for polypeptide hormones are present in cell membranes in extremely small quantities and comprise only a small fraction of a percent of the total membrane protein. Therefore, any method that can be used for their detection and characterization must have exquisite sensitivity and specificity. Antibodies provide reagents with these essential features. In this chapter we described a series of monoclonal antibodies to receptors for insulin and insulin-like growth factor-I and their use in characterizing these receptors.

Insulin-like growth factor-I, also called somatomedin-C, is a polypeptide hormone that is synthesized and secreted by several tissues in response to growth hormone. It mediates many of the effects of growth hormone, including stimulation of growth and sulfation of cartilage. It is structurally similar to insulin, having about 45% homology with the amino acid residues in insulin, including three disulfide bonds, which correspond to those of insulin (Rinderknecht and Humbel, 1978; Klapper *et al.*, 1983). Like proinsulin, it has a connecting peptide linking regions corresponding to the A and B chains. In addition, it has a short continuation in the region corresponding to the *C*-terminus of the A chain. With this degree of structural similarity, it is not surprising that both insulin and insulin-like growth factor-I bind to and activate each other's receptors, although with considerably lower affinity than each binds to its own receptor (Hintz *et al.*, 1972; Van Wyk *et al.*, 1980; Kasuga *et al.*, 1981; Massaqué and Czech, 1982; Bhaumick *et al.*, 1981).

The structures of the receptors for insulin and insulin-like growth factor-I are also quite similar. [For recent reviews, see Czech (1983), Czech *et al.* (1983) and Jacobs and Cuatrecasas (1983).] Both receptors are composed of two copies of an α subunit and two copies of a β subunit (Fig. 11.1). These have

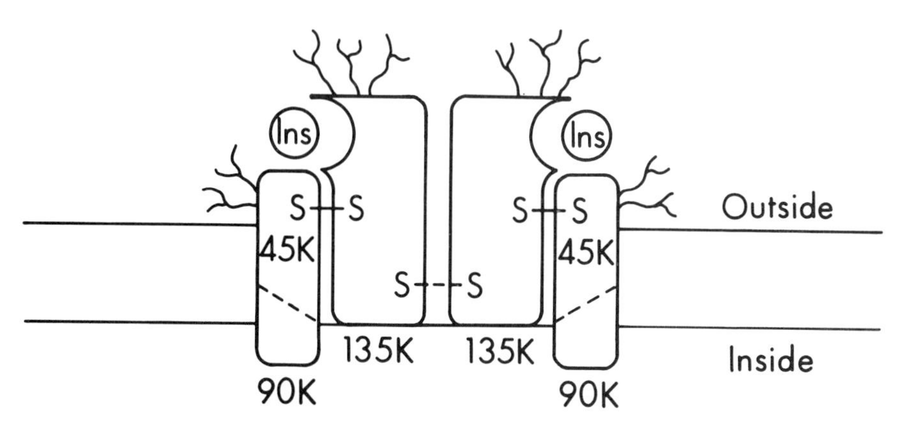

Fig. 11.1 Proposed subunit structure of receptors for insulin and insulin-like growth factor-I.

approximate molecular weights of 135 000 and 90 000. They form tetramers in which the subunits are linked to a variable degree by interchain disulfide bonds.

11.2 PRODUCTION AND INITIAL CHARACTERIZATION OF MONOCLONAL ANTIBODIES

11.2.1 Antibodies produced with purified placental receptor

We have produced three monoclonal antibodies by using as an antigen insulin receptor purified from human placenta by affinity chromatography on insulin–Sepharose (Kull *et al.*, 1982, 1983). Since the type I insulin-like growth factor receptor has substantial affinity for insulin, it might be anticipated that some of this receptor would also adsorb to the insulin–Sepharose column and co-purify with insulin receptors. Indeed, Bennett *et al.* (1981) have reported that this does occur, and we have found that approximately 30% of the insulin-like growth factor-I binding activity present in placental extracts adsorbs to insulin–Sepharose columns. However, no insulin-like growth factor-I binding activity was recovered when the column was eluted with urea. In spite of these negative results, it is possible that insulin-like growth factor-I receptor was present in the eluate but in a form denatured by urea and no longer capable of binding hormone. In view of the antibodies that were obtained when this material was used as an antigen, this seems likely.

SJL mice were immunized subcutaneously with approximately 3 μg of purified receptor emulsified in complete Freund's adjuvant. They were boosted three times at 3-week intervals with similar amounts of receptor in incomplete Freund's adjuvant. Three days prior to sacrifice, the mice were boosted with an intravenous injection of purified receptor. Their spleen and local lymph node cells were fused with FO myeloma cells and the hybrids seeded at low density with peritoneal macrophages as a feeder layer (Fazekas de St. Groth and Scheidegger, 1980). Viable clones were screened for their ability to immunoprecipitate solubilized insulin receptor or insulin-like growth factor-I receptor in an assay in which the receptors were labeled by binding the ^{125}I-labeled hormone (Kull *et al.*, 1982, 1983). Six clones were initially found to be positive. Each was positive for immunoprecipitation of both insulin receptor and insulin-like growth factor-I receptor. One clone died. Two others stopped producing antibody. The remaining three were serially subcloned five times by limiting dilution and grown in ascites fluid in Balb/c $\times$ SJL F_1 hybrids. We refer to these three clones and the antibodies they produce as αIR-1, αIR-2 and αIR-3.

Once the clones had been sufficiently expanded, it was necessary to establish their specificities. This was complicated since ^{125}I-insulin and ^{125}I-insulin-like growth factor-I will to some extent label both receptors. To overcome this complication, we examined the ability of various concentrations of unlabeled insulin and insulin-like growth factor-I to inhibit the immunoprecipitation of bound labeled insulin and insulin-like growth factor-I (Kull *et al.*, 1983). The concentration curve for this inhibition reflects the affinity of the unlabeled hormone for the immunoprecipitated receptor to which the labeled hormone was bound. Using this approach, we showed that all three antibodies immunoprecipitate both receptors, although αIR-1 is approximately ten times more effective in immunoprecipitating insulin receptors, whereas αIR-2 and αIR-3 are more than 100 times as effective in immunoprecipitating receptors for insulin-like growth factor-I.

To further establish the specificity of these antibodies, membrane proteins were non-specifically radioiodinated with lactoperoxidase. The membranes were then solubilized with Triton X-100 and immunoprecipitated with normal mouse serum, αIR-1, αIR-2 or αIR-3, and the immunoprecipitates analyzed by SDS-polyacrylamide gel electrophoresis. αIR-1 specifically immunoprecipitated two bands with molecular weights of 135 000 and 90 000, which correspond to the known α and β subunits of the insulin receptor (Kull *et al.*, 1983). αIR-2 and αIR-3 also specifically immunoprecipitated two bands. These, however, have somewhat different molecular weights than the corresponding bands present in insulin receptors and vary slightly depending on the tissue studied (Kull *et al.*, 1983). In placenta they have molecular weights of 132 000 and 96 000, whereas in IM-9 cells they have molecular weights of 136 000 and 96 000.

11.2.2 Antibody produced with IM-9 cells

A fourth monoclonal antibody was produced by Roth *et al.* (1982). They immunized Balb/c mice with IM-9 cells, a human lymphocyte cell line that has an abundance of both insulin and insulin-like growth factor-I receptors. Hybrids were screened for their ability to produce antibody that inhibited ^{125}I-insulin binding to IM-9 cells. One stable hybrid clone with this property was identified. The antibody produced also inhibited ^{125}I-insulin-like growth factor-I receptor binding, although a 300-fold higher concentration was required than to inhibit ^{125}I-insulin receptor binding (Roth *et al.*, 1983a). To examine its specificity, its ability to immunoprecipitate [^{35}S]methionine-labeled proteins from IM-9 cells was determined. Only the α and β subunits of the insulin receptor were specifically immunoprecipitated (Roth *et al.*, 1982).

11.3 PROPERTIES OF THE MONOCLONAL ANTIBODIES

The properties of these monoclonal antibodies and their interaction with the receptors are interesting, particularly when compared with the several polyclonal antisera that are available. All four monoclonal antibodies are IgG_1 K

(Roth *et al.*, 1982; Kull *et al.*, 1983). This is perhaps not surprising, since IgG_1 K is the most common class of mouse immunoglobulin.

Both αIR-2 and αIR-3 react with the α subunit (Kull *et al.*, 1983). This was established by dissociating lactoperoxidase-labeled receptors with SDS and dithiothreitol and determining which subunit was immunoprecipitated. αIR-1 failed to immunoprecipitate either subunit after this treatment (Kull *et al.*, 1983). Presumably the native structure of the receptor is required for its recognition. As a result, we were able to determine with which subunit it interacts.

Roth *et al.* (1983b) using a different approach demonstrated that the antibody they produced also reacts with the α subunit. They treated labeled receptor with elastase, which drastically degraded the β subunit but left the α subunit intact. Their antibody immunoprecipitated the residual α subunit.

Neither αIR-1, -2 or -3 significantly inhibit hormone binding, despite the fact that αIR-2 and αIR-3 interact with α, the subunit most directly involved in hormone binding.* In contrast, the antibody prepared by Roth *et al.* (1982) competitively inhibits hormone binding, and conversely, insulin inhibits the binding of this antibody to the receptor.

As previously discussed, all four antibodies recognize both insulin and insulin-like growth factor-I receptors, which suggests extensive immunochemical similarities between these receptors. Consistent with this, several polyclonal sera react with both receptors (Rosenfeld *et al.*, 1981; Jonas *et al.*, 1982; Armstrong *et al.*, 1983). Since three of the monoclonal antibodies recognize an epitope in the α subunit, at least part of this immunochemical similarity must reside in that subunit.

Preliminary studies indicate that αIR-1 does not mimic the biological effects of insulin. The antibody produced by Roth *et al.* (1982) also lacks agonist activity. Because it blocks the binding of insulin to its receptor, it acts as a competitive antagonist. This is in marked contrast to several polyclonal anti-insulin receptor sera, which all have agonist activity (see, for example, Kahn *et al.*, 1977; Jacobs *et al.*, 1978). The insulin-like activity of these polyclonal sera is related to their ability to cross-link receptors (Kahn *et al.*, 1978). Monovalent Fab′ fragments are devoid of agonist activity, but activity is recovered by cross-linking with anti-Fab′ antibody. It is interesting to speculate that the lack of agonist activity of the monoclonal antibodies may be related to the fact that since they recognize a single epitope, it may be difficult for them to form cross-linked aggregates larger than dimers.

Also in contrast with available polyclonal anti-insulin receptor sera, which have little species specificity, both αIR-1 and the antibody produced by Roth *et al.* (1982) are species specific. αIR-1 recognizes human receptor, but not receptor from horse, rat or mouse (Kull *et al.*, 1982). The antibody produced by Roth *et al.* (1982) also recognizes human receptor but not rat receptor. The

*Note added in proof: Recent studies indicate that αIR-3 does inhibit hormone binding.

species specificity of these antibodies is surprising in the light of their ability to cross-react with insulin-like growth factor-I receptors. Both insulin and insulin-like growth factor-I receptors are present in a number of mammalian and avian species. With respect to their hormone-binding properties, insulin-like growth factor-I receptors are quite similar in these different species but are clearly different from insulin receptors, which are also quite similar in different species. Therefore, it would be anticipated that receptors for insulin and insulin-like growth factor-I diverged earlier phylogenetically than the evolution of mammalian species, and that insulin receptors from different mammalian species would be immunochemically more closely related to each other than to insulin-like growth factor receptors, even in the same species. The results obtained with αIR-1 and the antibody produced by Roth *et al.* indicate that there are epitopes where this is not the case. This might provide useful clues as to the mechanism by which these receptors have evolved.

11.4 RECEPTOR PURIFICATION

One major advantage of the monoclonal approach to producing antibodies is that purified receptor is not required to obtain highly specific antibodies. In fact, as illustrated by the experience of Roth *et al.* (1982), whole cells can be a suitable immunogen. However, once antibodies have been produced, they can be utilized for receptor purification.

Roth and Cassell (1983) have prepared an immunoaffinity column with the monoclonal antibody they produced and used it to purify insulin receptor from solubilized IM-9 cells. Receptor was eluted from the column biospecifically with insulin. This method of elution is more specific and less likely to denature the receptor than those required to elute receptor from other antibody or insulin affinity columns, which have utilized urea, high salt or low pH (Jacobs *et al.*, 1977; Fujita-Yamaguchi *et al.*, 1983; Harrison and Itin, 1980). Perhaps the reason this method could be used with the monoclonal antibody is that it has a somewhat lower affinity for the receptor, and binds to a single epitope which is competitively inhibited by insulin. The receptor eluted by this procedure appears to be purified to homogeneity on the basis of SDS-polyacrylamide gel electrophoresis, which reveals two bands corresponding to the known α and β subunits of the insulin receptor. It binds insulin and has insulin-stimulated protein kinase activity (Roth and Cassell, 1983). Thus, it retains the two known functions of the insulin receptor.

We have been attempting to purify insulin-like growth factor-I receptors from human placenta with an immunoaffinity column prepared by coupling αIR-3 to CL-Sepharose 4B activated with cyanogen bromide. The column quantitatively adsorbs insulin-like growth factor-I receptors from solubilized placenta while the majority of insulin receptor is not retained. After washing, the small amount of insulin receptor that is retained can be eluted at pH 9.5.

Insulin-like growth factor-I receptor can then be eluted free of insulin receptor at pH 10.3. This procedure results in nearly quantitative recovery of the insulin-like growth factor-I binding activity applied to the column. Unfortunately, polyacrylamide gel electrophoresis of the eluted receptor indicates that it is still contaminated by a large number of protein bands which are major components of the solubilized placental extract. We are currently attempting to modify the affinity purification procedure to eliminate contamination or develop additional procedures for further purification.

11.5 IMMUNOPRECIPITATION OF NON-SPECIFICALLY LABELED RECEPTORS AS A USEFUL TOOL FOR STUDYING THEIR PROPERTIES

Because of their specificity, antibodies to receptors for insulin and insulin-like growth factor-I have been useful reagents for identifying non-specifically labeled receptor. In these studies, receptors for insulin or insulin-like growth factor-I, present in intact cells, membranes, or other complex mixtures of proteins, are labeled by some non-specific means, such as lactoperoxidase iodination, phosphorylation or biosynthetically with labeled precursors. The labeled receptors are then immunoprecipitated with specific antibodies, and analyzed by SDS-polyacrylamide gel electrophoresis and radioautography. Polyclonal anti-insulin receptor sera have been available for a number of years and have been used extensively to study the insulin receptor in this manner. The development of monoclonal antibodies specific for insulin-like growth factor-I receptor has made it possible to carry out similar studies of that receptor.

11.5.1 Receptor biosynthesis

Using polyclonal anti-insulin receptor sera, Kasuga *et al.* (1982a) and Deutsch *et al.* (1983) have shown that when cells are incubated with [^{35}S]methionine, the first immunoreactive peptide that becomes labeled has a molecular weight of approximately 200000. Only at later times are the 135000- and 90000-molecular weight α and β subunits of the mature receptor labeled. They suggested that this high-molecular weight peptide may be a biosynthetic precursor to the insulin receptor. To further evaluate this, and to determine if a similar precursor is involved in the biosynthesis of insulin-like growth factor-I receptors, we incubated cells with [^{35}S]methionine in the presence of monensin (Jacobs *et al.*, 1983a), a carboxylic ionophore that inhibits the post-translational maturation of membrane glycoproteins by blocking their transport through the Golgi apparatus, and inhibiting terminal glycosylation and proteolytic processing (for a review on monensin, see Tartakoff, 1983). The labeled cells were then solubilized and immunoprecipitated with αIR-1 and αIR-3. Monensin completely blocked the synthesis of mature forms of α and β subunits of both receptors, but caused the appearance of a major 180 000-molecular weight polypeptide and considerably smaller amounts of 115 000- and 89 000-molecular weight polypeptides. These were not seen in immunoprecipitates of cells that had not been treated with monensin (Jacobs *et al.*, 1983a).

These results raise two major questions. What is the relation of the 180 000-, 115 000- and 89 000-molecular weight polypeptides to the mature α and β subunits? Since when cells are cultured in the presence of monensin, both αIR-1 and αIR-3 immunoprecipitate biosynthetically labeled peptides of similar size, do receptors for insulin and insulin-like growth factor-I have common precursors? To answer these questions, cells treated with monensin were affinity labeled with ^{125}I-insulin (Jacobs *et al.*, 1983a). This method of labeling is of course specific for insulin receptors, and in particular for the α subunit. The cells were then solubilized and immunoprecipitated with αIR-1 or αIR-3. αIR-1 immunoprecipitated the labeled 180 000- and 115 000-molecular weight bands, which suggests that these are precursors or abnormally processed forms of the α subunit. αIR-3 failed to immunoprecipitate either labeled band. This indicates that the precursors for the insulin receptor are not the same as those for the insulin-like growth factor-I receptor even though they have similar molecular weights and cannot be distinguished by SDS-polyacrylamide gel electrophoresis (Jacobs *et al.*, 1983a).

As will be discussed more extensively in later sections of this chapter, in the presence of ATP, the β subunit of receptors for both insulin and insulin-like growth factor-I are phosphorylated. To determine the relation between the β subunits of the mature receptors and the forms found in the presence of monensin, IM-9 cells that had been cultured in the presence of monensin were solubilized, incubated with [^{32}P]ATP under conditions in which phosphorylation of the β subunits normally occur, and then immunoprecipitated with αIR-1 and αIR-3. Both antibodies specifically immunoprecipitated phosphorylated 180 000-molecular weight bands, which were not present if the cells had not been incubated with monensin. This suggests that the 180 000-molecular weight peptides are also precursors of the β subunits, and is consistent with the findings of Kasuga *et al.* (1982a) that peptide maps of the 180 000-molecular weight peptide precurors of the insulin receptor overlap with peptide maps of both the α and β subunits of the mature receptor.

The 180 000-molecular weight precursors of both receptors that are found in monensin-treated cells exist as disulfide-linked dimers, which when run on SDS-polyacrylamide gels without reduction migrate slightly more rapidly than the disulfide-linked tetramers of the mature receptors (Jacobs *et al.*, 1983a).

The evidence presented above, in conjunction with what is known about the effects of monensin and membrane glycoprotein synthesis in general suggests

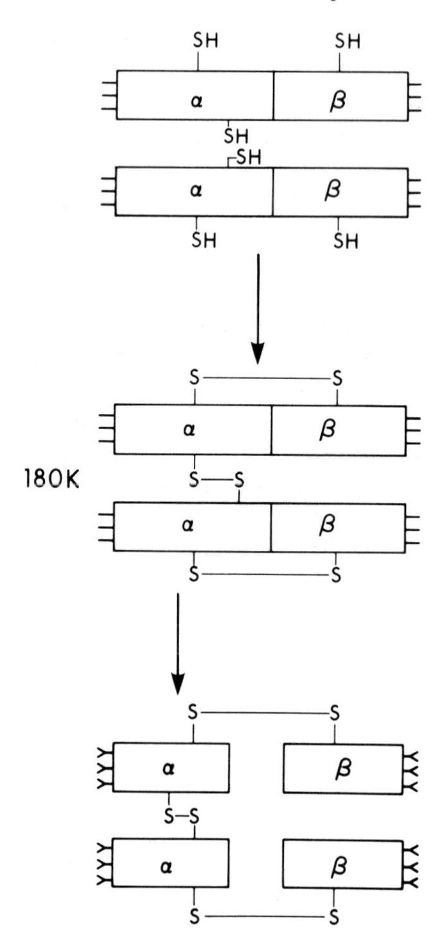

Fig. 11.2 Biosynthetic scheme for the biosynthesis of receptors for insulin and insulin-like growth factor-I.

the following scheme for synthesis of receptors for insulin and insulin-like growth factor-I (Fig. 11.2). Both receptors are synthesized as a single polypeptide chain on ribosomes attached to the rough endoplasmic reticulum. Co-translationally, these chains are core glycosylated and possibly a signal peptide is cleaved to generate the 180 000-molecular weight polypeptides seen in the presence of monensin. Quite early, these 180 000-molecular weight peptides form homodimers which are linked by interchain disulfide bonds, and also presumably form intrachain disulfide bonds linking regions correponding to the α and β subunits. The dimers are then transported to the Golgi apparatus where they are proteolytically cleaved at the region where the α and β subunits are connected, and terminally glycosylated to generate mature α and β

subunits forming a disulfide-linked tetramer. Monensin blocks the last two processes and results in the accumulation of 180 000-molecular weight precursors and also smaller amounts of 115 000- and 89 000-molecular weight forms which have been proteolytically processed but not terminally glycosylated. In the absence of monensin, the fully mature receptors are then transported from the Golgi apparatus to the cell membrane.*

11.5.2 Receptor phosphorylation

Kasuga *et al.*, and subsequently several other laboratories, have demonstrated that the β subunit of the insulin receptor is phosphorylated and that phosphorylation is enhanced by insulin (Roth and Cassell, 1983; Kasuga *et al.*, 1982b,c,d; Petruzelli *et al.*, 1982; Avruch *et al.*, 1982; Zick *et al.*, 1983; Shia and Pilch, 1983; Van Obberghen *et al.*, 1983). In intact cells, phosphorylation occurs on tyrosine, threonine and serine residues (Kasuga *et al.*, 1982c). However, in solubilized preparations, only tyrosine is phosphorylated (Kasuga *et al.*, 1982d; Petruzelli *et al.*, 1982; Avruch *et al.*, 1982; Shia and Pilch, 1983). The receptor itself is a tyrosine-specific protein kinase, whose activity is stimulated by insulin, and which undergoes autophosphorylation (Roth and Cassell, 1983; Shia and Pilch, 1983; Van Obberghen *et al.*, 1983; Kasuga *et al.*, 1983). Aside from autophosphorylation, insulin receptor catalyzes the tyrosine phosphorylation of several substrates in broken cell preparations (Petruzelli *et al.*, 1982; Stadtmauer and Rosen, 1983), but its other natural *in vivo* substrates are unknown. Several viral oncogene products and other growth factor receptors, including epidermal growth factor and platelet-derived growth factor, are also tyrosine-specific protein kinases (Hunter and Sefton, 1982; Cohen *et al.*, 1980; Ek and Heldin, 1982; Wishimura *et al.*, 1982). Thus, although currently the functional significance of insulin receptor protein kinase activity has not been established, it appers likely to be involved in mediating the actions of insulin, particularly its pleiotropic effects.

Monoclonal antibodies have been used to demonstrate that the insulin-like growth factor-I receptor is also phosphorylated on its β subunit, an effect which is enhanced by insulin-like growth factor-I (Jacobs *et al.*, 1983b). To do this, IM-9 cells were incubated with $[^{32}P]H_3PO_4$, solubilized and immunoprecipitated with αIR-1 or αIR-3. The β subunit of the insulin receptor, which was immunoprecipitated by αIR-1, and the β subunit of the insulin-like growth factor-I receptor, which was immunoprecipitated by αIR-3, were both phosphorylated in the basal state. The extent of phosphorylation of the β subunits of both receptors was increased if the cells were incubated with insulin or insulin-like growth factor-I. Each hormone was more effective at enhancing the

*Editor's note: see also a recent paper by Hedo *et al.* (1983) (*J. Biol. Chem.*, **258,** 10 020–10 026), which provides further evidence for the processing of α and β subunits of the insulin receptor from a common polypeptide precursor.

phosphorylation of its own receptor. Stimulation of phosphorylation of each receptor by the heterologous hormone probably results from the cross-reactivity of the hormones. However, another possibility is that each hormone can serve as a substrate for the other receptor.

Phosphorylation of insulin-like growth factor-I receptor that had been solubilized from IM-9 cells was also demonstrated by using [γ^{32}P]ATP as a substrate and immunoprecipitating with αIR-3 (Jacobs *et al.*, 1983b). Under these conditions, phosphorylation occurred exclusively on tyrosyl residues. If solubilized insulin-like growth factor-I receptors were first immunoprecipitated and then incubated with [γ^{32}P]ATP there was also phosphorylation of the β subunit (Jacobs *et al.*, 1983), indicating that the receptor itself or a protein tightly associated with it was the responsible kinase. Similar results were obtained with insulin-like growth factor-I receptor solubilized from human placenta (Pilch, 1983).

11.5.3 Phorbol esters and receptor phosphorylation

Protein kinase-C is a serine- or threonine-specific protein kinase whose activity is regulated by calcium, phosphatidylserine and diacylglycerol (Nishizuka and Takai, 1981). In several cells, diacylglycerol released in response to hormonal stimulation, presumably activates protein kinase-C, which in turn, it has been postulated, may mediate many of the actions of the hormone (Nishizuka and Takai, 1981). Recently, it has been shown that phorbol esters directly activate protein kinase-C by binding to the diacylglycerol site (Castagna *et al.*, 1982; Niedel *et al.*, 1983). Phorbol esters modulate a variety of cellular effects (Blumberg 1980, 1981). On the basis of currently available information, it is reasonable to assume that they do so through activation of protein kinase-C. One of these effects is to decrease the affinity of insulin for its receptor in intact cells (Grunberger and Gordon, 1982; Thomopoulos *et al.*, 1982). This occurs rapidly, within minutes after the addition of phorbol esters, suggesting that a direct effect of protein kinase-C on the receptor may be involved. To evaluate this, we determined the effect of phorbol esters on the state of phosphorylation of receptors for insulin and insulin-like growth factor-I (Jacobs *et al.*, 1983c).

The endogenous pool of ATP of IM-9 cells was labeled by incubating them with [^{32}P]H_3PO_4. The cells were then incubated in the presence or absence of phorbol esters. The cells were solubilized and immunoprecipitated with αIR-1 or αIR-3. Phorbol esters caused a rapid increase in the extent of phosphorylation of the β subunits of both the insulin receptor (immunoprecipitated by αIR-1), and the insulin-like growth factor-I receptor (immunoprecipitated by αIR-3) (Jacobs *et al.*, 1983c). Insulin added simultaneously with phorbol esters further increased the extent of phosphorylation of the insulin receptor, and its effects appeared to be additive to those of phorbol esters. Peptide maps indicated that phorbol esters stimulate phosphorylation of different sites on the

β subunit from those phosphorylated by insulin (Jacobs *et al.*, 1983c). This is not surprising, since protein kinase-C is a serine- and threonine-specific kinase, while the insulin receptor is a tyrosine-specific protein kinase.

It is not clear if the receptor phosphorylation seen in response to phorbol esters is directly mediated by protein kinase-C, or if phorbol esters stimulate a phosphorylation cascade, which only indirectly results in the phosphorylation of these receptors. However, the results do suggest that protein kinase-C may play an important role in the physiological regulation of receptors of insulin and insulin-like growth factor-I.

REFERENCES

Armstrong, G.D., Hollenberg, M.D., Bhaumic, B., Bala, R.M. and Maturo, J.M. III (1983), *Can. J. Biochem.*, **61**, 650–656.

Avruch, J., Nemenoff, R.A., Blackshear, P.J., Pierce, M.W. and Osathanondh, R. (1982), *J. Biol. Chem.*, **257**, 15162–15166.

Bennett, A., Daly, F.T. and Hintz, R.L. (1981), *Diabetes*, **30**, Suppl. 1, 55A.

Bhaumick, B., Goren, H.J. and Bala, R.M. (1981), *Horm. Metab. Res.*, **13**, 515–518.

Blumberg, P.M. (1980), *CRC Crit. Rev. Toxicol.*, 153–197.

Blumberg, P.M. (1981), *CRC Crit. Rev. Toxicol.*, 199–234.

Castagna, M., Takai, Y., Kaibuichi, K., Sano, K., Kikkawa, U. and Nishizuka, Y. (1982), *J. Biol. Chem.*, **257**, 7847–7851.

Cohen, S., Carpenter, G. and King, L., Jr. (1980), *J. Biol. Chem.*, **255**, 4834–4842.

Czech, M.P. (1983), *Cell*, **31**, 8–10.

Czech, M.P., Oppenheimer, C.L. and Massaqué, J. (1983) *Fed. Proc.*, **42**, 2598–2601.

Deutsch, P.J., Wan, C.F., Rosen, O.M. and Rubin, C.S. (1983), *Proc. Natl. Acad. Sci. U.S.A.*, **80**, 133–136.

Ek, B. and Heldin, C-H. (1982), *J. Biol. Chem.*, **257**, 10486–10492.

Fazekas de St. Groth, S. and Scheidegger, D. (1980), *J. Immunol. Methods*, **35**, 1–21.

Fujita-Yamaguchi, Y., Choi, S., Sakamoto, Y. and Itakura, K. (1983), *J. Biol. Chem.*, **258**, 5045–5049.

Grunberger, G. and Gordon, P. (1982), *Am. J. Physiol.*, **243**, E319–E324.

Harrison, L.C. and Itin, A. (1980), *J. Biol. Chem.*, **255**, 12066–12072.

Hintz, R.L., Cemmons, D.R., Underwood, L.E. and Van Wyck, J.J. (1972), *Proc. Natl. Acad. Sci. U.S.A.*, **69**, 2351–2353.

Hunter, T. and Sefton, B.M. (1982), in *The Molecular Actions of Toxins, Viruses and Interferon* (P. Cohen, and S. Van Heyningen, eds), Elsevier/North Holland, New York, pp. 337–370.

Jacobs, S. and Cuatrecasas, P. (1983), *Annu. Rev. Pharmacol. Toxicol.*, **23**, 461–479.

Jacobs, S., Shechter, Y., Bissell, K. and Cuatrecasas, P. (1977), *Biochem. Biophys. Res. Commun.*, **77**, 981–988.

Jacobs, S., Chang, K-J. and Cuatrecasas, P. (1978), *Science*, **200**, 1283–1284.

Jacobs, S., Kull, F.C., Jr. and Cuatrecasas, P. (1983a), *Proc. Natl. Acad. Sci. U.S.A.*, **80**, 1228–1231.

Jacobs, S., Kull, F.C., Jr., Earp, H.S., Svoboda, M.E., Van Wyk, J.J. and Cuatrecasas, P. (1983b) *J. Biol. Chem.*, **258**, 9581–9584.

Jacobs, S., Sahyoun, N.E., Saltiel, A.R. and Cuatrecasas, P. (1983c), *Proc. Natl. Acad. Sci. U.S.A.*, **80**, (in press).

Jonas, H.A., Baxter, R.C. and Harrison, L.C. (1982), *Biochem. Biophys. Res. Commun.*, **109**, 463–470.

Kahn, C.R., Baird, K., Flier, J.S. and Jarrett, D.B. (1977), *J. Clin. Invest.*, **60**, 1094–1106.

Kahn, C.R., Baird, K.L., Jarrett, D.B. and Flier, J.S. (1978), *Proc. Natl. Acad. Si. U.S.A.*, **75**, 4209–4213.

Kasuga, M., Van Obberghen, E., Nissley, S.P. and Rechler, M.M. (1981), *J. Biol. Chem.*, **256**, 5305–5308.

Kasuga, M., Hedo, J.A., Yamada, K.M. and Kahn, C.R. (1982a), *J. Biol. Chem.*, **257**, 10392–10399.

Kasuga, M., Karlson, F.A. and Kahn, C.R. (1982b), *Science*, **215**, 185–187.

Kasuga, M., Zick, Y., Blithe, D.L., Karlsson, F.A., Häring, H.U. and Kahn, C.R. (1982c), *J. Biol. Chem.*, **257**, 9891–9894.

Kasuga, M. Zick, Y., Blithe, D.L., Crettaz, M. and Kahn, C.R. (1982d), *Nature (London)*, **298**, 667–669.

Kasuga, M., Fujita-Yamaguchi, Y., Blithe, D.L. and Kahn, C.R. (1983), *Proc. Natl. Acad. Sci. U.S.A.*, **80**, 2137–2141.

Klapper, D.C., Svoboda, M.E. and Van Wyk, J.J. (1983), *Endocrinology*, **112**, 2215–2217.

Kull, F.C., Jr., Jacobs, S., Su, Y.-F. and Cuatrecasas, P. (1982), *Biochem. Biophys. Res. Commun.*, **106**, 1019–1026.

Kull, F.C., Jr., Jacobs, S., Su, Y.-F., Svoboda, M.E., Van Wyk, J.J. and Cuatrecasas, P. (1983), *J. Biol. Chem.*, **258**, 6561–6566.

Massaqué, J. and Czech, M.P. (1982), *J. Biol. Chem.*, **257**, 5038–5045.

Niedel, J.E., Kuhn, L.J. and Vanderbark, G.R. (1983), *Proc. Natl. Acad. Sci. U.S.A.*, **80**, 36–40.

Nishimura, J., Huang, J.S. and Deuel, T.F. (1982), *Proc. Natl. Acad. Sci. U.S.A.*, **79**, 4303–4307.

Nishizuka, Y. and Takai, Y. (1981), in *Protein Phosphorylation Cold Spring Harbor Conferences on Cell Proliferation* (O. Rosen and E.G. Krebs, eds.), Cold Spring Harbor Laboratory, New York, pp. 237–250.

Petruzelli, L.M., Ganguly, S., Smith, C.J., Cobb, M.H., Rubin, C.S. and Rosen, O.M. (1982), *Proc. Natl. Acad. Sci. U.S.A.*, **79**, 6972–6976.

Pilch, P.F. (1983), *Nature (London)*, (in press).

Rinderknecht, E. and Humbel, R.E. (1978), *J. Biol. Chem.*, **253**, 2769–2775.

Rosenfeld, R.G., Baldwin, D., Jr., Dollar, L.A., Hintz, R.L., Olefsky, J.M. and Rubenstein, A. (1981), *Diabetes*, **30**, 979–982.

Roth, R.A. and Cassell, D.J. (1983), *Science*, **219**, 299–301.

Roth, R.A., Cassell, D.J., Wang, K.Z., Maddux, B.A. and Goldfine, I.D. (1982), *Proc. Natl. Acad. Sci. U.S.A.*, **79**, 7312–7316.

Roth, R.A., Maddux, B., Wong, K.Y., Styne, D.M., Vliet, G.V., Humbel, R.E. and Goldfine, I.D. (1983a), *Endocrinology*, 1865–1867.

Roth, R., Cassell, D.J. and Misirow, M.L. (1983b), *Fed. Proc.*, **42**, 1790.

Shia, M.A. and Pilch, P.F. (1983), *Biochemistry*, **22,** 717–721.
Stadtmauer, L.A. and Rosen, O.M. (1983), *J. Biol. Chem.*, **258,** 6682–6685.
Tartakoff, A.M. (1983), *Cell,* **32,** 1026–1028.
Thomopoulos, P., Testa, U., Gourdin, M.-F., Hervey, C., Titeux, M. and Vainchenker, W. (1982), *Eur. J. Biochem.*, **29,** 389–393.
Van Obberghen, E., Rossi, B., Kowalski, A., Gazzano, H. and Ponzio, G. (1983), *Proc. Natl. Acad. Sci. U.S.A.*, **80,** 945–949.
Van Wyk, J.J., Svoboda, M.E. and Underwood, L.E. (1980), *J. Clin. Endocrinol. Metab.*, **50,** 206–208.
Zick, Y., Kasuga, M., Kahn, C.R. and Roth, J. (1983), *J. Biol. Chem.*, **258,** 75–80.

12 Monoclonal Antibodies against the Membrane Receptor for Epidermal Growth Factor: A Versatile Tool for Structural and Mechanistic Studies

J. SCHLESSINGER, I. LAX, Y. YARDEN, H. KANETY and T. LIBERMANN

Acknowledgements

This work was supported by grants from the National Institute of Health (CA-25820), Stiftung Volkswagenwerk, U.S. Israel Binational Science Foundations and Israel Cancer Research Fund.

Note added in proof

In collaboration with the laboratory of Dr M. Waterfield, we have recently sequenced several peptides which were derived from the EGF receptor from human cells. It was shown that each of six peptides derived from the human EGF receptor very closely matched a part of the deduced sequence of the V-erb-B transforming protein of avian erythroblastosis virus (AEV) (Downward *et al.*, 1984).

The V-erb-B oncogene is a member of the *src* gene family. Our results support the view that the transforming protein encoded by V-erb-B acquired the cellular gene sequence of a truncated EGF receptor lacking the external-binding domain but retaining the transmembranal domain and a domain involved in stimulating cell proliferation (Downward *et al.*, 1984).

Monoclonal Antibodies to Receptors: Probes for Receptor Structure and Function
(*Receptors and Recognition*, Series B, Volume 17)
Edited by M. F. Greaves
Published in 1984 by Chapman and Hall, 11 New Fetter Lane, London EC4P 4EE

12.1 INTRODUCTION

12.1.1 Epidermal growth factor (EGF), its biological activity and interaction with target cells

The proliferation of eukaryotic cells is regulated by several polypeptide hormones and growth factors which are present in body fluids in low concentrations. One of the best characterized growth factors is epidermal growth factor (EGF). EGF is a small protein of molecular weight 6045. It acts on various epithelial cells, fibroblasts and other cell types *in vitro* and *in vivo* (for reviews see Carpenter and Cohen, 1979; Schlessinger *et al.*, 1983a,b). The binding of EGF to its surface receptor leads to the activation of various early and delayed cellular processes leading to cell proliferation. Early effects include an increase in ion flux (Rozengurt and Heppel, 1975; Moolenaar *et al.*, 1982), and changes in cell morphology (Chinkers *et al.*, 1979) and in the organization of the cytoskeleton (Schlessinger and Geiger, 1981).

Several methods were used to follow the fate of EGF after binding to target cells (for review, see Schlessinger *et al.*, 1983a). The picture which emerges from these studies is as follows. EGF binds to randomly distributed surface receptors which rapidly rotate and translate in the plasma membrane of the cell (Schlessinger *et al.*, 1978a,b; Hillman and Schlessinger, 1982; Zidovetzki *et al.*, 1981). The binding of EGF to its surface receptors leads to aggregation of the hormone–receptor complexes, mainly in coated areas but also in non-coated areas on the plasma membrane (Haigler *et al.*, 1979). The occupied EGF receptors in coated pits pinch off and form coated vesicles. The coated vesicles lose their clathrin coat and appear in an organelle called the endosome. The pH within the endosomes has been shown to be 5 ± 0.2 (Tycko and Maxfield, 1982). At this pH, EGF molecules dissociate from the receptor. Nevertheless, in contrast with other ligands which are internalized via receptor-mediated endocytosis (e.g. transferrin) both EGF and its receptor seem to be degraded by lysosomal enzymes (Carpenter and Cohen, 1976). The rapid degradation of receptor molecules leads to their down-regulation; receptor molecules cannot be recycled and reutilized. However, it is not clear yet whether all the receptors are degraded or whether a fraction of the internalized receptor can be recycled and reutilized (Schlessinger *et al.*, 1983b).

12.1.2 The EGF receptor–kinase system

One of the first responses mediated by EGF is the phosphorylation of various membrane proteins, including EGF receptor itself (Carpenter, *et al.*, 1978, 1979; Chen *et al.*, 1982). The EGF-sensitive kinase is tyrosine-specific, cyclic nucleotide-independent and appears to be an integral part of the receptor molecule itself (Ushiro and Cohen, 1980; Buhrow *et al.*, 1982).* The EGF receptor–kinase was identified by numerous studies as a membrane glycoprotein of 170000 daltons (Linsley and Fox, 1979; Carpenter *et al.*, 1979; Cohen *et al.*, 1980).

EGF induces the phosphorylation of several cellular proteins including a 36000-dalton protein (Hunter and Cooper, 1981; Erikson *et al.*, 1981) which is one of the main substrates of pp60^{src} (i.e. the transforming protein of Rous sarcoma virus). Moreover, an antibody generated against a synthetic peptide homologous with the amino acid sequence of the phosphorylation site of pp60^{src} immunoprecipitates a functional EGF receptor–kinase from human A431 cells (Lax *et al.*, 1984). This and other observations (Chinkers and Cohen, 1981; Kudlow *et al.*, 1981) suggest molecular similarity between the kinase activities and the phosphorylation sites of EGF receptor and pp60^{src}.

12.2 MONOCLONAL ANTIBODIES AGAINST EGF RECEPTOR

We have generated monoclonal antibodies against different domains on the receptor molecule (Table 12.1). Initially we used human epidermoid carcinoma A431 cells as immunogen (Schreiber *et al.*, 1981). These cells bear an unusually high number of EGF receptors ($\sim 2 \times 10^6$ receptors per cell, Fabricant *et al.*, 1977). Mice were immunized with A431 cells. Spleen cells from these mice were fused with non-secreting murine myelomas. The hybrid cells were grown in selection medium and several screening assays were used in order to detect and subsequently select the anti-EGF receptor antibodies with the required properties.

12.2.1 Monoclonal antibodies which block the binding of ^{125}I-labeled EGF to EGF receptor

We have screened for anti-receptor antibodies which specifically block the binding of ^{125}I-labeled EGF to EGF receptor of A431 cells. Several hybridoma clones secreting such antibodies were found. One clone denoted 2G2 secretes IgM antibodies which have the following properties (Schreiber *et al.*, 1981a).

2G2-IgM binds to cells which bear EGF receptor in proportion to the number of EGF receptors on these cells. 2G2-IgM binds to EGF receptor on cells from various species including human, dog, rat and mouse cells; it does not bind to avian cells. 2G2-IgM does not bind to cells which do not bear EGF receptors, namely various lymphoid cells and transformed cells.

2G2-IgM mimics various effects of EGF. It activates the EGF-sensitive kinase, induces morphological changes in A431 cells and stimulates DNA synthesis and proliferation of human foreskin fibroblasts. Hence, 2G2-IgM

*Recently reviewed by Carpenter (1983).

Table 12.1 Various antibodies against EGF receptor

Type of antibody	Cross-reactivity	Effect of EGF binding	Biological activity	References
Monoclonal 29(1)-IgG_1	Human, mouse	None	None	Yarden and Schlessinger (1983) and data in this chapter
Monoclonal TL5-IgG	Human, bovine, dog, rat, mouse	None	None	Schreiber *et al.* (1983) and data in this chapter
Monoclonal 2G2-IgM	Human, rat, mouse	Yes	Yes	Schreiber *et al.* (1981a) Schreiber *et al.* (1983) Hapgood *et al.* (1983)
Monoclonal R_1-IgG_{2b}	Only human cells	None	None	Waterfield *et al.* (1982)
Monoclonal 528-IgG	Human cells	Yes	Yes	Kawamoto *et al.* (1983)
Monoclonal G49-IgG	Only human A431 cells	Yes	No data	Gregoriou and Rees (1983)
Polyclonal anti-membranes prepared from A431 cells	Human, bovine, dog, rat, mouse	Yes	None	Yarden and Schlessinger (1983) Haigler and Carpenter (1980)
Polyclonal anti-shed vesicles from A431 cells	Human, mouse	No data	No data	Yarden *et al.* unpublished data
Polyclonal antibodies against pure denatured EGF receptor	Human, mouse	None	No data	Yarden *et al.* unpublished data

Editor's note: some additional publications have recently appeared describing polyclonal or monoclonal antibodies against the EGF receptor: Richert *et al.* (1983); Carlin *et al.* (1982a,b, 1983); Carpenter *et al.* (1983).

acts as a full agonist of EGF, inducing both the early and delayed effects mediated by the growth factor (Schreiber *et al.*, 1981a; Schreiber *et al.*, 1983; Hapgood *et al.*, 1983). The biological activity of 2G2-IgM parallels the biological activity of EGF. In fibroblasts, where EGF induces DNA synthesis and cell proliferation, 2G2-IgM acts also as a mitogen. However, in rat pituitary cells (GH_3) EGF partially blocks cell proliferation, changes the morphology of the cells (i.e. the round cells become flat and elongated) and induces the expression of the prolactin gene (Johnson *et al.*, 1980a). Monoclonal antibody 2G2-IgM has the same effects on these cells (Hapgood *et al.*, 1983).

Following binding and internalization, a small fraction of EGF molecules accumulates in the nucleus of GH_3 cells. The inhibition of hormone degradation by chloroquine increases the amount of EGF accumulated in the nucleus (Johnson *et al.*, 1980b). These results, together with the observation that EGF induces changes in the structure of chromatin in isolated nuclei of GH_3 cells, suggest that the interaction between EGF and specific sites on the nucleus mediates the expression of specific genes (Johnson and Eberhardt, 1982). Similar mechanisms were proposed for the initiation of the delayed effects mediated by insulin (Goldfine *et al.*, 1977) and nerve growth factor (Andres *et al.*, 1977; Yankner and Shooter, 1979).

Hapgood *et al.* (1983) have shown that 2G2-IgM binds specifically to the rat pituitary GH_3 cells and blocks the binding of ^{125}I-EGF to these cells. Moreover, like EGF, the monoclonal antibody induces the synthesis of prolactin and morphological changes in these cells.

The EGF receptor molecule from GH_3 cells was identified as a 170000-dalton protein associated with a tyrosine-specific kinase. It is similar to but not identical with the receptor from A431 cells (Hapgood *et al.*, 1983).

The capacity of 2G2-IgM to induce various 'EGF-like' effects provides an important insight into the mechanism of action of EGF. It indicates that EGF receptor when properly triggered contains all the biological attributes necessary for the induction of various effects of EGF. The appropriate change(s) in the receptor molecule induced either by EGF or by the monoclonal antibodies (i.e. conformational change, phosphorylation, proteolytic cleavage?) leads to the activation of the pleiotropic effects of the growth factor. As EGF and 2G2-IgM are structurally unrelated it can be concluded that the internalized EGF molecule or its fragment is not essential as a 'second messenger'.

Initiation of the various responses could occur as a consequence of direct interaction between the activated receptor and intracellular sites or via a putative intermediate molecule which acts as an internal stimulus of the activated EGF receptor.

12.2.2 The role of antibody valence

Monovalent Fab′ fragments of 2G2-IgM (2G2-Fab′) were prepared from purified intact IgM by a short trypsin-cleavage procedure (Schreiber *et al.*, 1983). The affinity of the monovalent 2G2-Fab′ fragment towards EGF receptor is approximately 20-fold smaller than the apparent affinity of the intact IgM antibody. Like the intact antibody, 2G2-Fab′ competes for the binding of ^{125}I-EGF towards EGF receptor on human fibroblasts. The inhibition is concentration-dependent and correlates with receptor occupancy on the cells. Unlike the binding of the intact antibody which cannot be blocked by EGF, the binding of 2G2-Fab′ is inhibited by saturating concentrations of the growth factor. The monovalent Fab′ fragments do not induce receptor clustering and fail to induce DNA synthesis in fibroblasts. However, 2G2-Fab′ activates the EGF-sensitive kinase, inducing phosphorylation of EGF receptor and of other membrane proteins (Schreiber *et al.*, 1983). The addition of 2G2Fab′ followed by anti-mouse Ig antibodies induced receptor clustering and restored the capacity of the monovalent antibody to stimulate DNA synthesis in fibroblasts. Hence, in many respects 2G2-Fab′ is similar to cyanogen bromide-cleaved EGF (Yarden *et al.*, 1982; Schreiber *et al.*, 1981b, 1983). Both ligands activate the EGF-sensitive kinase and fail to induce receptor clustering and DNA synthesis. Moreover, both stimulate DNA synthesis only when cross-linked on the cell surface with antibodies against them (Schechter *et al.*, 1978; Schreiber *et al.*, 1983). Hence, protein phosphorylation induced either by EGF or by the non-mitogenic ligands is apparently not a sufficient signal either for receptor clustering or for the induction of DNA synthesis. Hence, the activation of DNA synthesis is not a consequence of a single step involving receptor occupancy and induction of protein phosphorylation. Rather, several biochemical signals are generated during the events of EGF receptor clustering, internalization and processing of the receptor molecule (Yarden *et al.*, 1982; Schreiber *et al.*, 1983; Schlessinger *et al.*, 1983a,b).

12.2.3 Generation of monoclonal antibodies used for the isolation of a functional EGF receptor–kinase system

We have generated and characterized several monoclonal antibodies of which two are currently employed for the isolation of large quantities of EGF receptor for structural studies. They are two hybridomas clones denoted TL5-IgG and 29(1)-IgG_1 (Table 12.1).

Generation of monoclonal antibodies against EGF receptor which immunoprecipitate a functional EGF receptor–kinase system

Mice were immunized with fixed A431 cells as previously described by Schreiber *et al.* (1981a). The antiserum from one mouse contained high titers of antibodies which reacted with cell-surface components of A431 cells and precipitated the EGF receptor from these cells. The precipitated EGF receptor was associated with a functional kinase activity which underwent autophosphorylation upon addition of [γ-^{32}P]ATP to the immunoprecipitate (Fig. 12.1(a),B). The phosphorylated product co-migrated with the 170000–150000-dalton proteins of the EGF receptor which had been phosphorylated upon the addition of 500 ng of EGF/ml to a membrane preparation of A431 cells (Carpenter *et al.*, 1979) (Fig. 12.1(a),D). Similar results were obtained when polyclonal rabbit antibodies against membranes from A431 cells were used to immunoprecipitate the EGF receptor, followed by the addition of

[γ-^{32}P]ATP to the immunoprecipitate (Fig. 12.1(a),A). Normal mouse antiserum (Fig. 12.1(a),C) and normal rabbit antiserum (data not shown) failed to precipitate detectable amounts of EGF receptor from A431 cells.

This immunoprecipitation/phosphorylation procedure was consequently employed in this laboratory as an assay for the selection of monoclonal antibodies which can immunoprecipitate a *functional EGF receptor–kinase system*.

Three days after a booster injection, spleen cells of the immunized mouse were fused with NSO cells, a non-secreting murine myeloma (Galfre *et al.*, 1977; Galfre and Milstein, 1981). Hybrid cells were distributed into 240 wells of Costar trays and grown in selection medium containing hypoxanthine, aminopterin and thymidine (HAT). Growth of hybrid cells was observed in all wells. Hybrid cultures secreting mouse Ig were selected with a solid-phase radioimmunoassay: 212 hybrid cultures secreted more than 1 μg of mouse Ig per ml of medium per 10^6 hybrid cells per day.

Positive Ig-secreting hybridoma cultures were screened for their binding capacity to a variety of cells as previously described by Schreiber *et al.* (1981a). The initial screening protocol includes binding to A431 cells and other cells which bear EGF receptors, and lack of binding to the lymphomas Molt and Daudi which are devoid of EGF receptors. The 82 hybridoma cultures selected in the first screening step were subsequently screened for their capacity to secrete Ig which specifically precipitates a functional EGF receptor–kinase system (Fig. 12.1(b)), without necessarily interfering with the binding of EGF to the receptor. Since we used protein A–Sepharose rather than a second antibody for the immunoprecipitation procedure, we have favored antibodies from the IgG subclasses. Six out of 82 hybridoma cultures

---

Fig. 12.1 *(Opposite)* Screening of monoclonal antibodies against the EGF receptor of A431 cells (a) Radioautography of SDS-PAGE analysis of immunoprecipitates of EGF–receptor kinase complexes by sera of immunized mice. Aliquots of solubilized A431 membranes were incubated with rabbit polyclonal anti-A431 membrane IgG (A), antiserum from a mouse immunized with A431 cells (B) or normal mouse serum (C) bound to protein A–Sepharose as described in Lax *et al.* (1984). The immunoprecipitate was phosphorylated by [γ-^{32}P]ATP as described in Lax *et al.* (1984). Immunoprecipitates with rabbit anti-A431 membrane IgG fraction (A), mouse anti-A431 cell serum (B) or normal mouse serum (C) were used. A431 membranes (D) were phosphorylated according to standard procedures (Chinkers *et al.*, 1979) and used as marker for the EGF receptor. (b) Immunoprecipitation of EGF–receptor kinase complex by hybridoma culture medium. Aliquots of solubilized A431 cells were incubated with hybridoma culture medium bound to protein A–Sepharose; the immunoprecipitate was phosphorylated by [γ-^{32}P]ATP and analyzed by SDS-PAGE and radioautography, as described in Lax *et al.* (1984). A, Total extract of A431 cells. B to N, Radioautography of phosphorylated proteins in the immunoprecipitates with different hybridoma culture media.

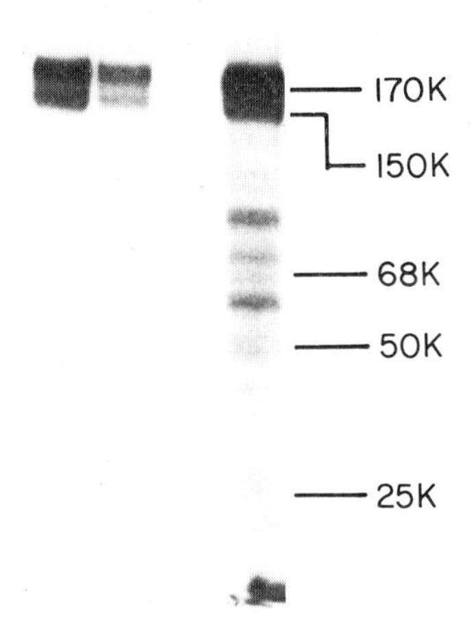
A
B
C
D
170K
150K
68K
50K
25K

(a)

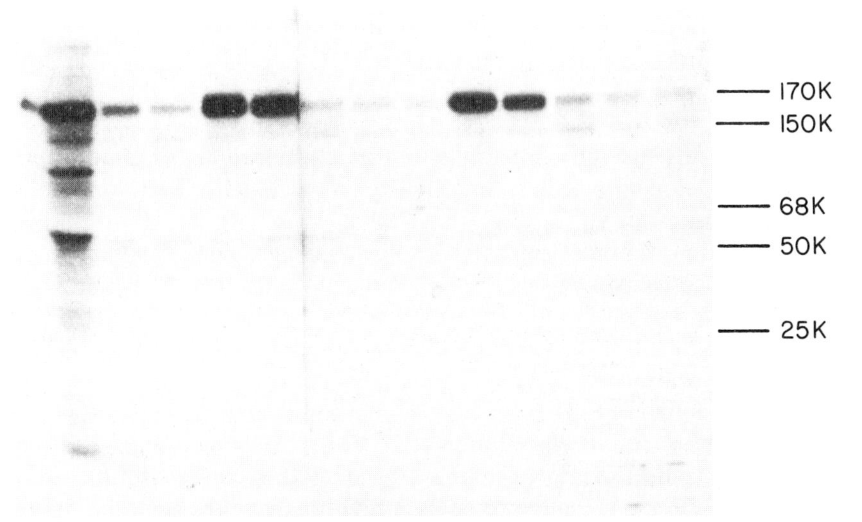
A
B
C
D
E
F
G
H
I
K
L
M
N
170K
150K
68K
50K
25K

(b)

screened secreted antibodies which precipitated a functional EGF receptor–kinase system. Fig. 12.1(b) depicts the immunoprecipitation/autophosphorylation assay for 12 different Ig-secreting hybrid cultures. A highly phosphorylated protein co-migrating with phosphorylated EGF receptor in membranes from A431 cells is observed in lanes D, E, I and K of Fig. 12.1(b). None of these cultures secreted antibodies which inhibited binding of ^{125}I-EGF to EGF receptor on A431 cells. Antibodies secreted from culture N did inhibit the binding of ^{125}I-EGF to EGF receptor, and EGF inhibits antibody binding to A431 cells. However, these antibodies did not give a positive result in the immunoprecipitation/phosphorylation assay (Fig. 12.1(b), lane N). Two of the positive hybridomas (Fig. 12.1(b),E) were cloned in soft agar from single-cell colonies. One clone that secreted high quantities of IgG_3 antibodies (denoted TL5-IgG) was grown in cultures and in the form of ascites in mice. Monoclonal antibody TL5-IgG binds in a saturable fashion to A431 cells (Fig. 12.2), human foreskin fibroblasts, mouse 3T3 fibroblasts and rat NRK pituitary cells. It does not bind to Molt or Daudi cells which do not bear EGF receptors.

The second clone which secreted IgG_1 antibody (denoted 29(1)-IgG) was also grown in culture and in the form of ascites in mice. 29(1)-IgG binds to human and mouse cells. It does not affect the binding of TL5-IgG to EGF receptor nor does the binding of TL5-IgG influence the binding of 29(1)-IgG to EGF receptor. Hence the two antibodies bind to different domains of the EGF receptor. Monoclonal antibody 29(1)-IgG is currently used for the isolation of large quantities of EGF receptor using immunoaffinity chromatography (Yarden *et al.*, 1984).

EGF enhances the affinity of monoclonal anti-EGF receptor TL5-IgG towards EGF receptors on intact A431 cells

TL5-IgG does not interfere with the binding of EGF to the receptor molecule. We have further investigated the effect of EGF on the binding of TL5-IgG to EGF receptor on A431 cells. The equilibrium binding of ^{125}I-TL5-IgG to A431 cells at 4 °C is shown in Fig. 12.2(a). Preincubation of the cells at 4 °C with EGF did not modify the apparent affinity or the concentration of TL5-IgG bound at saturation. However, when the cells were preincubated at 37 °C for 20 min with 10 nM EGF prior to antibody binding at 4°C, both apparent affinity and concentration of bound antibody at saturation were significantly altered (Fig. 12.2(a)). Scatchard analysis (Fig. 12.2(b)) indicates that the binding of EGF results in an increase in the apparent binding constant $K_{app.}$ of TL5-IgG from 7×10^7 M^{-1} to 3×10^8 M^{-1} and a reduction in number of bound antibody molecules at saturation from 1.2×10^6 molecules per cell to 8.1×10^5 molecules per cell. The number of EGF receptors was measured independently with ^{125}I-EGF as 1.4×10^6 per cell. Preincubation at 37 °C for 20 min with 10 nM EGF caused an internalization of approximately 10% of EGF receptor. Hence part of the effect of EGF on the number of TL5-IgG-binding sites could be the consequence of EGF-mediated receptor internalization. Two not necessarily exclusive interpretations can be offered for the EGF-induced increase in the affinity of TL5-IgG. First, EGF at 37 °C may induce clustering of its receptor (for reviews, see Schlessinger *et al.*, 1983a,b) which favors bivalent binding of TL5-IgG instead of the usual, predominantly monovalent mode of binding of monoclonal antibodies to surface receptors (Mason and Williams, 1980). This, in turn, should increase the affinity of TL5-IgG towards EGF receptor. Second, EGF at 37 °C may bring about a conformational change in the receptor molecule so that the antigenic determinants recognized by TL5-IgG become more accessible, facilitating antibody binding without any net stoichiometry change.

Monoclonal antibody against EGF receptor TL5-IgG immunoprecipitates a functional EGF receptor–kinase system

Fig. 12.3 shows [^{35}S]methionine-labeled EGF receptor isolated from A431 and 3T3 cells and immunoprecipitated with TL5-IgG bound to protein A–Sepharose and analyzed by SDS-polyacrylamide gel electrophoresis. The EGF receptor from A431 cells co-migrated (Fig. 12.3, B, D) with the EGF receptor from 3T3 cells (see Fig. 12.3, F), both appearing as a single band with apparent molecular weight 170000. Control antibodies failed to immunoprecipitate EGF receptor (Fig. 12.3, C, E, G). Binding and immunoprecipitation studies

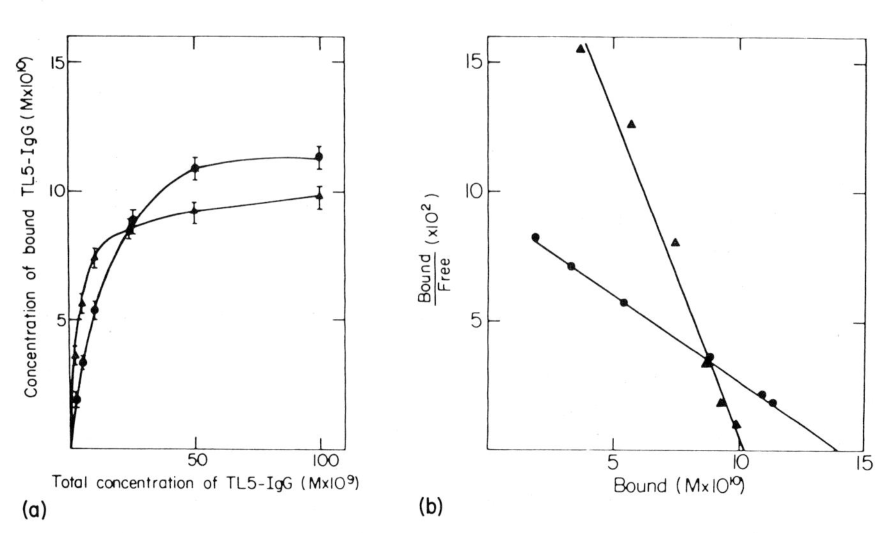

Fig. 12.2 Effect of EGF on the equilibrium binding of TL5-IgG to cells. (a) A431 cells were preincubated at 37 °C for 20 min with either buffer (●) or 10 nM EGF (▲) before incubation at 4 °C for 30 min with TL5-IgG. (b) Scatchard plot analysis of the binding curves from (a).

indicate that TL5-IgG recognizes an antigenic determinant on the EGF receptor common to human, rat and mouse cells. The molecular weight of the EGF receptor from different cells and tissues of various mammalian species is also very similar (Libermann *et al.*, unpublished results), indicating that the gene(s) for EGF receptor is (are) highly conserved. In other experiments, solubilized proteins from [^{35}S]methionine-labeled A431 cells were first incubated with Blue Sepharose which was reported to bind EGF receptor from human placenta (Hollenberg, 1979). Bound proteins were eluted with 1 M NaCl and the EGF receptor was immunoprecipitated from the eluate with the TL5-IgG antibodies (Fig. 12.3, D).

We have used TL5-IgG to investigate the link between the EGF receptor and the EGF-sensitive kinase. Fig. 12.4, A and B, show SDS-PAGE analysis of immunoprecipitated EGF receptor from A431 cells which had been phosphorylated in the precipitate. Fig. 12.4, A′ and B′, depict two-dimensional,

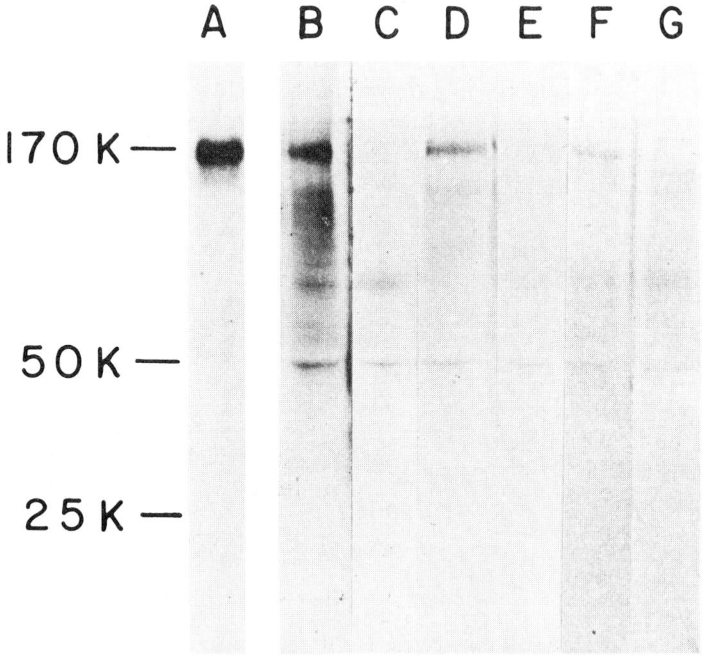

Fig. 12.3 Immunoprecipitation of biosynthetic labeled EGF receptor by TL5-IgG. A431 or 3T3 cells were biosynthetically labeled with [^{35}S]methionine, lysed with 1% Triton X-100 and subjected to immunoprecipitation. Radioautography of 5–15% acrylamide gradient SDS-PAGE analysis of the following immunoprecipitates: precipitated EGF receptor from A431 cells phosphorylated in the precipitate (A); precipitate with purified TL5-IgG (B, D); control antibody (C, E); precipitate from [^{35}S]methionine-labeled A431 cells before (B, C) and after (D, E) Blue Sepharose chromatography; precipitate with TL5-IgG ascites (F) and control ascites (G) from [^{35}S]methionine-labeled 3T3-cells.

tryptic peptide analysis of the phosphopeptides of EGF receptor. The data clearly show that the immunopurified EGF receptor is associated with a functional protein kinase activity. As noted above, TL5-IgG does not interfere with the binding of EGF to its receptor (Schreiber *et al.*, 1983). It was therefore possible to examine whether EGF stimulated the autophosphorylation of EGF receptor in the immunoprecipitate. The addition of 500 ng of EGF/ml to the immunoprecipitate induced a 2–5-fold increase in the phosphorylation of EGF receptor (Fig. 12.4, A, B). Similar results were obtained when EGF was added during the immunoprecipitation reaction (data not shown). Comparison of the two-dimensional tryptic phosphopeptide maps of the isolated EGF receptor in the presence (Fig. 12.4, A′) or absence (Fig. 12.4, B′) of EGF indicates that EGF increases the level of phosphate in all the tryptic peptides obtained from

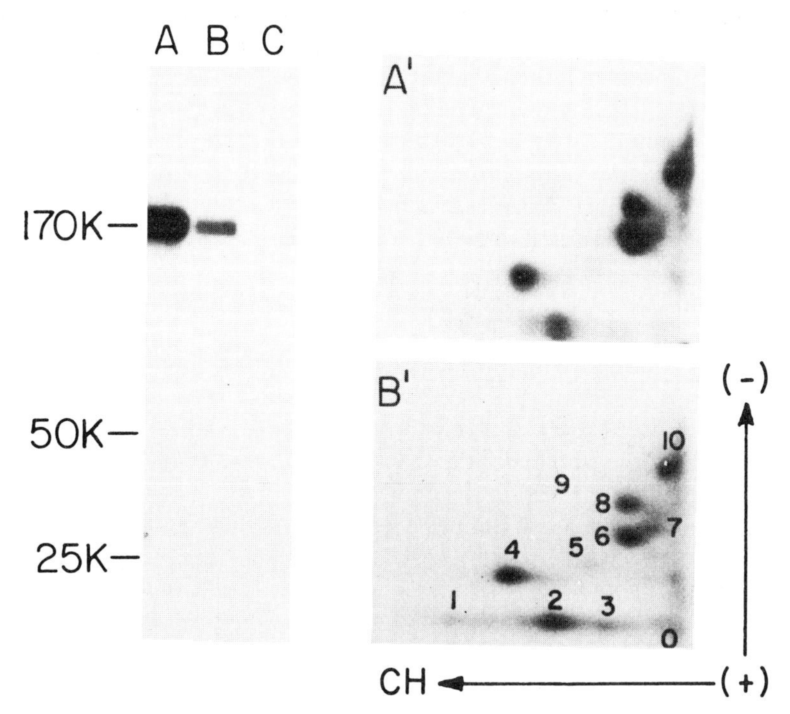

Fig. 12.4 Immunoprecipitation of functional EGF-receptor–kinase complexes with monoclonal antibody TL5-IgG. Aliquots of solubilized A431 cells were incubated with either ascites fluid of TL5-IgG or control ascites bound to protein A–Sepharose. The immunoprecipitate was phosphorylated by [γ-^{32}P]ATP in the presence or absence of EGF as described in Lax *et al.* (1984). Radioautography of SDS-PAGE analysis of phosphorylated immunoprecipitate with TL5-IgG (A and B) or with control ascites (C) and in the presence (A) or absence (B) of EGF. A′ and B′, two-dimensional tryptic analysis of phosphopeptides derived from the 170 000-dalton EGF-receptor band in A and B respectively.

the immunoprecipitated receptor. Ten distinct phosphopeptides overlap with the tryptic phosphopeptides observed in the EGF receptor of A431 cells phosphorylated *in vitro* (data not shown).

Monoclonal anti-EGF receptor antibody TL5-IgG immunoprecipitates EGF receptor which is photoaffinity-labeled with [γ-^{32}P]-8-azido-ATP

Photoaffinity labeling of ATP- or ADP-binding sites of proteins by exposure to [γ-^{32}P]-8-azido-ATP followed by ultraviolet irradiation was previously used successfully for specific labeling of ATP-binding sites on various proteins (Haley *et al.*, 1979; Owens and Haley, 1978; Czarnubi *et al.*, 1979). We have used this approach for the photoaffinity labeling of the putative ATP-binding site on EGF receptor. Fig. 12.5, A, depicts SDS-PAGE analysis of EGF receptor from A431 cells immunoprecipitated with TL5-IgG and photoaffinity-labeled with [γ-^{32}P]-8-azido-ATP. Both EGF receptor and the light and heavy chains of the antibody were labeled (Fig. 12.5, A). Preincubation for 15 minutes at 60 °C (Fig. 12.5, B) or treatment with 2 mM *N*-ethylmaleimide (NEM) for 60 minutes at room temperature abolished the photolabeling of the receptor molecule but did not affect the photolabeling of the light and heavy chains of the antibody. These treatments are known to inactivate the protein kinase of EGF receptors (Buhrow *et al.*, 1982), indicating that the photolabeling of the receptor is specific.

In a second experiment, shed membrane vesicles from A431 cells were photoaffinity labeled. The major photolabeled proteins are EGF receptor and bovine serum albumin (BSA) which is present in the reaction mixture at a concentration of 1 mg/ml (Fig. 12.5, C). Heat inactivation (Fig. 12.5, E) or treatment with NEM (data not shown) decreased the labeling of the receptor but did not affect the labeling of BSA and the other membrane proteins, again demonstrating the specificity of the photolabeling of the receptor. The labeled vesicles were solubilized and the EGF receptor was immunoprecipitated with TL5-IgG (Fig. 12.5, G). The major photolabeled protein is the 170 000-dalton EGF receptor together with several degradation products which were not observed when the immunoprecipitated receptor was photolabeled (Fig. 12.5, I). The antibodies did not precipitate photolabeled receptor from heat-inactivated vesicles (Fig. 12.5, E), and [γ-^{32}P]-8-azido-ATP did not label proteins from the membrane vesicles when the reaction was performed in the dark (Fig. 12.5, D, F, H, K). Buhrow *et al.* (1982) have labeled an ATP/ADP-binding site on EGF receptor in membrane vesicles from A431 cells with the affinity reagent 5′-*P*-fluorosulfonylbenzoyladenosine.

Studies of the EGF receptor in intact A431 cells and in A431-derived membrane vesicles employing monoclonal anti-receptor antibodies and chemical labeling techniques have therefore established that the kinase activity is an integral part of the receptor, as evidenced by (a) retention of such activity in the TL5-IgG immunoprecipitate and (b) the demonstration of a nucleotide-

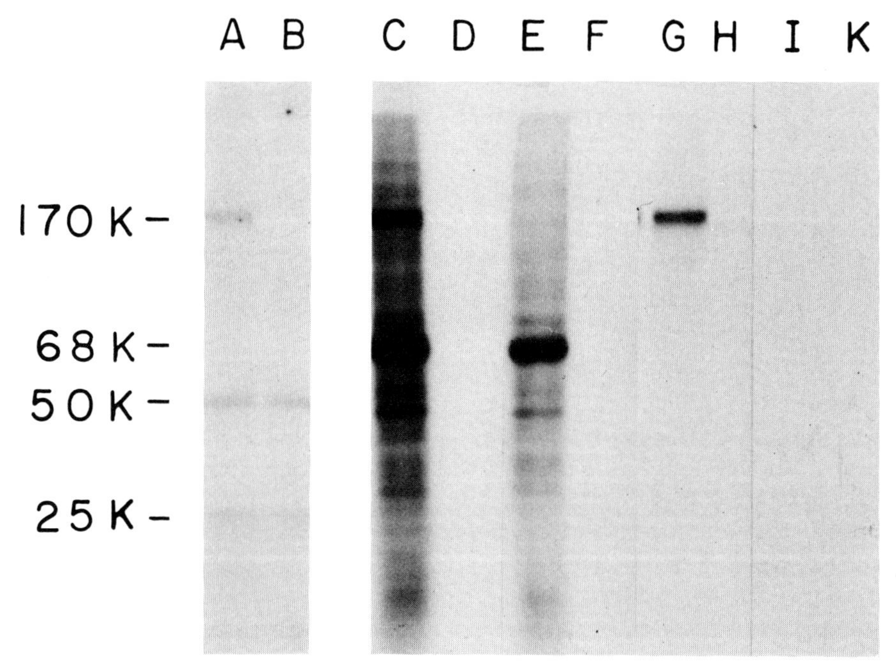

Fig. 12.5 Photoaffinity labeling of EGF receptor with [γ-^{32}P]-8-azido-ATP. Immunoprecipitates either from A431 cells (A, B) or membrane vesicles from A431 cells (C–F) were treated with 3μCi of [γ-^{32}P]-8-azido-ATP in the presence (A, B, C, E) or absence (D, F) of u.v. irradiation. Subsequently the EGF receptor was immunoprecipitated from the labeled vesicles (G–K). Immunoprecipitate from A431 cells (A) and from heat-inactivated (15 min at 60°C) sample (B). Membrane vesicles exposed (G) or not exposed (H) to u.v. irradiation and heat-inactivated membrane vesicles (15 min at 60°C) exposed (E) or not exposed (F) to u.v. irradiation. Immunoprecipitated EGF receptor from solubilized membrane vesicles exposed (G) or not exposed (H) to u.v. irradiation and heat-inactivated vesicles exposed (I) or not-exposed (K) to u.v. irradiation.

binding site essential for autophosphorylation. That stimulation of autophosphorylation and increase in the apparent affinity of TL5-IgG for receptor following EGF binding supports our proposed receptor model (Schlessinger *et al.*, 1983a); this postulates that the EGF receptor is allosterically modulated by hormone binding through induced conformational changes, and that these are manifested by enhanced kinase activity and modification of antigenic determinants.

12.3 STRUCTURAL ANALYSIS OF EGF RECEPTOR USING MONOCLONAL ANTIBODIES

12.3.1 Purification of EGF receptor by immunoaffinity chromatography

The isolation of large quantities of purified EGF receptor will greatly facilitate structural analysis of its functional domains, including the kinase and phosphorylation sites. We have therefore developed an efficient immunoaffinity procedure for this purpose (Yarden *et al.*, 1984). EGF receptor from A431 cells was immobilized on a 29(1)-IgG–Sepharose column. The column was then washed extensively with loading buffer and the immunoaffinity-purified receptor was phosphorylated by the addition of [γ-^{32}P]ATP

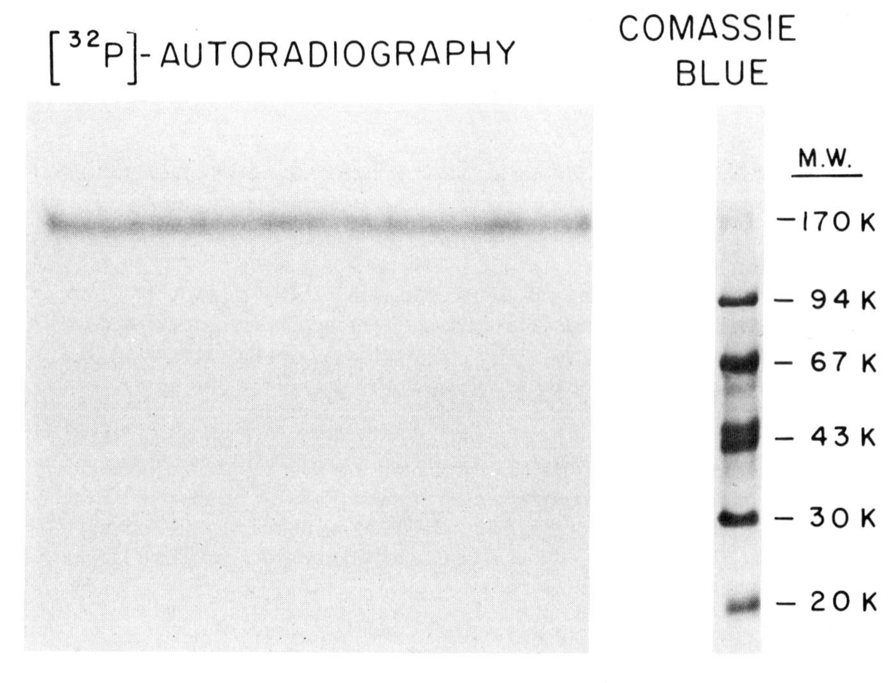

Fig. 12.6 Purification of EGF receptor by immunoaffinity chromatography followed by preparative SDS/polyacrylamide gel electrophoresis. EGF receptor extracted from A431 cells was immobilized on a 29(1)-IgG–Sepharose column. The column was washed extensively and the immunoaffinity-purified receptor was phosphorylated by the addition of [γ-^{32}P]ATP together with $MnCl_2$. After 10 minutes at 40 °C the column was washed and the receptor was eluted at pH 2.5. The eluted phospholabeled receptor was applied on a preparative gel after being heated in electrophoresis sample buffer. Approximately 5% of the receptor preparation was loaded on a separate lane after mixing with low-molecular-weight markers (Pharmacia). This separate lane was fixed, stained with Coomassie Blue and compared with the ^{32}P-radioautogram of the unfixed gel.

(20 μCi, 5000 Ci/mmol) and 3 mM $MnCl_2$. After 10 min at 4 °C the receptor was eluted with a solution composed of 50 nM glycine–HCl buffer at pH 2.5, 0.1% Triton X-100, 150 mM NaCl and 10% glycerol. The peak fractions were pooled according to radioactive content.

For the analysis of the primary structure of the receptor molecule (in collaboration with Dr M. Waterfield, ICRF, London), the eluted receptor, after being heated for 3 min at 95 °C in sample buffer, was applied to a preparative polyacrylamide gel (3 × 120 × 150 mm). The ^{32}P-labeled receptor was easily located on the unfixed gel (see Fig. 12.6). The receptor was finally electro-eluted from the gel and further used in primary structure studies and as an immunogen (Yarden *et al.*, 1984).

Using this procedure it is possible to purify approximately 150 μg of receptor molecules (Yarden *et al.*, unpublished work) from 2×10^9 cultured A431 cells.

12.3.2 The biosynthesis of EGF receptor

The biosynthetic pathway of EGF receptor has been studied by pulse–chase experiments. A431 cells grown in 35 mm dishes were washed with methionine-free Dulbecco's minimal essential medium (DMEM) and then starved for 30 min at 37 °C with 300 μl of methionine-free DMEM containing 10% dialysed fetal calf serum (FCS). Then, 70 μl of [^{35}S]methionine (1220 Ci/mmol) were added for 15 min at 37 °C. The medium was removed and replaced with 2 ml of DMEM supplemented with 10% FCS. After increasing time periods (0, 15, 30, 60, 120 min) the dishes were placed on ice, washed three times with ice-cold PBS and immediately scraped in 200 μl of solubilization buffer. Fig. 12.7 depicts the radioautogram of a SDS/polyacrilamide gel containing labeled EGF receptor immunoprecipitated with TL5-IgG antibodies.

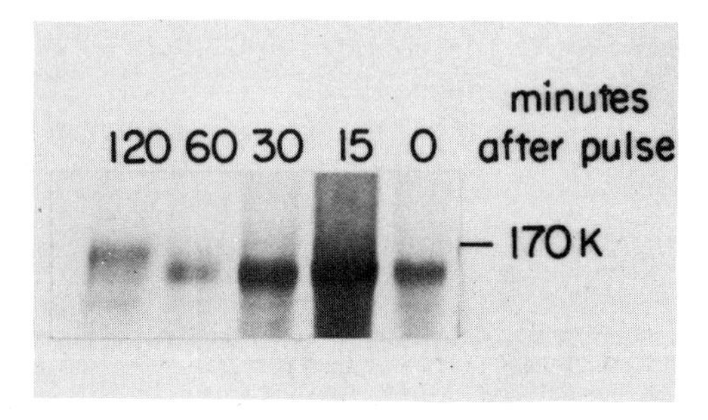

Fig. 12.7 Pulse–chase experiments using EGF receptor from A431 cells. A431 cells were labeled with [^{35}S]methionine for 15 minutes at 37 °C and then chased with DMEM containing 10% FCS for increasing periods (0, 15, 30, 60 and 120 minutes). Then the cells were scraped from the dish in solubilization buffer and the EGF receptor was immunoprecipitated using TL5-IgG antibodies. The labeled receptors were analyzed by SDS-PAGE followed by radioautography.

The EGF receptor can be detected after a pulse of 15 min as a precursor polypeptide with an approximate molecular weight of 140000. It is fully converted to the 170000-dalton polypeptide approximately 2 hours after the pulse. Mayes *et al.* (1984) have shown that in the presence of tunicamycin a 135000-dalton form of the receptor is precipitated with monoclonal antibodies, while in the presence of monensin a 160000-dalton polypeptide is detected. The 170000–175000-dalton species apparently represents the fully glycosylated receptor.

12.3.3 Valency of EGF receptor *in situ*

EGF receptor clustering induced by EGF or by anti-receptor antibodies plays an important role in the transduction of the signals mediated by the growth factor. It is therefore of great importance to study the valency of the receptor molecule towards its ligands, the various aggregation states of the receptor molecule and molecular interactions which control these processes (Schlessinger, 1979, 1980). For these studies, we have used the monoclonal antibody TL5-IgG and its monovalent Fab′ fragment. As previously indicated TL5-IgG does not interfere with the binding of EGF to the receptor molecule, fails to induce receptor clustering and does not possess any intrinsic biological activity (Schreiber *et al.*, 1983).

A431 cells were incubated with 10 μg of rhodamine-TL5 (R-TL5)/ml and 10 μg of fluorescein-TL5 (F-TL5)/ml for 1 h at 4 °C and then for an additional 10 min at 37 °C. Both R-TL5- and F-TL5-labeled receptors appear homogeneously distributed on the cell surface (see Fig. 12.8). The lateral mobility of the fluorescently labeled receptors was measured by the fluorescence photobleaching recovery (FPR) method (for review, see Schlessinger and Elson, 1982). The lateral diffusion coefficient of the fluorescently labeled receptor was $D = (5–7)\times 10^{-10}$ cm^2/s and the mobile fraction was $R(\%) = 50–70$ at 23 °C. Similar values of D and $R(\%)$ were determined for EGF receptor labeled with R-TL5 (Hillman and Schlessinger, 1982).

It is noteworthy that the bivalent TL5-IgG antibodies do not induce the formation of visible EGF receptor clusters. This is to be expected for a monoclonal antibody as the majority of the molecules will be bound univalently (Mason and Williams, 1980) and the remaining doubly bound molecules can only group two antigenic determinants. On treatment of A431 cells prelabeled with R-TL5 with anti-rhodamine antibodies for 10 min at 37°C numerous fluorescent patches were observed on the cell surface. FPR experiments show that the cross-linked receptors are essentially immobile ($D < 5\times 10^{-12}$ cm^2/s, $R(\%) < 5\%$). Patching of the receptors was not observed when cells prelabeled with F-TL5 were incubated with anti-rhodamine antibodies, the mobility of the labeled receptors remaining unaffected.

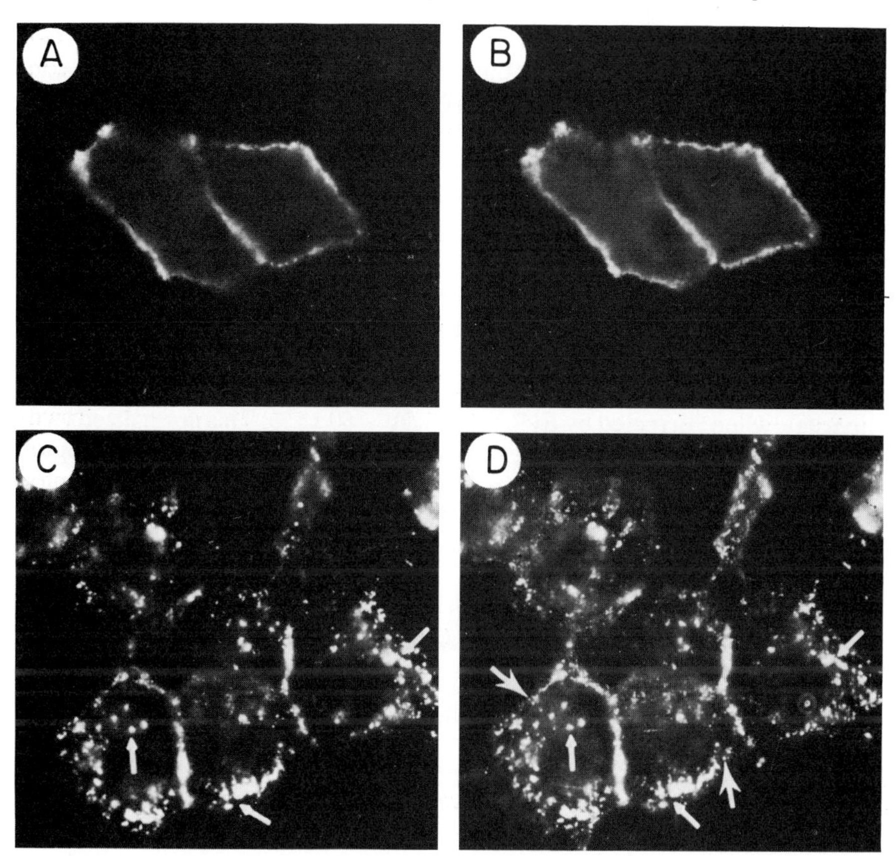

Fig. 12.8 The distribution of fluorescently labeled TL5-IgG on A431 cells. A431 cells were labeled with a mixture of 6×10^{-8} M R-TL5-IgG and F-TL5-IgG for 60 minutes at 40 °C. Some of the samples were fixed with 3% formaldehyde (A, B) and mounted for fluorescence microscopy. Other samples were further incubated for 10 minutes at 37 °C with anti-rhodamine antibodies (C, D). Panels A and C depict fluorescence micrographs obtained with optical filters selective for the observation of fluorescein emission. Panels B and D depict fluorescence micrographs obtained with optical filter for selective observation of rhodamine emission. Overlapping patches of R-TL5 and F-TL5 are marked by small arrows (see in C and D). Patches which are observed in panel D but not in C are marked by large arrows. Magnification ×732.

Two experiments employing doubly labeled A431 cells were performed to determine the effective aggregation state of EGF receptor as detected by TL5 binding. In the first experiment A431 cells were labeled with equal amounts of R-TL5 and F-TL5 and incubated for 10 minutes at 37 °C with anti-rhodamine antibody. A large fraction of both markers subsequently underwent rapid aggregation on the cell surface and many visible patches containing both F-TL5 and R-TL5 were observed. This result is consistent with aggregation of EGF receptor prior to effector binding. At saturating levels of the second antibody, the mobile fraction of F-TL5-labeled receptors should provide a measure of the aggregation state of the receptor *in situ*. Thus, if the receptor is monovalent towards its ligand, 100% of the bound F-TL5 antibody will be mobile, a dimeric receptor will induce 50% immobilization, and a trimer or tetramer will cause 25% or 12.5% immobilization respectively. The mobilities of fluorescein- and rhodamine-TL5 were measured separately using the argon laser lines at 488 nm and 514 nm respectively. The addition of anti-rhodamine antibodies immobilized most of the R-TL5 antibody as expected, while $\sim 20 \pm 5\%$ of the F-TL5 antibody remained mobile with $D = (5-7) \times 10^{-10}\ cm^2/s$. The normalized, mobile fraction of F-TL5 was $30 \pm 10\%$ of the total, when corrected by $R(\%)$ (R-TL5) $\sim 60 \pm 5\%$. This normalized value is clearly most consistent with the dimer or trimer as the predominant species of EGF receptor prior to the addition of EGF.

In the second experiment, A431 cells were labeled with 1 μg of R-TL5/ml, so that less than 50% of the receptors were occupied. The cells were then incubated with anti-rhodamine antibody for 10 min at 37 °C, fixed with 3% formaldehyde for 10 min at 23 °C and then incubated for a further 30 min at 23 °C with excess F-TL5. On fluorescence microscopy, the rhodamine label appeared only in patches while fluorescein was found either in patches ('co-patched' with R-TL5) or diffusely distributed over the cell surface. Hence, when partially labeled receptors are induced to form patches by a second cross-linking antibody against the labeling substrate, there is simultaneous co-patching of unlabeled receptors which are then available for further labeling *in situ*. This again indicates that the receptor is multivalent towards the monoclonal antibody, and, by implication, towards its natural effector. These studies of receptor valence in the absence of EGF can detect only a weighted average; a distribution of microaggregates is most likely present because of encounters between monomeric and oligomeric species. The addition of EGF induces further clustering and subsequent internalization of the occupied receptors (Haigler *et al.*, 1978; Hopkins and Bethroyd, 1981; Schlessinger *et al.*, 1983a,b).

12.3.4 Generation of monoclonal antibodies against pure EGF receptor

In an immunoaffinity run, as described in Section 12.3.1, we routinely isolate 150 μg of pure EGF receptor from A431 cells. This material was also used as an immunogen and for subsequent screening of newly generated monoclonal antibodies.

Mice were immunized with 10 μg of pure receptor electroeluted from a preparative polyacrylamide gel. The antisera from the immunized mice contained high titers of antibodies which reacted with the cell surface of A431 cells and immunoprecipitated both ^{125}I-labeled pure, denatured EGF receptor and a functional EGF receptor–kinase (Lax *et al.*, unpublished result).

Three days after booster injection, spleen cells of an immunized mouse were fused with NSO cells. Hybrid cells were distributed into 500 wells of Costar trays and grown in HAT selection medium. A fast and specific screening procedure was developed which utilizes ^{125}I-labeled pure EGF receptor to screen for antibodies against the receptor molecule (Lax *et al.*, unpublished result). Using this screening procedure it is possible to detect monoclonal antibodies which recognize different structural determinants on the pure, denatured receptor. Characterization of such monoclonal antibodies is now in progress.

12.4 OTHER ANTIBODIES AGAINST EGF RECEPTOR

12.4.1 Polyclonal antibodies against EGF receptor

Several polyclonal antibodies against EGF receptor of different form and purity have been generated and used in recent years (see Table 12.1).

Polyclonal antibodies against membranes prepared from A431 cells

In several studies rabbit polyclonal antibodies against membranes (prepared according to Thom *et al.*, 1977) from A431 cells were used. These antibodies recognize the membrane receptor of EGF, immunoprecipitate a functional EGF receptor–kinase and block the binding of ^{125}I-EGF to its receptor (Haigler and Carpenter, 1980; Hapgood *et al.*, 1983); however, they lack intrinsic biological activity. These antisera contain antibodies against other molecular components of A431 membranes. Nevertheless, as EGF receptor from A431 cells is an excellent immunogen polyclonal antisera may be enriched for antibodies against EGF receptor and therefore can be used as an efficient reagent for the immunoprecipitation of receptor. These antibodies recognize EGF receptor from human, bovine, dog, rat and mouse cells (Table 12.1).

Polyclonal antibodies against shed vesicles prepared from A431 cells

Polyclonal antibodies were prepared against shed vesicles from A431 cells, which contain a native form of EGF receptor (Cohen *et al.*, 1982). These antibodies recognize both native and denatured EGF receptor and efficiently immunoprecipitate EGF receptor from human and mouse (Webb *et al.*, 1983; Hortsch *et al.*, 1983) cells (Table 12.1).

Polyclonal antibodies against pure denatured EGF receptor

EGF receptor was purified by immunoaffinity chromatography using a monoclonal 29(1)-IgG_1–Sepharose column, and the pure, denatured 170000-dalton polypeptide electroeluted from a preparative polyacrylamide gel was employed as an immunogen. The polyclonal, rabbit antibodies recognize both the native and denatured forms of EGF receptor, do not interfere with the binding of EGF to A431 cells, and efficiently immunoprecipitate EGF receptor from human and mouse cells (Table 12.1).

Cross-reactivity of the polyclonal antibodies against the pure, denatured EGF receptor with receptors from other species has not yet been examined. We are currently characterizing the biological properties of these reagents.

12.4.2 Analysis of structural domains on EGF receptor using various antibodies

A model describing EGF receptor as an 'allosteric receptor' composed of various functional domains was proposed by Schlessinger *et al.* (1983a,b). According to this model, EGF receptor involves various functional domains or sites: (i) the combining site for EGF, which serves as an allosteric regulator of the receptor molecule; (ii) the tyrosine-specific, cyclic nucleotide-independent protein kinase; (iii) the locus of phosphorylation sites at tyrosine and possibly also serine residues which most likely reside in the cytoplasmic portion of the receptor; (iv) an aggregation site which facilitates the receptor–receptor and/or receptor–coated-pit interaction; (v) a modulation site for interaction with other species known to modulate the binding and activity of EGF (e.g. the tumor promoter phorbol ester (TPA) and vasopressin: for review, see Schlessinger *et al.*, 1983); (vi) the site of attachment of EGF-receptor-associated carbohydrate.

It is proposed that the binding of EGF to the combining site induces conformational change in the receptor which leads allosterically to alterations in activity or accessibility of other sites.

A long-term goal of this laboratory is the preparation of various antibodies against different domains of EGF receptor and the application of immunological and biophysical methods to the problem of structural assignment of the various sites and the elucidation of their respective biological roles.

To this end it would be particularly advantageous to possess monoclonal antibodies against determinants of the cytoplasmic side of the EGF receptor, where the kinase and phosphorylation site are presumably situated. Because of the high immunogenicity of the external portion of the EGF receptor from A431 cells it is, however, likely that the monoclonal and polyclonal antibodies presently available are primarily, if not exclusively, directed against external moieties.

An alternative approach involves utilization of antibodies against synthetic peptides from other proteins possessing functions which are similar to those of the EGF receptor and which are perhaps structurally similar in a local sense. Thus, for example, the EGF receptor–kinase appears similar in many respects to the kinase activity of the transforming protein of Rous sarcoma virus, pp60^{src}. A synthetic peptide derived from the phosphorylation site of pp60^{src} is phosphorylated by the EGF-sensitive kinase (Pike *et al.*, 1982). Moreover, antibodies against pp60^{src} are phosphorylated by preparations containing EGF receptor from A431 cells (Chinkers and Cohen, 1981; Kudlow *et al.*, 1981). However, antibodies against pp60^{src} fail to precipitate the EGF receptor–kinase from A431 and other cells.

We have recently shown (Lax *et al.*, 1984) that antibodies generated against a synthetic peptide corresponding to the phosphorylation site of pp60^{src} will immunoprecipitate the EGF receptor–kinase. In addition, this peptide blocks specifically the autophosphorylation of EGF receptor. These results indicate that at least one domain on EGF receptor is antigenically related to a specific domain on pp60^{src}.

Interestingly the antibodies against the synthetic peptide do not bind to intact A431 cells, suggesting that they may indeed bind to the cytoplasmic portion of EGF receptor. If so, the EGF receptor must span the plasma membrane, with the phosphorylation site located, as proposed, on the cytoplasmic face (Schlessinger *et al.*, 1983b).

12.5 FUTURE PROSPECTS

Immunological reagents based upon the various monoclonal and polyclonal antibodies generated against the EGF receptor have afforded assignment of various structural domains on the receptor molecule and should continue to advance our understanding of the mechanism of action of EGF.

The determination of the primary structure of the EGF receptor using protein microsequencing methods and recombinant-DNA technology will, in addition to providing valuable structural information, greatly facilitate the immunological approaches outlined in this review. Thus, it should be possible to prepare various synthetic peptides corresponding to a number of potentially critical structural domains of the receptor molecule; antibodies generated against these peptides might then be used to locate and functionally characterize specific domains on the receptor *in situ*. It is anticipated that such strategies will ultimately provide a comprehensive view of the relationship of structure to function for the EGF receptor system as well as for other membrane receptors.

REFERENCES

Andres, R.Y., Jeng, I. and Bradshaw, R.A. (1977), *Proc. Natl. Acad. Sci. U.S.A.*, **74,** 2785–2789.

Buhrow, S.A., Cohen, S. and Staros, J.V. (1982), *J. Biol. Chem.*, **257,** 4019–4022.

Carpenter, G. (1983), *Mol. Cell. Endocrinol.*, **31,** 1–19.

Carpenter, G. and Cohen, S. (1976), *J. Cell Biol.*, **71,** 159–171.

Carpenter, G. and Cohen, S. (1979), *Ann. Dev. Biochem.*, **48,** 193–216.

Carpenter, G., King, L., Jr. and Cohen, S. (1978), *Nature (London)*, **276,** 409–410.

Carpenter, G., King, L., Jr. and Cohen, S. (1979), *J. Biol. Chem.*, **254**, 4884–4891.
Carpenter, G., Stoscheck, C.M., Preston, Y.A. and Deharro, J.E. (1983), *Proc. Natl. Acad. Sci. U.S.A.*, **80**, 5627–5630.
Chinkers, M. and Cohen, S. (1981), *Nature (London)*, **290**, 516–519.
Chinkers, M., McKanna, T.J.A. and Cohen, S. (1979), *J. Cell Biol.*, **83**, 260–265.
Cohen, S., Carpenter, G. and King, L., Jr. (1980), *J. Biol. Chem.*, **255**, 4834–4842.
Cohen, J., Hiroshi, U., Christa, S. and Michael, C. (1982), *J. Biol. Chem.*, **257**, 1523–1531.
Czarnubi, J., Geahler, H. and Haley, B. (1979), *Methods Enzymol.*, **LVI**, 642–653.
Downward, J., Yarden, Y., Mayes, E., Scrace, E., Totty, N. Stockwell, P., Ullrich, A., Schlessinger, J. and Waterfield M.D. (1984), *Nature*, **307**, 521–526.
Erikson, E., Shealy, P.J. and Erikson, R.L. (1981), *J. Biol. Chem.*, **256**, 11381–11384.
Fabricant, R.N., DeLarco, J.E. and Todaro, G.J. (1977), *Proc. Natl. Acad. Sci. U.S.A.*, **74**, 565–568.
Galfre, G. and Milstein, C. (1981), *Methods Enzymol.*, **73**, 3–46.
Galfre, G., Howe, S.C., Milstein, C., Butcher, G.W. and Howard, J.C. (1977), *Nature (London)*, **266**, 550–552.
Goldfine, I.D., Smith, G.J., Wong, K.Y. and Jones, A.L. (1977), *Proc. Natl. Acad. Sci. U.S.A.*, **74**, 1368–1372.
Gregoriou, M. and Rees, A.R. (1983), *Cell Biol. Int. Rep.*, **17**, 539–540.
Haigler, H.T. and Carpenter, G. (1980), *Biochim. Biophys. Acta*, **598**, 314–325.
Haigler, H.T., Ash, J.F., Singer, S.J. and Cohen, S. (1978), *Proc. Natl. Acad. Sci. U.S.A.*, **75**, 3317–3321.
Haigler, H.T., McKanna, J.A. and Cohen, S. (1979), *J. Cell Biol.*, **81**, 382–395.
Hapgood, J., Libermann, T.A., Lax, I., Yarden, Y., Schreiber, A.B., Naor, Z. and Schlessinger, J. (1983), *Proc. Natl. Acad. Sci. U.S.A.*, **80**, 6451–6455.
Hillman, G. and Schlessinger, J. (1982), *Biochemistry*, **21**, 1667–1672.
Hollenberg, M.D. (1979), *Vitam. Horm.*, **37**, 69–110.
Hopkins, C.R. and Boothroyd, B. (1981), *Eur. J. Cell Biol.*, **24**, 259–265.
Hortsch, M., Schlessinger, J., Gootwine, E. and Webb, C.G. (1983), *EMBO J.*, **2**, 1937–1941.
Hunter, T. and Cooper, J.A. (1981), *Cell*, **24**, 741–752.
Johnson, L.K. and Eberhardt, N.L. (1982), *J. Cell. Biochem. Suppl.*, **6**, 139.
Johnson, L.K., Baxter, J.D., Vlodavsky, Y. and Gospodarowicz, D. (1980a), *Proc. Natl. Acad. Sci. U.S.A.*, **77**, 394–398.
Johnson, L.K., Vlodavsky, I., Baxter, J.D. and Gospodarowicz, D. (1980b), *Nature (London)*, **287**, 340–343.
Kawamoto, T., Sato, J.D., Le, A., Polikoff, J., Sato, J.H. and Mendelsohn, J. (1983), *Proc. Natl. Acad. Sci. U.S.A.*, **80**, 1337–1341.
Kudlow, J.E., Buss, J.E. and Gill, G.N. (1981), *Nature (London)*, **290**, 519–521.
Lax, I., Bar-Eli, M., Libermann, T.A., Yarden, Y. and Schlessinger, J. (1984), submitted.
Linsley, P.S. and Fox, C.F. (1980), *J. Supramol. Struct.*, **14**, 511–525.
Mason, D.W. and Williams, A.F. (1980), *Biochem. J.*, **187**, 1–20.
Mayes, E.L.V. and Waterfield, M.D. (1984), *EMBO J.*, (in press).
Moolenaar, W.H., Yarden, Y., deLaat, S.W. and Schlessinger, J. (1982), *J. Biol. Chem.*, **257**, 8502–8506.
Owens, J.R. and Haley, B.E. (1978), *J. Supramol. Struct.*, **9**, 57–68.

Pike, L.J., Gallis, B., Casuellie, J.E., Bornstein, P. and Krebs, E.G. (1982), *Proc. Natl. Acad. Sci. U.S.A.*, **79,** 1443–1447.

Richert, N.D., Willingham, M.C. and Pastan, I. (1983), *J. Biol. Chem.*, **258,** 8902–8907.

Rozengurt, E. and Heppel, L.A. (1975), *Proc. Natl. Acad. Sci. U.S.A.*, **72,** 4492–4495.

Schlessinger, J. (1979), in *Physical Chemical Aspects of Cell Surface Events in Cellular Regulation* (C. De Lisi and R. Blumenthal, eds), Elsevier Press, New York, pp. 89–111.

Schlessinger, J. (1980), *Trends Biochem. Sci.*, **5,** 210–214.

Schlessinger, J. and Elson, E.L. (1982), in *Biophysics, Methods of Experimental Physics,* **20** (E. Ehrenstein and H. Lear, eds), Academic Press, New York, pp. 197–226.

Schlessinger, J. and Geiger, B. (1981), *Exp. Cell Res.*, **134,** 273–279.

Schlessinger, J., Schechter, Y., Willingham, M.C. and Pastan, I. (1978a), *Proc. Natl. Acad. Sci. U.S.A.*, **75,** 2659–2663.

Schlessinger, J., Schechter, Y., Cuatrecasas, P., Willingham, M.C. and Pastan, I. (1978b), *Proc. Natl. Acad. Sci. U.S.A.*, **75,** 5353–5358.

Schlessinger, J., Schreiber, A.B., Libermann, T.A., Lax, I., Avivi, A. and Yarden, Y. (1983a) *Cell Membr. Methods Rev.*, **1,** 117–144.

Schlessinger, J., Schreiber, A.B., Levi, A., Lax, I., Libermann, T. and Yarden, Y. (1983b), *CRC Crit. Rev. Biochem.*, **14,** 93–111.

Schreiber, A.B., Lax, I., Yarden, Y., Eshhar, Z. and Schlessinger, J. (1981a), *Proc. Natl. Acad. Sci. U.S.A.*, **78,** 7535–7539.

Schreiber, A.B., Yarden, Y. and Schlessinger, J. (1981b), *Biochem. Biophys. Res. Commun.*, **101,** 517–523.

Schreiber, A.B., Libermann, T.A., Lax, I., Yarden, Y. and Schlessinger, J. (1983), *J. Biol. Chem.*, **258,** 846–853.

Schechter, Y., Hernaez, L., Schlessinger, J. and Cuatrecasas, P. (1978), *Nature (London)*, **278,** 835–838.

Thom, D., Powell, A.J., Lloyd, C.W. and Rees, D.A. (1977), *Biochem. J.*, **168,** 187–194.

Tycko, B. and Maxfield, F.R. (1982), *Cell,* **28,** 643–651.

Ushiro, H. and Cohen, S. (1980), *J. Biol. Chem.*, **255,** 8363–8365.

Waterfield, M.D., Mayes, E.L.V., Stroobant, P., Bennet, P.L.P., Young, S., Goodfellow, P.N., Banting, G. and Ozanne, B. (1982), *J. Cell Biochem.*, **20,** 149–161.

Webb, C.G., Hortsch, M., Gootwine, E. and Schlessinger, J. (1983), *Cell Biol. Int. Rep.*, **7,** 531–532.

Yankner, B.A. and Shooter, E.M. (1979), *Proc. Natl. Acad. Sci. U.S.A.*, **76,** 1268–1273.

Yarden, Y., Harari, I. and Schlessinger, J. (1984), submitted.

Yarden, Y., Schreiber, A.B. and Schlessinger, J. (1982), *J. Cell. Biol.*, **92,** 687–693.

Zidovetzki, R., Yarden, Y., Schlessinger, J. and Jovin, T.M. (1981), *Proc. Natl. Acad. Sci. U.S.A.*, **78,** 6981–6985.

13 Biochemistry and Molecular Biology of Receptors: Applications of Antibodies

C. SCHNEIDER

Glossary of terms and abbreviations

cDNA Single-stranded DNA copied from mRNA using reverse transcriptase

Protein A Protein from *Staphylococcus aureus* (usually Cowan Strain I); useful for its property of binding to the Fc region of immunoglobulin molecules

TR Transferrin receptor

HAT Hypoxanthine, aminopterin and thymidine (ingredients in culture medium for selecting against enzyme-deficient mutants)

Acknowledgements

I wish to thank Dr Markku Kurkinen for introducing me to the cDNA cloning technology, Miss Ushi Asser for her helpful technical assistance, Mr Robert Sutherland for his help in the biochemical characterization of the antigens, Dr J.G. Williams and Dr David Ish-Horowicz for helpful comments and suggestions, and Dr M.F. Greaves for advice and encouragement. Finally, I wish to thank Mrs Jackie Needham for her patience and persistence in typing this manuscript.

Monoclonal Antibodies to Receptors: Probes for Receptor Structure and Function
(*Receptors and Recognition*, Series B, Volume 17)
Edited by M. F. Greaves
Published in 1984 by Chapman and Hall, 11 New Fetter Lane, London EC4P 4EE

13.1 INTRODUCTION

The advent of monoclonal antibodies is having a major impact on the study of receptors as illustrated in other chapters of this book. Here I briefly review some technical aspects of the use of monoclonal antibodies for the biochemical purification and molecular cloning of receptors using, as an example, the transferrin receptor studied in this laboratory (Sutherland *et al.*, 1981; Schneider *et al.*, 1982a,b, 1983a,b) and others (see Chapter 10) and recently reported studies on the cloning of acetylcholine receptors (see below).

13.2 IMMUNOAFFINITY PURIFICATION OF RECEPTOR PROTEIN

Once a monoclonal antibody against a particular receptor molecule has been characterized, the latter can be purified to homogeneity provided that the antibody has a reasonable affinity or good immunoprecipitability of the antigen. For this purpose, it is necessary to construct an immobilized matrix to which the antibody is covalently bound and which retains the maximal binding activity. This approach has been used with monoclonal antibodies bound to Sepharose to isolate receptors for acetylcholine (Lennon *et al.*, 1980; Momoi and Lennon, 1982), insulin (Roth and Cassell, 1983), low-density lipoprotein receptors (Schneider, W.J. *et al.*, 1982), β-adrenergic receptors (see Chapter 6 by Fraser) and, recently, epidermal growth factor (see Schlessinger *et al.*, Chapter 12). Monoclonal antibodies to other cell-surface components have also been used to prepare immunoadsorbent columns (see Dalchau and Fabre, 1982, for review). Affinity chromatography, using insolubilized ligands recognized by the receptors' binding sites or antibodies to hormones which can bind hormone–receptor complexes (Heinrich *et al.*, 1980), provide alternative strategies for purification (reviewed in Jacobs and Cuatrecasas, 1981).

Conventional methods of immobilizing antibodies or other affinity ligands on solid matrices usually employ cyanogen bromide-activated Sepharose (Jacobs and Cuatrecasas, 1981). The binding of antibodies to cyanogen bromide-activated Sepharose often, however, generates affinity columns with low activity because of the multisite attachment and orientation of the immunoglobulin molecule which reduces the efficiency of antibody–antigen interaction. Protein A from *Staphylococcus aureus* has the property of binding to the Fc portion of immunoglobulin molecules. A matrix employing protein A–Sepharose, therefore, offers optimal spatial orientation of antibodies (Gersten and Marchalonis, 1978). Protein A on Sepharose may also function as a 'spacer' molecule to decrease steric hindrance effects of the matrix (Jacobs and Cuatrecasas, 1981). Such an affinity matrix has been successfully constructed and used in this laboratory (Schneider *et al.*, 1982a). The method consists of binding a monoclonal antibody directly to protein A–Sepharose 4B or indirectly via rabbit anti-mouse immunoglobulin (depending on whether the particular mouse monoclonal antibody binds to protein A). This is followed by covalently cross-linking the complex with dimethyl pimelimidate (a bivalent cross-linker). The matrix derived by this procedure has a stable and optimal orientation of the antibody molecules, thus allowing high-efficiency antigen

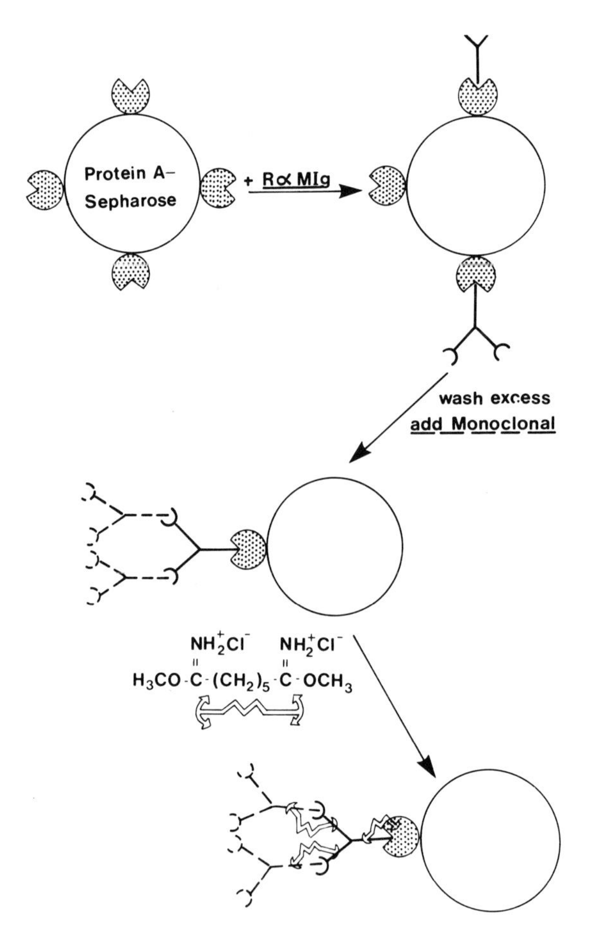

Fig. 13.1 Affinity matrices are made either as one-layer systems when a monoclonal antibody binds directly to protein A or as two-layer systems using rabbit anti-mouse Ig and the mouse monoclonal (as exemplified in the figure). After washing the unbound antibody the complex is covalently cross-linked using dimethyl pimelimidate (a bivalent cross-linker) (Schneider *et al.*, 1982a). The method is fast and permits microscale preparation of matrices for analysis of antigen as well as large-scale one-step purification of membrane components from cell lysates. RαMIg is rabbit anti-mouse immunoglobulin.

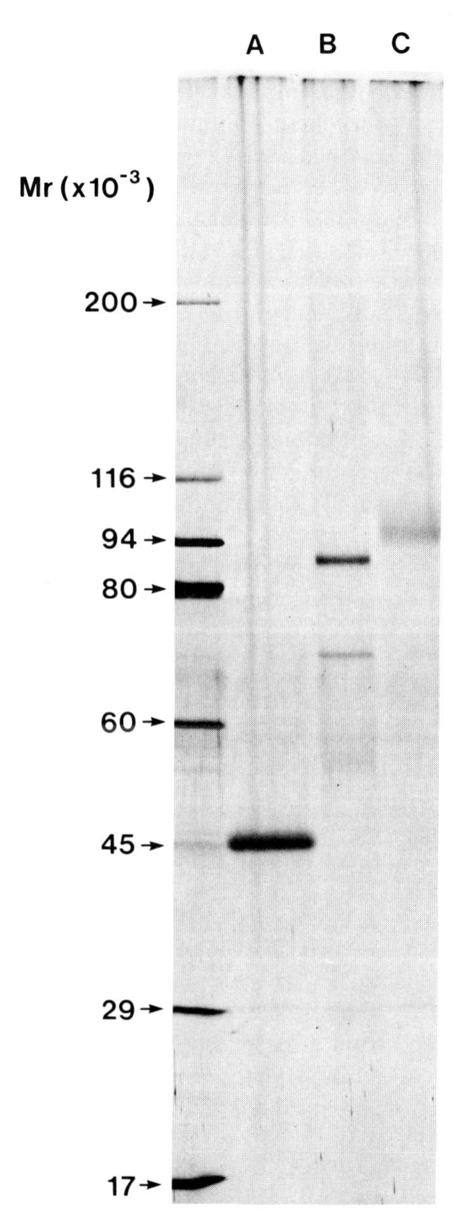

Fig. 13.2 SDS-polyacrylamide gel electrophoresis of antigens isolated from all extracts by a one-step immunoaffinity procedure. Track A. HLA-AB heavy chain (45 000 daltons) eluted from monoclonal antibody W6/32 matrix (β_2-microglobulin not resolved under these conditions). Track B. Transferrin receptor (90 000 daltons) plus trace of bovine serum albumin (69 000 daltons) eluted from monoclonal antibody OKT9 matrix. Track C. cALLA antigen (100 000 daltons) eluted from monoclonal antibody J5 matrix. Gels were run on a polyacrylamide gradient from 6 to 12%. Standard molecular weights are represented on the left. The gel was visualized by staining for total protein using the silver nitrate method (Schneider *et al.*, 1982a).

binding with no significant antibody leakage (Fig. 13.1). Up to 12 mg of monoclonal antibody can be covalently bound per ml of protein A–Sepharose 4B, thus limiting the non-specific interactions and allowing a one-step purification of the antigen from a crude cell lysate. This type of column has proved to be very useful for the isolation of the human transferrin receptor, and other membrane proteins (Fig. 13.2).

A potential drawback for functional studies of the protein eluted from an antibody column is the use of harsh elution conditions to break the antigen–antibody complex. This usually causes denaturation of receptor (the antibody retaining its binding activity). When the monoclonal antibody reacts with an epitope associated with the receptor binding site it may be possible to elute with free hormone under milder conditions as shown by Roth and Cassell (1983) for a monoclonal anti-insulin receptor antibody affinity column.

Because of the relatively small volume of the above-mentioned matrix (1 ml of the matrix contains about 12 mg of bound antibody), it can be easily assembled on to a gel-filtration support (Sephadex G50) to completely and immediately remove by centrifugation the denaturation buffer used for detaching the antigen from the column, the protein coming through in the 'void volume'.

13.3 MOLECULAR CLONING OF RECEPTOR GENES

The various approaches available at present to clone genes for receptor proteins are illustrated in Fig. 13.3; they can be grouped into two main categories.

(i) The cDNA approach, which consists of copying an mRNA population into cDNA, cloning that into *Escherichia coli* and screening with 'specific' probes for the sequence of interest. The cDNA clones obtained can then be used to pull out the gene from a 'cosmid' or lambda genomic library. This approach has been successfully used to clone genes for cell-surface glycoproteins of the major histocompatibility loci (reviewed by Orr, 1982). Several groups have recently reported the isolation and sequencing of cDNA clones of the acetylcholine receptor α, β and δ subunits from *Torpedo* electric organs (Ballivet *et al.*, 1982; Giraudat *et al.*, 1982; Sumikawa *et al.*, 1982; Noda *et al.*, 1982, 1983) and the α subunit from mouse muscle (Merlie *et al.*, 1983).

(ii) The heterologous DNA transfection approach, which consists of transfecting, for instance, human DNA into mouse cells, and screening with a species-specific monoclonal antibody for the positive, stable transfectants. These can be enriched by fluorescence cell sorting (see Chapter 10 by Trowbridge and Newman; see also Kavathas and Herzenberg, 1983), amplified and the DNA used to construct a genomic library. The human specific colonies are then screened taking advantage of the particular species-specific repetitive sequences of human DNA.

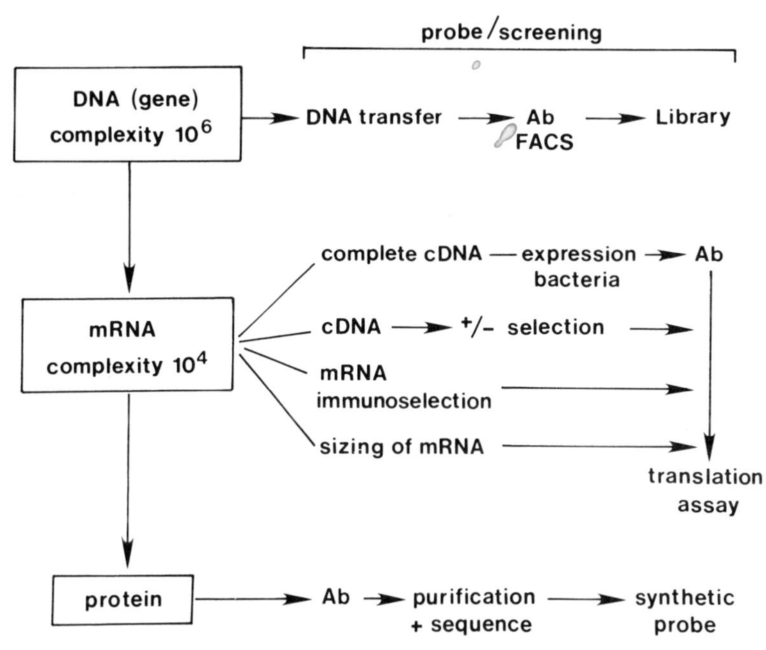

Fig. 13.3 Summary of the current approaches for cloning the gene for a receptor protein. The typical eukaryotic genome contains sufficient DNA to generate 10^6 DNA fragments of a size suitable for genomic cloning. The screening procedure thus has to be extremely selective if the starting point is of such complexity (see Section 13.2.2). In contrast there are 10 000–30 000 different mRNA sequences in a typical eukaryotic cell (Williams and Lloyd, 1979) present at widely varying abundances. If the abundance of the particular receptor mRNA is 0.1–0.01% of the mRNA population, a 10 000-member cDNA library will represent the particular sequence. The various screening possibilities used at present include (a) direct screening with an antibody for the expressed protein in the bacteria; (b) differential screening with cDNA probes made from mRNA of cells containing the particular mRNA sequence at >0.1%, since the sensitivity of the *in situ* colony hybridization will not detect probes present at less than 0.1%; (c) immunoselection of mRNA from polysomes, giving a high enrichment of the particular mRNA adequate for screening the cDNA library against the non-enriched mRNA population; (d) enrichment of the particular mRNA sequence by sizing, through sucrose gradient centrifugation or agarose gels. This could give the adequate enrichment of 10 times over the low abundance level (0.01%) present at the beginning. If some of the sequence of the receptor protein is known, synthetic oligonucleotide can be made to (a) use as a hybridization probe or (b) prime a total mRNA population for the enzymatic synthesis of cDNA (Chan *et al.*, 1979; Noyes *et al.*, 1979).

Obviously the two approaches are very different, and the differences influence the requirement of each method for particular qualities of the antibody, which is the essential probe for both methods.

13.3.1 The cDNA strategy: identifying appropriate antibodies

To use this strategy it is essential to have an antibody which can recognize the protein as synthesized *in vitro*. Establishing the *in vitro* biosynthesis of a certain protein to be cloned is an essential step in the cDNA approach and will in fact represent the assay from beginning to end for following through the cloning procedure. The possible limitation of monoclonal antibodies in this respect can be exemplified in the case of the transferrin receptor.

Several monoclonal antibodies against the transferrin receptor (see Chapter 10 by Trowbridge and Newman) were not able to immunoprecipitate the receptor synthesized *in vitro*. We presumed that the failure of each individual monoclonal antibody to immunoprecipitate the *in vitro* product reflects monospecificity for epitopes which are highly conformation-dependent and not available on the primary translation product synthesized *in vitro*.

In this situation the restricted specificity of monoclonal antibodies is clearly a disadvantage. This problem can, however, be circumvented by deliberately screening for monoclonal antibodies which immunoprecipitate the *in vitro* synthesized product or by using mixtures of monoclonals with specificity for different epitopes of the same receptor protein. In the latter instance, exemplified by recent studies on the p97, a transferrin-related protein melanoma cell-surface antigen (Brown *et al.*, 1982), the use of several different monoclonal antibodies, each binding with weak affinity, may produce a precipitable complex (K.E. Hellstrom, personal communication).

In general the *in vitro* synthesized product may be more similar sterically to a denatured (unfolded) form of the receptor protein than to the native, processed form (Soreq *et al.*, 1982; see also Fuchs *et al.*, Chapter 8). An alternative strategy is therefore to deliberately immunize animals with denatured receptor to produce polyclonal or monoclonal antibodies which recognize non-conformation-dependent antigenic determinants (see Chapter 8 by Fuchs *et al.* on the acetylcholine receptor; see also Tzartos *et al.*, 1981, and papers describing the cloning of acetylcholine receptor referred to above). We raised a polyclonal antibody against the denatured transferrin receptor (Schneider *et al.*, 1983a). The receptor was purified by monoclonal antibody (OKT9) affinity chromatography as described above, further reduced and alkylated and analyzed by preparative SDS-PAGE. The antiserum generated in rabbits to immunopurified TR was monospecific and capable of reacting with the *in vitro* synthesized molecule (Schneider *et al.*, 1983a). Another advantage of such a monospecific antiserum is that a fraction of its reactivity will recognize the nascent receptor molecule present on polysomes engaged in its synthesis. This allows for the immunoselection of the mRNA from the polysomes engaged in the synthesis of the particular protein.

The important advantage of this approach is that it becomes possible to isolate mRNA species that are present at low frequency. Thus Korman *et al.* (1982) successfully isolated mRNA polysomes specific for the cell-surface protein HLA-DR which is present at 0.01 to 0.05%. The monoclonal antibody they used was produced against protein purified under conditions that probably selected for non-conformational determinants. Similarly the transferrin receptor mRNA could be isolated, although it represents a very-low-abundance mRNA species (0.01%; Schneider *et al.*, 1983b).

Once an mRNA population is shown to contain the particular mRNA coding for the protein recognized by the antiserum or monoclonal antibody, the RNA can be used for the next step. RNA is copied into DNA by reverse transcriptase, double-stranded DNA is made from this template and ligated to a suitable vector containing an antibiotic resistance marker. Competent *E. coli* are then transformed with the plasmid constructed, and colonies growing on antibiotic plates are then used for screening. A description of the methods and vectors used to make cDNA libraries is beyond the scope of this chapter so the reader is referred to Williams (1981) and Maniatis *et al.* (1982).

The most critical step for identification of the recombinant clones of interest is the judicious choice of the probe to be used. The normally employed 'primary screen' is the '*in situ* hybridization', devised by Grunstein and Hogness (1975), whereby bacterial colonies are replicated on an inert support (Millipore filters), lyesed *in situ* with alkali and then hybridized to a radioactively labeled probe. The most commonly used method to label mRNA is to make ^{32}P-labeled cDNA via reverse transcriptase in the presence of radiolabeled nucleotides, generating a probe of very high specific activity. If the mRNA sequence of interest constitutes at least 0.25–0.5% of the total mRNA population in the tissue or cell line where it is most abundant, the ^{32}P-probe made from such an mRNA will identify the clones in the colony hybridization assay. If, in contrast, the mRNA sequence is present at less than 0.1%, it will not be detected as a signal in the *in situ* hybridization assay (Williams and Lloyd, 1979). Taking advantage of this fact, a differential screening procedure can be used. This approach has been successfully used to clone the acetylcholine receptor by screening a cDNA library made from *Torpedo* electric organ mRNA (AChR mRNA – 0.1–0.4%) with a positive, hot cDNA probe from the same mRNA source versus a negative cDNA probe prepared from spleen or liver mRNA (AChR mRNA $< 0.1\%$). The mRNA sequence of interest must constitute at least 0.25–0.5% of the mRNA population in which it is most abundant, if it is to be recognized by this procedure. We anticipate that the majority of receptor proteins will generally be in the low-abundance class of mRNAs (0.01%), and if the 'exceptional' cell system presenting the mRNA at a level greater than 0.01% is not found, the probe mRNA has to be enriched over the detection threshold.

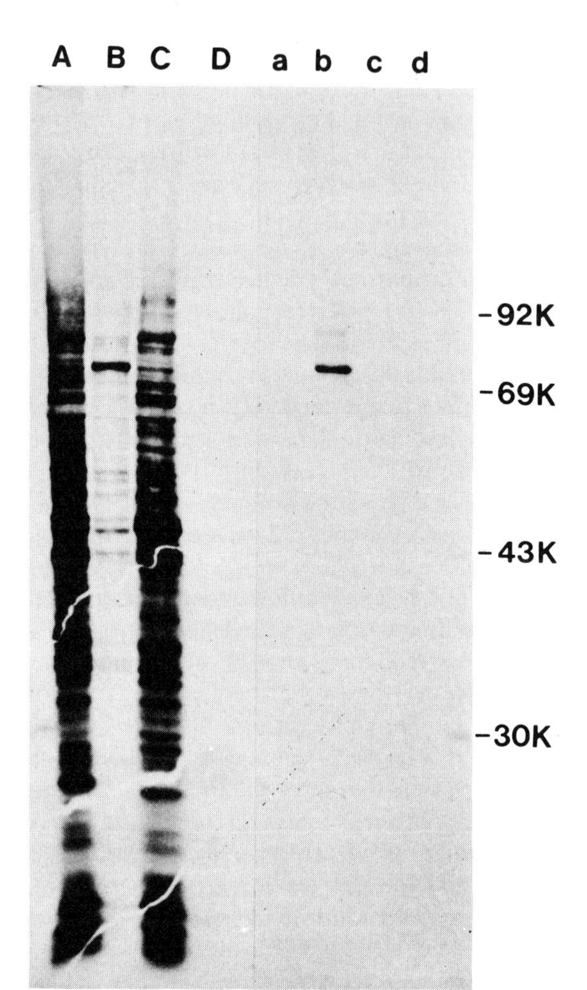

Fig. 13.4 Immunoselection of TR-mRNA. An IgG fraction of rabbit anti-TR serum was added to polysomes extracted from HeLa cells; the resulting polysome–antibody complexes were isolated by adsorption to protein A–Sepharose. After extensive washes, the RNA was specifically eluted from the protein A–Sepharose column with Mg chelators. The mRNA was then purified by oligo(dT)-cellulose chromatography and translated *in vitro*. When immunopurified mRNA was translated, the pattern of total translation products from total mRNA (lane A) revealed a dramatic increase in a 78 000-dalton primary translation product (lane B). This major product was precipitable by anti-TR serum (lane b). Lane C represents the total translation product from poly(A)+ RNA isolated from polysomes that failed to bind to protein A–Sepharose; lane c is the corresponding immunoprecipitation with anti-TR. Lane D, total translation products with no RNA added; lane d, corresponding immunoprecipitation. The analysis of [^{35}S]methionine-labeled proteins was by 7–14% gradient SDS-PAGE and fluorographic exposure time was 3 hours (Schneider *et al.*, 1983b).

The most reliable method to construct such probes is by polysome immunoselection. Polysomes isolated by conventional methods are made to react with an IgG fraction of the antiserum. The antibody recognizing the nascent polypeptide complex (on the polysome) is bound to protein A–Sepharose 4B. The RNA bound is eluted by disrupting the polysomes with EDTA, then passed over an oligo(dT)-cellulose column to select for mRNA. This is translated *in vitro* and the products analyzed by SDS-PAGE.

As can be seen in Fig. 13.4, the immunoselected mRNA for the transferrin receptor gives a major band when the total translation products are analyzed by SDS-PAGE. This major product was precipitable by anti-TR serum. The purification of the TR-mRNA by this method is at least 1000-fold. The immunopurified TR-mRNA was used to make [^{32}P]cDNA for a 'positive' probe and the polysomal RNA which failed to bind to the protein A–Sepharose column was used as a template for making [^{32}P]cDNA to use as a 'negative' probe. Because the immunoselected mRNA still contained contaminating, abundant messenger species, most positive colonies are common to the two probes (Fig. 13.5). The circled areas in Fig. 13.5 identify colonies reacting with the positive probe only; these were shown by positive hybrid selection (see below) to be TR-clones (Schneider *et al.*, 1983b). This differential screening procedure has now been used successfully in many laboratories and enormously simplifies the screening work.

A very promising technique for isolating cDNA clones of interest involves cloning the cDNA into plasmids that promote expression of the cDNA in *E. coli*. The resultant cDNA 'expression libraries' are then screened for the appropriate translation products using an immunological assay. This has been proved successful in isolating cDNA clones for the relatively high abundant mRNA (0.1–1%) coding for chicken tropomyosin (Helfman *et al.*, 1983). In this case also, use of a polyclonal serum was essential; a monoclonal antibody may in fact detect a site near the *N*-terminus of the protein, which would mean only a small fraction of the appropriate cDNA clones would make an antigen that a particular monoclonal antibody would detect. However, again a mixture of several monoclonal antibodies recognizing antigenic determinants throughout the length of the protein could be substituted for polyclonal sera. Although promising and very direct, this approach has not been proved efficient for isolating cDNA clones of low-abundance mRNA species ($< 0.01\%$). The chances that (a) the inserted cDNA in the plasmid is in the correct reading frame for translational activity and (b) that the cDNA is long enough to be in the coding region for a definitive antigenic site recognized by the antibody make the system highly inefficient so that a very large number of clones ($> 100\,000$) has to be screened to be really sure to 'fish' the clone of interest. Nonetheless, the technology needed to bypass these points is being developed (Young and Davis, 1983) and soon we will know if it can be applied successfully to clone low-abundance mRNA.

Another very successful screening procedure is the use of synthetic oligonucleotides as hybridization probes themselves or as primers for the enzymatic synthesis of cDNA (Chan *et al.*, 1979; Noyes *et al.*, 1979). This approach has recently been exploited for cloning the acetylcholine receptor (α, β and δ

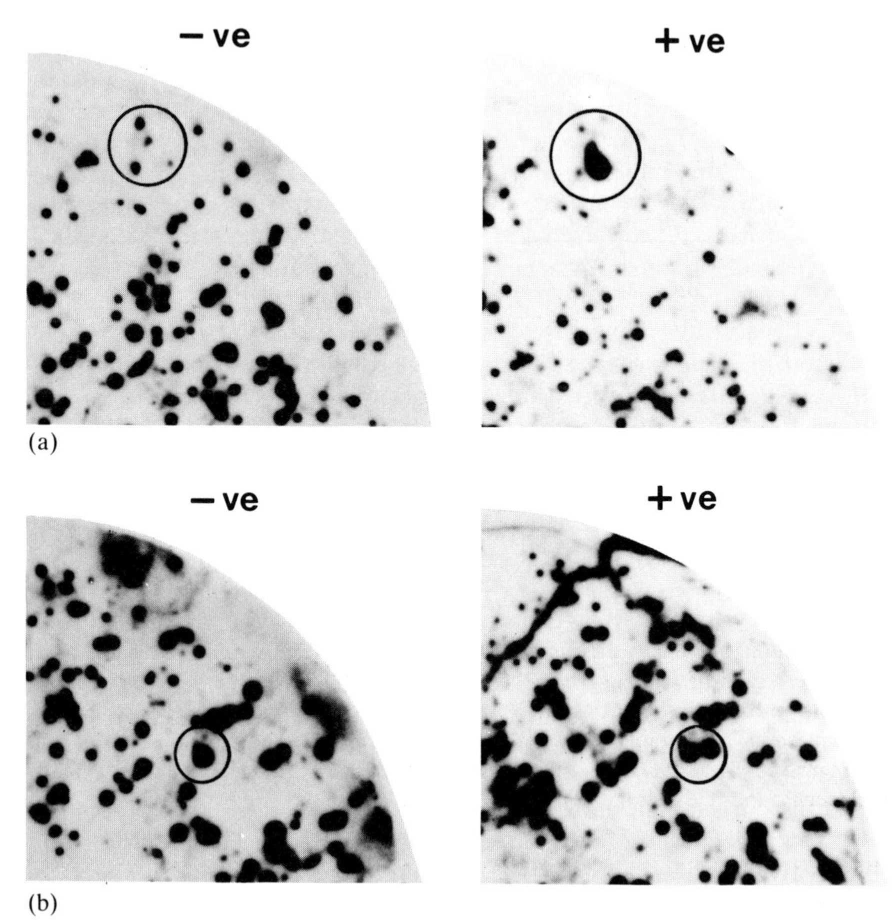

Fig. 13.5 Screening of a placental cDNA library for TR-specific clones. Nitrocellulose filters containing the transformed colonies grown in the presence of ampicillin were replica-plated and fixed. Immunopurified mRNA was used to make [^{32}P]cDNA for a 'positive' probe and the polysomal RNA which failed to bind to the protein A–Sepharose column was used as template for making [^{32}P]cDNA to use as a 'negative' probe. Because the immunoselected mRNA still contained contaminating, abundant messenger species, most positive colonies are common to the two probes (–ve versus +ve). This fact is actually useful in matching the radioautographs in search for the signals reacting only with the immunoselected probe and not with the negative probe (circled areas). The few differential signals are picked and the plasmid DNA isolated and used for hybrid selection (see Fig. 13.6). (a) and (b) are two different sets of clones (Schneider *et al.*, 1983b).

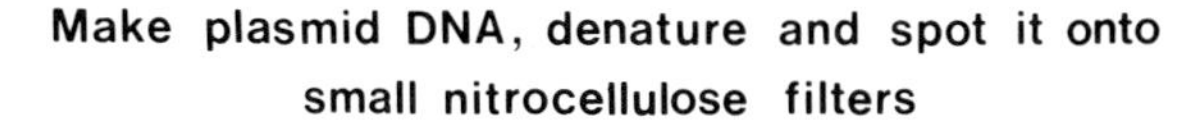

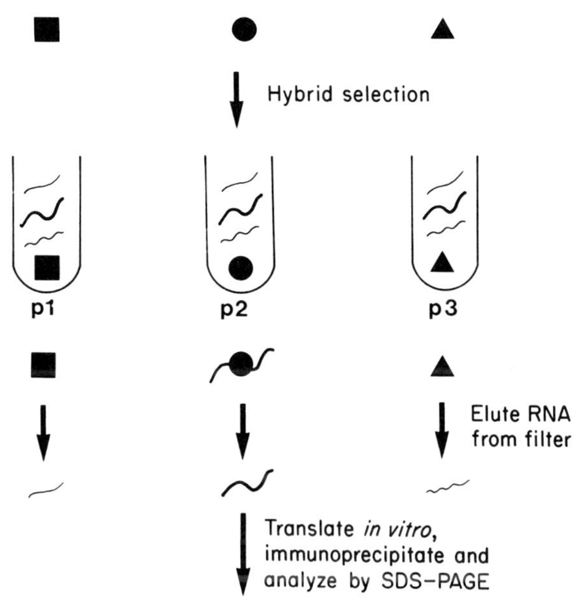

Fig. 13.6 Specific mRNAs can be purified from total cellular mRNA by hybridization to plasmid DNA, containing the complementary inserted sequence, that has been denatured and immobilized on nitrocellulose filters. RNA recovered from the filter hybridization reactions is translated *in vitro* to produce a protein that can be immunoprecipitated and analyzed by SDS-PAGE. When TR plasmid DNA isolated from the positive (circled) clones in Fig. 13.5 was tested in this assay together with some other plasmid DNA preparations, only the TR plasmid DNA showed increase in a band co-migrating with the TR primary translation product which would be precipitated with the specific anti-TR serum (Schneider *et al.*, 1983b).

subunits) genes and took advantage of sequence data available on the receptor proteins to synthesize the appropriate oligonucleotide probe (Noda *et al.*, 1982, 1983; Sumikawa *et al.*, 1982). This approach is, however, most applicable in high-technology laboratories and industries, where co-operation between protein chemists, DNA chemists and cloning facilities is maximal and is less suitable or practical for the 'do it yourself' laboratory for which this chapter is primarily intended.

To provide definitive evidence that cDNA clones isolated represent sequences complementary to the original mRNA and coding for the receptor protein of interest, hybrid selection is used*. This technique consists of binding

*Editor's note: or nucleotide sequencing when the receptor protein sequence is already known at least in part (e.g. acetylcholine receptor; Raftery *et al.*, 1980; Noda *et al.*, 1982).

the denatured plasmid DNA to small nitrocellulose filters. The plasmid DNA on the filter is then allowed to hybridize with total RNA and at the end of the hybridization the filter is washed and the RNA bound to it is eluted and translated *in vitro* (Fig. 13.6). The total products of the *in vitro* translation are then immunoprecipitated with the appropriate antibody and analyzed by SDS-PAGE and radioautography as shown in Fig. 13.6 for the transferrin receptor cDNA clones. If the plasmid DNA contains sequences that are complementary to the specific mRNA the protein product of it will be enriched with respect to plasmid DNA that does not contain that sequence. Monoclonal antibodies to denatured acetylcholine receptor have been successfully used in this context to identify cDNA clones (references above).

Once the cDNA clones have been identified and characterized, they can be used for obtaining the sequence of the coding regions or exons of the gene and also as a probe to screen a genomic library to obtain the entire gene.

13.3.2 The heterologous DNA transfection strategy: advantages and requirements of monoclonal antibodies

This approach consists of transfecting human DNA, usually by calcium phosphate precipitation, into mouse cells (see Fig. 13.7) along with a selectable marker, e.g. thymidine kinase (TK) if recipient cells are TK^-. The transfectants are then selected for the presence of the *TK* gene by growth in HAT medium.

Primary screening involves the selection of the transfectants that react with a species-specific monoclonal antibody directed against the antigen coded for by the gene of interest. An essential requirement for this technique is to have a species-specific monoclonal antibody reacting with a membrane antigen *in situ* and that can therefore be used as a basis for selection (e.g. by cell sorting; see Kavathas and Herzenberg, 1983). Therefore the technique is limited to cell-surface molecules and requires polyclonal or monoclonal antibodies binding to exposed determinants of the native receptor molecules.

Once the positive, stable transfectants are identified, they are enriched by cell sorting (i.e. using a fluorescence activated cell sorter) and expanded (see Chapter 10 by Trowbridge and Newman). DNA isolated from these transfectants is then prepared and used to make a genomic library. The recombinants are then screened for human specific sequences by using labeled total human DNA very rich in repetitive sequences that are human-specific. The recombinants showing reactivity with this probe are then characterized. This approach is currently being used to isolate the human transferrin receptor gene (Trowbridge and Newman, Chapter 10 and F. Ruddle, personal communication). Usually the investigators have to go back to screen a cDNA bank to obtain the coding region of the gene, since 'walking along' the gene is in many case a protracted and tedious exercise.

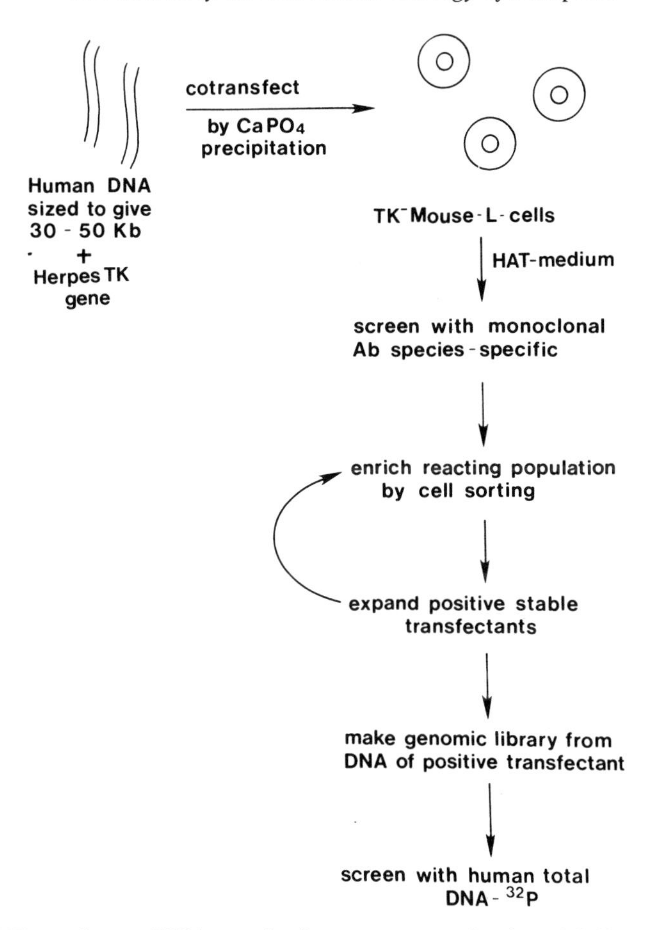

Fig. 13.7 Heterologous DNA transfection strategy; see Section 13.2.2.

During the next few years we can anticipate that the approaches outlined here will be successfully applied to many cell-surface receptor molecules. The anticipated availability of gene sequence data should provide us with new insights into receptor structure, inter-relationships and evolution.

REFERENCES

Ballivet, M., Patrick, J., Lee, J. and Heinemann, S. (1982), *Proc. Natl. Acad. Sci. U.S.A.*, **79**, 4466–4470.

Brown, J.P., Hewick, R.M., Hellstrom, K.E., Doolittle, R.F. and Dreyer, W.J. (1982), *Nature (London)*, **296**, 171–173.

Chan, S.J., Noyes, B.E., Agarwal, K.L. and Steiner, D.F. (1979), *Proc. Natl. Acad. Sci. U.S.A.*, **76**, 5036–5040.

Dalchau, R. and Fabre, J.W. (1982), in *Monoclonal Antibodies in Clinical Medicine* (A.J. McMichael and J.W. Fabre, eds), Academic Press, London, pp. 519–556.

Gersten, D.M. and Marchalonis, J.J. (1978), *J. Immunol. Methods*, **24**, 305–309.

Giraudat, J., Devillers-Thiery, A., Auffray, C., Rougeon, F. and Changeux, J.P. (1982), *EMBO J.*, **1**, 713–717.

Grunstein, M. and Hogness, D. (1975), *Proc. Natl. Acad. Sci. U.S.A.*, **72**, 3961–3965.

Heinrich, J., Pilch, P.F. and Czech, M.P. (1980), *J. Biol. Chem.*, **255**, 1732–1737.

Helfman, D., Feramisco, J., Fiddes, T., Thomas, P. and Hughes, S. (1983), *Proc. Natl. Acad. Sci. U.S.A.*, **80**, 31–35.

Jacobs, S. and Cuatrecasas, P. (1981), in *Membrane Receptors. Methods for Purification and Characterization* (S. Jacobs and P. Cuatrecasas, eds), Chapman and Hall, London, pp. 61–86.

Kavathas, P. and Herzenberg, L.A. (1983), *Proc. Natl. Acad. Sci. U.S.A.*, **80**, 524–528.

Korman, A.J., Knudsen, P.J., Kaufman, J.F. and Strominger, J.L. (1982), *Proc. Natl. Acad. Sci. U.S.A.*, **79**, 1844.

Lennon, V.A., Thompson, M. and Chen, J. (1980), *J. Biol. Chem.*, **255**, 4395–4398.

Maniatis, T., Fritsch, E.F. and Sambrook, J. (1982), *Molecular Cloning: A Laboratory Manual*, Cold Spring Harbor Laboratory, New York.

Mendex, B., Valenzuela, P., Martial, J.A. and Baxter, J.D. (1980), *Science*, **209**, 695–697.

Merlie, J.P., Sebbane, R., Gardner, S. and Lindstrom, J. (1983), *Proc. Natl. Acad. Sci. U.S.A.*, **80**, 3845–3849.

Momoi, M.Y. and Lennon, V.A. (1982), *J. Biol. Chem.*, **257**, 12 757–12 764.

Noda, M., Takahashi, H., Tanabe, T., Toyosato, M., Furutani, Y., Hirose, T., Asai, M., Inayama, S., Miyata, T. and Numa, S. (1982), *Nature (London)*, **299**, 793–797.

Noda, M., Takahashi, H., Tanabe, T., Toyosato, M., Kikyotani, S., Hirose, T., Asai, M., Takashima, H., Inayama, S., Miyata, T. and Numa, S. (1983), *Nature (London)*, **301**, 251–255.

Noyes, B.E., Mevarich, M., Stein, R. and Agarwal, K.L. (1979), *Proc. Natl. Acad. Sci. U.S.A.*, **76**, 1770–1774.

Orr, H.T. (1982) in *Histocompatibility Antigens: Structure and Function* (P. Parham and J. Strominger, eds), Chapman and Hall, London, pp. 3–51.

Raftery, M.A., Hunkapillar, M.W., Strader, C.D. and Hood, L.E. (1980), *Science*, **208**, 1454–1456.

Roth, R.A. and Cassell, D.J. (1983), *Science*, **219**, 299–301.

Schneider, C., Newman, R.A., Asser, U., Sutherland, D.R. and Greaves, M.F. (1982a), *J. Biol. Chem.*, **257**, 10 766–10 769.

Schneider, C., Sutherland, R., Newman, R. and Greaves, M.F. (1982b), *J. Biol. Chem.*, **257**, 8516–8522.

Schneider, C., Asser, U., Sutherland, D.R. and Greaves, M.F. (1983a), *FEBS Lett.*, **158**, 259–264.

Schneider, C., Kurkinen, M. and Greaves, M. (1983b), *EMBO J.*, **2**, 2259–2263.

Schneider, W.J., Beisiegel, U., Goldstein, J.L. and Brown, M.S. (1982), *J. Biol. Chem.*, **257**, 2664–2673.

Soreq, H., Bartfeld, D., Parvari, R. and Fuchs, S. (1982), *FEBS Lett.*, **139**, 32–36.

Sumikawa, K., Houghton, M., Smith, J.C., Bell, L., Richards, B.M. and Barnard, E.A. (1982), *Nucleic Acids Res.*, **10,** 5809–5822.
Sutherland, R., Delia, D., Schneider, C., Newman, R., Kemshead, J. and Greaves, M. (1981), *Proc. Natl. Acad. Sci. U.S.A.*, **78,** 4515–4519.
Tzartos, S.J., Rand, D.E., Einarson, B.L. and Lindstrom, J.M. (1981), *J. Biol. Chem.*, **256,** 8635–8645.
Williams, H.G. (1981), *Genetic Engineering* (R. Williamson, ed.), Vol. 1, Academic Press, New York, p. 2.
Williams, J.G. and Lloyd, M.M. (1979), *J. Mol. Biol.*, **129,** 19–35.
Young, R. and Davis, R. (1983), *Proc. Natl. Acad. Sci. U.S.A.*, **80,** 1194–1198.

Index